T0215045

Communications
in Computer and Information Science   1336

Jianhua Qian · Honghai Liu ·
Jiangtao Cao · Dalin Zhou (Eds.)

Robotics and Rehabilitation Intelligence

First International Conference, ICRRI 2020
Fushun, China, September 9–11, 2020
Proceedings, Part II

Editors
Jianhua Qian
Liaoning Shihua University
Fushun, China

Jiangtao Cao 🆔
Liaoning Shihua University
Fushun, China

Honghai Liu 🆔
University of Portsmouth
Portsmouth, UK

Dalin Zhou 🆔
University of Portsmouth
Portsmouth, UK

ISSN 1865-0929   ISSN 1865-0937 (electronic)
Communications in Computer and Information Science
ISBN 978-981-33-4931-5   ISBN 978-981-33-4932-2 (eBook)
https://doi.org/10.1007/978-981-33-4932-2

This Springer imprint is published by the registered company Springer Nature Singapore Pte Ltd.
The registered company address is: 152 Beach Road, #21-01/04 Gateway East, Singapore 189721, Singapore

Preface

The First International Conference on Robotics and Rehabilitation Intelligence (ICRRI 2020) was held at Liaoning Shihua University, China. ICRRI is a newly-established international conference sponsored by Springer, IEEE SMCS Japan Chapter, and IEEE SMCS Portsmouth Chapter, focusing on the advanced development of Intelligence in Robotics and Rehabilitation Engineering. ICRRI 2020 covered both theory and applications in robotics, rehabilitation, and computational intelligence systems, and was successful in attracting a total of 188 submissions addressing state-of-the-art development and research covering topics related to Human-robot Interaction, Robotic Vision, Multi-agent Systems and Control, Robot Intelligence and Learning, Robot Design and Control, Robot Motion Analysis and Planning, Medical Robot, Robot Locomotion, Mobile Robot and Navigation, Biomedical Signal Processing, Artificial Intelligence in Rehabilitation, Computational Intelligence in Rehabilitation, Systems Modeling and Simulation in Rehabilitation, Neural and Rehabilitation Engineering, Wearable Rehabilitation Systems, Rehabilitation Education, Biomedical and Health Informatics, Policy on Healthcare Innovation and Commercialization, Advanced Control Theory and Applications, Artificial Intelligence, Big Data, and Optimization Methods. Following the rigorous reviews of the submissions, a total of 60 papers (31.9% acceptance rate) were selected to be presented in the conference during September 9–11, 2020. We hope that the published papers of ICRRI 2020 will prove to be technically constructive and helpful to the research community. We would like to express our sincere acknowledgment to the attending authors and the distinguished plenary speakers. Acknowledgment is also given to the ICRRI 2020 Program Committee members for their efforts in the rigorous review process. Special thanks are extended to Jane Li in appreciation of her contribution in the technical support throughout ICRRI 2020. Last but not least, the help from Jane Li and Celine Chang of Springer is appreciated for the publishing.

We hope that the readers will find this volume of great value for reference.

September 2020

Jianhua Qian
Honghai Liu
Dalin Zhou
Jiangtao Cao

Organization

General Chairs

Jianhua Qian	Liaoning Shihua University, China
Naoyuki Kubota	Tokyo Metropolitan University, Japan
Honghai Liu	University of Portsmouth, UK

General Co-chairs

Jiangtao Cao	Liaoning Shihua University, China
Junyou Yang	Shenyang University of Technology, China
Dalin Zhou	University of Portsmouth, UK

Program Chairs

Yonghui Yang	University of Science and Technology Liaoning, China
Qiang Zhao	Liaoning Shihua University, China

Special Session Chairs

Xiaofei Ji	Shenyang Aerospace University, China
Gongfa Li	Wuhan University of Science and Technology, China
Hongyi Li	Bohai University, China
Hongwei Gao	Shenyang Ligong University, China

Publication Chairs

Yinfeng Fang	Hangzhou Dianzi University, China
Kairu Li	Shenyang University of Technology, China
Chengli Su	Liaoning Shihua University, China
Guoliang Wang	Liaoning Shihua University, China

Publicity Chairs

Kaspar Althoefer	Queen Mary University of London, UK
Jangmyung Lee	National Pusan University, South Korea
Qiang Liu	Liaoning Shihua University, China
Nick Savage	University of Portsmouth, UK
Dalai Tang	Inner Mongolia University of Finance and Economics, China

Award Chairs

Nanshu Lu	The University of Texas at Austin, USA
Gaoxiang Ouyang	Beijing Normal University, China
Yuichiro Toda	Okayama University, Japan

Secretaries

Yue Wang	Liaoning Shihua University, China
Taoyan Zhao	Liaoning Shihua University, China

Contents – Part II

Robot Design and Control

Robotic Vision and Machine Intelligence

Optimization Method in Monitoring

Advanced Process Control in Petrochemical Process

Rehabilitation Intelligence

Contents – Part I

Electric Drive and Power System Fault Diagnosis

Robust Stability and Stabilization

Intelligent Method Application

Intelligent Control and Perception

Smart Remanufacturing and Industrial Intelligence

Intelligent Control of Integrated Energy System

Smart Healthcare and Intelligent Information Processing

IFHS Method on Moving Object Detection in Vehicle Flow

Rui Zhao, Haiping Wei, and Hongfei Yu[✉]

School of Computer and Communication Engineering, Liaoning Shihua
University, Fushun 113000, China
yuhfln@163.com

Abstract. Moving target detection is widely used in the field of intelligent monitoring. This paper proposes an Improved Frame Difference Method Fusion Horn-Schunck Optical Flow Method (IFHS) algorithm to detect moving targets in video surveillance. First, a three-frame difference method is performed on the image to obtain a difference image. Second, the differential image is tracked according to the HS optical flow method to obtain an optical flow vector. Finally, the moving target is obtained by fusing the optical flow vector information of the two difference results. The problem of incomplete contours of moving targets detected in video surveillance by the separate optical flow method and the three-frame difference method is solved. The experimental simulation and experimental applications in different scenarios prove the robustness and accuracy of the method.

Keywords: Moving object detection · IFHS · Three-frame difference method · Horn-Schunck optical flow method

1 Introduction

With the development of artificial intelligence and the emergence of intelligent robots, people will rely on machines to complete a large number of complex and tedious tasks. Computer vision is a study of related theories and technologies to try to build artificial intelligence systems that acquire "information" from images or multidimensional data. Moving object detection is a branch of image processing and computer vision. It refers to the process of reducing redundant information in time and space in video by computer vision to effectively extract objects that have changed spatial position. It has great significance in theory and practice, and has been concerned by scholars at home and abroad for a long time.

There are three main methods of traditional moving object detection: the first is the inter-frame difference method [1–4]. It subtracts the pixel values of two images in two adjacent frames or a few frames apart in the video stream, and performs thresholding on the subtracted image to extract the moving areas in the image. This method is more

This work was supported by National Natural Science Foundation of China (Grant No. 61702247) and the Natural Science Foundation of Liaoning Province of China (Grant No. 2019-ZD-0052).

J. Qian et al. (Eds.): ICRRI 2020, CCIS 1336, pp. 3–19, 2020.
https://doi.org/10.1007/978-981-33-4932-2_1

robust, but the detected moving target contour is incomplete, which is suitable for scenes with slow background changes and fast object movement.

The second is the background subtraction method [5, 6]. Background subtraction is a method for detecting moving objects by comparing the current frame in the image sequence with the background reference model. Its performance depends on the background modeling technology used. Its disadvantage is that it is particularly sensitive to dynamic background changes and external unrelated events.

The last method is optical flow method [7–9]. In the optical flow method, the optical flow of each pixel is calculated by using the optical flow. Then the threshold value segmentation is performed on the optical flow field to distinguish the foreground and background to obtain the moving target area. Optical flow not only carries the motion information of moving objects, but also carries rich information about the three-dimensional structure of the scene. It can detect moving objects without knowing any information about the scene. This method is less robust and susceptible to light, and the basic assumptions are difficult to meet.

The above methods can be applied to sports scenes in most fields. According to the movement background, the detection of moving objects can be divided into two cases: detection in static scenes and dynamic scenes. A static scene means that the camera and the motion scene remain relatively still. A dynamic scene is a sequence of video images captured by a camera mounted on a moving object. The above-mentioned methods have developed very rapidly in the past few years, and there are also some theoretical applications worth improving in various application scenarios. The main contribution of this paper is to propose an IFHS algorithm to improve the traditional algorithm and solve the following two shortcomings: The first is that the optical flow method alone is harsh and has poor robustness. The second is that the traditional frame-to-frame difference method has incomplete target contours for fast moving targets. This algorithm effectively improves the accuracy and efficiency of moving target detection in video surveillance. This paper is detecting video surveillance information and is therefore performed in a static scenario.

2   Inter-frame Difference Method

2.1   Three-Frame Difference Method

Commonly used frame difference methods are two-frame difference method [10, 11] and three-frame difference method [12, 13]. The result of the two-frame difference method may appear that the overlapping part of the target in the two-frame image is not easy to detect, and the center of the moving target is prone to voids. For slow moving targets, missed tests may occur. The three-frame difference method can overcome the double shadow problem that is easily generated by the difference between frames, and also suppresses the noise to a certain extent [14]. Therefore, this paper chooses the three-frame difference method. Figure 1 is a comparison of the matrix difference method of the first frame and the 26th frame.

The three-frame difference method is to change two adjacent frames in a video sequence to obtain characteristic information of a moving target. This method first uses three adjacent frames as a group to perform the difference between two adjacent frames.

Next, Binarize the difference results are binarized. In order to remove isolated noise points in the image and holes generated in the target, it is necessary to perform a morphological filtering process on the obtained binary image, to perform a closing operation [15]. After the image is enlarged and corroded, it is a logical AND operation. Therefore, the area of the moving target can be better obtained.

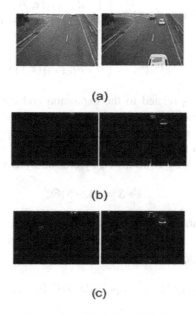

(a)

(b)

(c)

Fig. 1. Comparison images of frames 1 and 26: (a) Original image; (b) Image of two frame difference method; (c) Image of the three-frame difference method.

2.2   The Process of Three-Frame Difference Method

1) Let the image sequence of n frames be expressed as:

$$\{f0(x,y),\ldots,f(k)(x,y),\ldots,f(n-1)(x,y)\}$$

$f(k)(x,y)$ represents the k-th frame of the video sequence. Select three consecutive images in a video sequence:

$$f(k-1)(x,y),\ f(k)(x,y),\ f(k+1)(x,y)$$

Calculate the difference between two adjacent frames of images, the binarization results are expressed by Eqs. (1) and (2)

$$d_{\{k-1,k\}}(x,y) = \left| f_{\{k\}}(x,y) - f_{\{k-1\}}(x,y) \right| \tag{1}$$

$$d_{\{k,k+1\}}(x,y) = \left| f_{\{k+1\}}(x,y) - f_{\{k\}}(x,y) \right| \tag{2}$$

2) For the two difference images obtained, by setting an appropriate threshold the binarized images $b_{(k-1,k)}(x,y)$ and $b_{(k,k+1)}(x,y)$ can be obtained as shown in Eqs. (3) and (4).

$$b_{(k-1,k)}(x,y) = \begin{cases} 1 & d_{\{k-1,k\}}(x,y) \geq T \\ 0 & d_{\{k-1,k\}}(x,y) < T \end{cases} \tag{3}$$

$$b_{(k,k+1)}(x,y) = \begin{cases} 1 & d_{\{k,k+1\}}(x,y) \geq T \\ 0 & d_{\{k,k+1\}}(x,y) < T \end{cases} \tag{4}$$

3) The closed operation is related to the expansion and corrosion operations. The closed operation can be used to fill small cracks and discontinuities in the foreground object. And small holes, while the overall position and shape are unchanged. Closing is the dual operation of opening and is denoted by A•B; it is generated by dilating A by B, followed by erosion by B with Eq. 5.

$$A \bullet S = (A \oplus S) \ominus S \tag{5}$$

4) For each pixel (x,y), the above two binary images $b_{(k-1,k)}(x,y)$ with $b_{(k,k+1)}(x,y)$ perform logical and operation after closed operation to get $B_{(k-1,k)}(x,y)$. The result of the logical AND operation is represented by Eq. 6.

$$B_{(k-1,k)}(x,y) = b_{(k-1,k)}(x,y) \otimes b_{(k,k+1)}(x,y) \tag{6}$$

3   Optical Flow Method

3.1   Optical Flow Field

When the human eye observes a moving object, the scene of the object forms a series of continuously changing images on the retina of the human eye. This series of continuously changing information continuously "Flow" through the retina (the image plane), like a light "Flow", so called optical flow. Optical flow expresses the change of the image. Because it contains information about the movement of the target, it can be used by the observer to determine the movement of the target. In space, motion can be described by a motion field. However, on an image plane, the motion of an object is often reflected by the gray distribution of different images in the image sequence. Therefore, the motion field in space is transferred to the image and is expressed as optical flow field [16, 17].

The movement of a two-dimensional image is a projection of the three-dimensional object movement on the image plane relative to the observer. Ordered images can estimate the instantaneous image rate or discrete image transfer of two-dimensional images. The projection of 3D motion in a 2D plane is shown in Fig. 2.

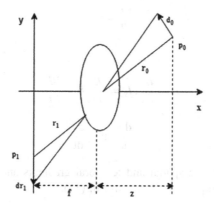

Fig. 2. Projection of 3D motion in a 2D plane.

3.2   Horn-Schunck Optical Flow Method

There are two types of constraints for the optical flow method:

1) The brightness does not change.
2) Time continuous or motion is "small motion".

Let $I(x, y, t)$ be the illuminance of the image point at time $t + \delta t$. If $u(x, y)$ and $v(x, y)$ are the x and y components of the optical flow at this point, suppose that point moves to $(x + \delta x, y + \delta y)$ and the brightness remains unchanged: $\delta x = u\delta t, \delta y = v\delta t$.

$$I(x + u\delta t, y + v\delta t, t + \delta t) = I(x, y, t) \tag{7}$$

Equation (7) is a constraint condition [18, 19]. This constraint cannot uniquely solve for u and v, so other constraints are needed, such as constraints such as continuity everywhere in the sports field. If the brightness changes smoothly through x, y, t the left side of Taylor series Eq. (7) can be extended to Eq. (8).

$$I(x, y, t) + \delta x \frac{\partial I}{\partial x} + \delta y \frac{\partial I}{\partial y} + \delta t \frac{\partial I}{\partial t} + \varepsilon \tag{8}$$

Where ε is the second and higher order terms for δx, δy, δt. The $I(x, y, t)$ on both sides of the above formula cancel each other, divide both sides by δt, and take the limit $\delta t \rightarrow 0$ to get the Eq. (9).

$$\frac{\partial I}{\partial x}\frac{dx}{dt} + \frac{\partial I}{\partial y}\frac{dy}{dt} + \frac{\partial I}{\delta t} = 0 \tag{9}$$

Equation (9) is actually an expansion of Eq. (10).

$$\frac{dI}{dt} = 0 \tag{10}$$

Assume:

$$I_x = \frac{\partial I}{\partial x}, I_y = \frac{\partial I}{\partial y}, I_t = \frac{\partial I}{\partial t}$$

$$u = \frac{dx}{dt}, \quad v = \frac{dy}{dt}$$

the relationship between the spatial and temporal gradients and the velocity components is obtained from Eq. (9) as shown in Eq. (11).

$$I_x u + I_y v + I_t = 0 \tag{11}$$

Finally, the constraint equation of light is shown in the following Eq. (12).

$$\nabla I \cdot v + I_t = 0 \tag{12}$$

Horn-Schunck optical flow [20–23] algorithm introduces global smoothing constraints to do motion estimation in images. The motion field satisfies both the optical flow constraint equation and the global smoothness. According to the optical flow constraint equation, the optical flow error is shown in the following Eq. (13).

$$e^2(x) = (I_x u + I_y v + I_t)^2 \tag{13}$$

among them $x = (x,y)^T$ For smooth changing optical flow, the sum of the squares of the velocity components is shown in Eq. (14).

$$s^2(x) = \iint \left[\left(\frac{\partial u}{\partial x}\right)^2 + \left(\frac{\partial u}{\partial y}\right)^2 + \left(\frac{\partial v}{\partial x}\right)^2 + \left(\frac{\partial v}{\partial y}\right)^2 \right] dxdy \tag{14}$$

Combining the smoothness measure with a weighted differential constraint measurement, where the weighting parameters control the balance between the image flow constraint differential and the smoothness differential:

$$E = \iint \{ e^2(x) + \alpha s^2(x) \} dxdy \tag{15}$$

The parameter α in the equation is a parameter that controls the smoothness, and the value of α is proportional to the smoothness. Use the variation method to transform the above equation into a pair of partial differential equations:

$$\alpha \nabla^2 u = I_x u + I_x I_y v + I_x I_t \tag{16}$$

$$\alpha \nabla^2 v = I_x I_y u + I_y^2 v + I_y I_t \tag{17}$$

The finite difference method is used to replace the Laplace in each equation with a weighted sum of the local neighborhood image flow vectors, and iterative methods are used to solve the two difference Eqs. (16) and (17).

The Horn-Schunck optical flow method uses the minimization function of (18) to find the differential of E with respect to u and v to obtain the optical flow vectors (21) and (22).

$$E = \sum_i \sum_j \left(e^2(x, y) + \alpha s^2(x, y) \right) \tag{18}$$

Find:

$$\left(I_x u + I_y v + I_t \right) I_x + \alpha(u - \bar{u}) = 0 \tag{19}$$

$$\left(I_x u + I_y v + I_t \right) I_y + \alpha(v - \bar{v}) = 0 \tag{20}$$

$\bar{u}$ and $\bar{v}$ are the averages in their neighborhoods, respectively. From the above two equations, u and v can be found. The optical flow is

$$u^{n+1} = \bar{u}^n - I_x \frac{I_x \bar{u}^n + I_y \bar{v}^n + I_t}{\alpha + I_x^2 + I_y^2} \tag{21}$$

$$v^{n+1} = \bar{v}^n - I_y \frac{I_x \bar{u}^n + I_y \bar{v}^n + I_t}{\alpha + I_x^2 + I_y^2} \tag{22}$$

The Horn-Schunck optical flow method is also called a differential method because it is a gradient-based optical flow method. This type of method is based on the assumption that the brightness of the image is constant. Calculate the two-dimensional velocity field. Therefore, the direction and contour of the moving object will be detected more completely.

The comparison between the detection results of the HS optical flow method and the original is shown in Fig. 3. The detected effect is affected by lighting, so moving targets are not particularly noticeable.

(a)

(b)

Fig. 3. Optical flow detection images of frames 1 and 26: (a) the original image, (b) the optical flow detection image.

4   IFHS Algorithm

The IFHS (Improved Frame Difference Method Fusion HS Optical Flow Method) algorithm is an improvement that combines the improved three-frame difference method with the HS optical flow method to detect motion information in video surveillance.

4.1   Mathematical Morphological Filtering and Denoising

Digital images are subject to noise pollution during the process of acquiring and transmitting, resulting in some black and white point noise, as shown in Fig. 4. Therefore, we need to perform a neighborhood operation on the pixels of the image by filtering to achieve the effect of removing noise. In the process of mathematical morphological image denoising [24], the filtering denoising effect can be improved by appropriately selecting the new installation and dimension of the structural elements. In the cascading process of multiple structural elements, the shape and dimensions of the structural elements need to be considered.

Original image    Noise image

Fig. 4. Original image and noise image

Mathematical morphology processes images by using certain structural elements to measure and extract the corresponding shapes in the images, and uses set theory to achieve the goal of analyzing and identifying the images, while maintaining the original information in the images. Assume that the structural elements are A_{mn}, n represents a shape sequence, and m represents a dimensional sequence. The process of $A_{mn} = \{A_{11}, A_{12}, \ldots A_{21}, \ldots, A_{nm}\}$ is to digitally filter and denoise digital images. According to the characteristics of noise, structural elements from small to large dimensions are adopted, and more geometric features of digital images are maintained. Therefore, a cascade filter is selected. The elements of the same shape structure are used to filter the image according to the dimensional arrangement from small to large. The series filtering composed of structural elements of different shapes is then connected in parallel. Finally, calculate and plot the Peak Signal to Noise Ratio (PSNR) value curve to show the denoising effect. PSNR is the ratio of the maximum signal power to the signal noise power. The larger the PSNR value between the two images, the more similar it is. The figure below shows the original image and the noise image as shown in Fig. 4. The effect of four series noise reduction is shown in Fig. 5.

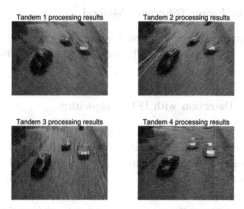

Fig. 5. Effect of four series noise reduction

The parallel denoising effect is shown in Fig. 6. The image of the denoising effect by the PSNR curve is shown in Fig. 7.

Fig. 6. Parallel denoising effect diagram

It is proved by experiments that de-noising through series and parallel filters is better than series filtering alone. So we use series and parallel denoising.

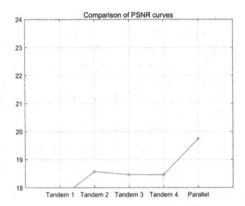

Fig. 7. PSNR curve image

4.2   Improved Three-Frame Difference Method

Aiming at the shortcomings of the traditional frame difference method, this paper uses series-parallel filtering and three-frame difference method to perform image processing. The specific process is shown in Fig. 8.

4.3   Moving Object Detection with IFHS Algorithm

Differentiate the k-frame images in the input video sequence with their $k - 1$ and $k + 1$ frames, respectively, and use the HS optical flow method to calculate the optical flow vectors of the two images; then set a threshold to remove the length less than this threshold optical flow vector; initialize $k = 3$; When a new frame enters, perform a difference operation on the previous frame and the next frame, use the appropriate threshold to binarize the new difference image and the previous difference image, and then Perform logical AND operation on the two binarized images, and iteratively accumulate the obtained moving targets in sequence. The specific operation is shown in Fig. 9.

In this paper, the algorithm operates on a computer with Inter i5-9300H,8g memory and 2.4 GHz. It is implemented by Matlab R2016a and its own toolbox. The data set involved in the algorithm is publicly provided on the Internet. The URL is as follows: http://cmp.felk.cvut.cz/data/motorway/. The "Toyota Motor Europe (TME) Motorway Dataset" is composed by 28 clips for a total of approximately 27 min (30000 + frames) with vehicle annotation. Annotation was semi-automatically generated using laser-scanner data. Image sequences were selected from acquisition made in North Italian motorways in December 2011. This selection includes variable traffic situations, number of lanes, road curvature, and lighting, covering most of the conditions present in the complete acquisition. The dataset comprises: Image acquisition: stereo, 20 Hz frequency, 1024 × 768 grayscale losslessly compressed images, 32° horizontal field of view, bayer coded color information (in OpenCV use CV_BayerGB2GRAY and CV_ BayerGB2BGR color conversion codes; please note that left camera was rotated upside down, convert to color/grayscale BEFORE flipping the image). A checkboard

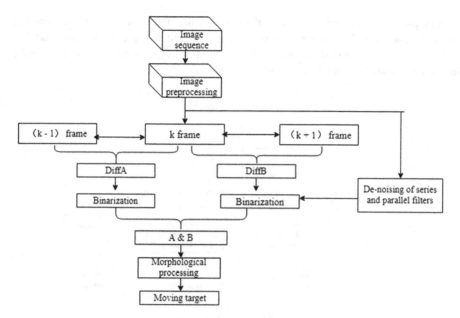

Fig. 8. Improved three-frame difference method flowchart

calibration sequence is made available. Laser-scanner generated vehicle annotation and classification (car/truck). A software evaluation toolkit (C ++ source code). The data provided is timestamped, and includes extrinsic calibration.

The dataset has been divided in two sub-sets depending on lighting condition, named "daylight" (although with objects casting shadows on the road) and "sunset" (facing the sun or at dusk). For each clip, 5 s of preceding acquisition are provided, to allow the algorithm stabilizing before starting the actual performance measurement. The data has been acquired in cooperation with VisLab (University of Parma, Italy), using the BRAiVE test vehicle.

In this paper, in order to fuse the advantages of various algorithms together to make up for their shortcomings, we combine the improved three-frame difference method and the HS optical flow method into a new method (Improved Frame Difference Method Fusion Horn-Schunck Optical Flow) to detect vehicle motion information in the video. For the sake of comparison, we take frame 1 and frame 26 as examples. It is simulated on the running software of matlab. After the video is subjected to the three-frame difference method, the difference result is tracked and detected by the HS optical flow method. First calculate the magnitude of the optical flow vector. Because optical flow is a complex matrix, you need to multiply by conjugate. Second, you need to calculate the average value of the optical flow amplitude to represent the speed threshold. Finally, this paper use threshold segmentation to extract moving objects, next filter to remove noise. The mathematical morphological filtering denoising was mentioned above. After multiple comparisons in the experiment, as shown in the detection result of Fig. 7, the comparison of the PSNR curves of the series filtering denoising and the parallel filtering denoising. This paper uses series, parallel, and filtering to denoise, and Xia compensates for the loss of the image during the transmission process, so as to make the detection target contour clearer in the following.

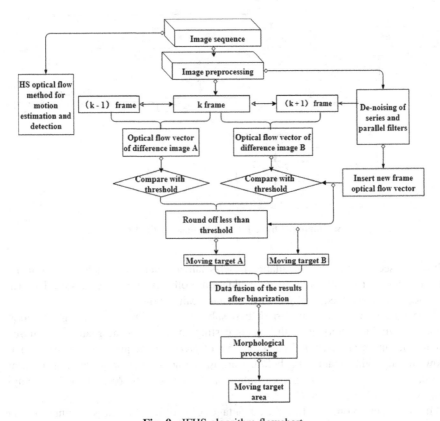

Fig. 9. IFHS algorithm flowchart

Remove the road by morphological erosion, and fill the car hole by morphological closure. In this way, the vehicle information in the video can be more completely retrieved. The image results of the specific method combined detection are shown in Fig. 10.

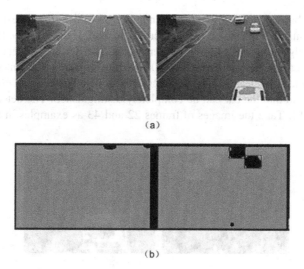

(a)

(b)

Fig. 10. IFHS algorithm detection image: (a) Original image (b) IFHS algorithm detection image

4.4 Vehicle Flow

This paper has an image of 120 × 160 pixels. We add a white virtual bar on its 30th line to make a "ruler" to measure the content. Then we also need to add a car border to show the tracked and show the tracked car. Because the vehicle needs to have two frames to pass the frame, in order to clearly show the frame effect, we take the middle frame 6 and 22 as an example. At the same time, we called the toolbox in matlab to calculate the area and border of the car. An image with a white ruler and a border in the image is shown in Fig. 11. The border of the car passed the white bar. We judged that it was passing by the car.

Fig. 11. Framed image of the car at frame 6 and 22.

Next, you need to calculate the percentage of the area of the car to the area of the drawn border. After many experimental prediction tests, we found that more than 40% of the area of the frame is a vehicle. As shown in the figure, we determine whether the vehicle is a vehicle based on whether the percentage is greater than 40%. This paper uses the white bar as a ruler to count the number of vehicles. This experiment uses highway video information, so no one will cross the road. If there is a pedestrian on the road, the new algorithm will draw a border around the detected object and the percentage of its area detected will not reach 40%. White bars are not counted. In order to make the identification clearer, the parameters should be adjusted from 40% to 45% in the experiment, so that the car can be completely distinguished. The detected image is shown in Fig. 12. Take the images of frames 22 and 43 as examples in this paper.

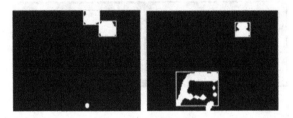

Fig. 12. An image with a border around the detected image

This paper selects a certain number of frames based on the speed of the vehicle and considers the vehicle to be in the process of moving from video surveillance to entry. Here in this video, the selected frame number is 20. This paper takes the maximum value of the vehicle count in 10 frames out of 20 frames as the number of vehicles we think is currently passing through the road section, and then add up to get the video traffic volume.

Due to the three-frame difference method, the video image has a total of 118 frames, with 20 frames as a part, and the result is rounded up. Then six partial loops are accumulated to get the number of vehicles in the video. The specific counting results are shown in Table 1.

Table 1. Video vehicle computer results

TIME	1	2	3	4	5	6
NUM	1	3	4	7	9	10

As can be seen from Table 1, each process of vehicle technology counts a total of 6 times, and the video passes through a total of 10 vehicles.

5   Evaluation

In order to test the effectiveness of the algorithm in this paper, the F-measure method [25] is used to evaluate the effectiveness of the proposed new moving target detection algorithm. F-Measure is the tp (true positives) detected by Precision and Recall weighted harmonic average fp (false positives) fn (false negatives) tn (true negatives).

$$Precision = \frac{tp}{(tp+fp)} \tag{23}$$

$$Recall = \frac{tp}{(tp+fn)} \tag{24}$$

$$f1 = \frac{2PR}{P+R} \tag{25}$$

$f1$ combines the results of P and R.When F1 is high, the comparison shows that the experimental method is ideal. The following Table 2 compares the experimental results.

Table 2. Analysis and comparison of different video detection F1 of each algorithm

Video	HS optical flow method	Three-frame difference method	IFHS algorithm
Video 1	0.61	0.73	0.82
Video 2	0.63	0.75	0.83
Video 3	0.66	0.79	0.81
Video 4	0.59	0.70	0.91
$f1$ average	0.62	0.74	0.84

From the standpoint of video alone, this article uses four videos for comparison. From the results of separate tests, it can still be proven that the accuracy of the algorithm in this paper is higher than the accuracy of the two algorithms alone.

6   Conclusion and Future Work

After testing in this paper, the algorithm has obvious advantages in detecting vehicle information in video than the two algorithms alone, which can prove the accuracy and robustness of the algorithm. The experimental results show that the IFHS method has a good guiding effect on vehicle flow detection. The test results in Tables 1 and 2 show that the algorithm can well compensate for image holes between frames. The matching optical flow information obtains the movement trajectory to achieve the positioning algorithm of the vehicle during the movement. It can reduce the number of false recognitions, speed up the detection speed, reduce the complexity of the computer, meet the needs of traffic flow statistics, and ensure the accuracy of the number of statistics. Through several comparative iterative experiments, we find that the method is both effective and stable.

In the future, this algorithm can intelligently monitor the information in video surveillance, saving human resources. In order to provide basic guarantees for intelligent transportation systems. In the future, research on monitoring and identifying vehicle information at intersections in real time and managing traffic lights at intersections may continue. If a large amount of traffic is detected, the changes in lighting will accelerate, otherwise it will slow down, thereby promoting the development and popularization of intelligent transportation systems.

References

1. Zhang, T., Jiang, P., Zhang, M.: Inter-frame video image generation based on spatial continuity generative adversarial networks. Signal Image Video Process. 13(8), 1487–1494 (2019). https://doi.org/10.1007/s11760-019-01499-0
2. Zhao, D.-N., Wang, R.-K., Lu, Z.-M.: Inter-frame passive-blind forgery detection for video shot based on similarity analysis. Multimed. Tools Appl. 77(19), 25389–25408 (2018)
3. Li, H., Xiong, Z., Shi, Z.: HSVCNN: CNN-based hyperspectral reconstruction from RGB videos. In: 2018 25th IEEE International Conference on Image Processing (ICIP). IEEE (2018)
4. He, Z.-J., Yang, C.-L., Tang, R.-D.: Research on structural similarity based inter-frame group sparse representation for compressed video sensing. Acta Electron. Sinica 46(3), 544–553 (2018)
5. Mabrouk, L., Huet, S., Houzet, D.: Efficient adaptive load balancing approach for compressive background subtraction algorithm on heterogeneous CPU–GPU platforms. J. Real-Time Image Process. (1) (2019)
6. Smeureanu, S., Ionescu, R.T.: Real-time deep learning method for abandoned luggage detection in video. In: European Signal Processing Conference (2018)
7. Wu, D., Hao, H., Wang, L.: Intelligent monitoring system based on Hi3531 for recognition of human falling action. In: 2018 International Conference on Intelligent Transportation, Big Data & Smart City (ICITBS) (2018)
8. Shah, N., Píngale, A., Patel, V.: An adaptive background subtraction scheme for video surveillance systems. In: IEEE International Symposium on Signal Processing & Information Technology. IEEE (2018)
9. Chen, H., Nedzvedz, A., Nedzvedz, O.: Correction to: wound healing monitoring by video sequence using integral optical flow. J. Appl. Spectrosc. 86, 1146 (2020)
10. Li, Y., Jia, H., Chen, S.: Single-shot time-gated fluorescence lifetime imaging using three-frame images. Opt. Express 26(14), 17936 (2018)
11. Li, B., Zan, S., Jing, Q., et al.: Detection and tracking method for maneuvering targets using improved gradient-based optical flow field. In: 2019 Chinese Control And Decision Conference (CCDC) (2019)
12. Mitiche, A., Mansouri, A.-R.: On convergence of the Horn and Schunck optical-flow estimation method. IEEE Trans. Image Process. 13(6), 848–852 (2004)
13. Han, X., Gao, Y., Lu, Z., et al.: Research on moving object detection algorithm based on improved three frame difference method and optical flow (2016)
14. Zhang, M., Zhou, X., Wang, C.: Noise suppression threshold channel estimation method using RC and SRRC filters in OFDM systems. In: 2018 IEEE 18th International Conference on Communication Technology (ICCT). IEEE (2018)
15. Batabyal, T., Acton, S.T.: Elastic Path2Path: automated morphological classification of neurons by elastic path matching. In: 2018 25th IEEE International Conference on Image Processing (ICIP). IEEE (2018)
16. Hu, X., Li, D., Zeng, X., et al.: A visualization method of facial expression deformation based on pore-scale facial feature matching. In: 2019 Chinese Control Conference (CCC). IEEE (2019)

17. Petrou, Z.I., Xian, Y., Tian, Y.: Towards breaking the spatial resolution barriers: an optical flow and super-resolution approach for sea ice motion estimation. Isprs J. Photogram. Remote Sens. **138**, 164–175 (2018)
18. Hur, J., Roth, S.: Iterative residual refinement for joint optical flow and occlusion estimation (2019)
19. Xiang, X., Zhai, M., Zhang, R., et al.: A CNNs-based method for optical flow estimation with prior constraints and stacked U-Nets. Neural Comput. Appl. (2018)
20. Wang, H., Zheng, J., Pei, B.: A robust optical flow calculation method based on wavelet. Dianzi Yu Xinxi Xuebao/J. Electron. Inf. Technol. **40**(12), 2945–2953 (2018)
21. Leila, C., Salah, C., Karima, B.: Fast motion estimation algorithm based on geometric wavelet transform. Int. J. Wavelets Multiresolut. Inf. Process. **17**(2) (2019)
22. Peng, Y., Chen, Z., Wu, Q.M.J., et al.: Traffic flow detection and statistics via improved optical flow and connected region analysis. Signal Image Video Process. **12**(1), 99–105 (2018)
23. Ali, I., Alsbou, N., Jaskowiak, J., et al.: Quantitative evaluation of the performance of different deformable image registration algorithms in helical, axial, and cone-beam CT images using a mobile phantom. J. Appl. Clin. Med. Phys. **19**(2), 62 (2018)
24. Shivarama Holla, K., Jidesh, P., Bini, A.A.: Multiple-coil magnetic resonance image denoising and deblurring with nonlocal total bounded variation. Iete Tech. Rev. 1–6 (2019)
25. Altschuler, M., Bloodgood, M.: Stopping active learning based on predicted change of f measure for text classification. In: 2019 IEEE 13th International Conference on Semantic Computing (ICSC). IEEE (2019)

Power Control Algorithm of D2D Communication Based on Non-cooperative Game Theory

Yanqiu Li$^{(\boxtimes)}$ and Chengyin Ye

Liaoning Shihua University, Fushun 113001, China
leeyanqiul63@163.com

Abstract. With the rapid development of the next generation mobile network, terminal equipment and data traffic are growing rapidly, and the developing trend of the communication network is presents intelligence and large capacity. D2D (device-to-device) communication sharing spectrum resources with cellular users is being developed rapidly, for it can effectively solve the problem of large-scale user terminal access, improve the utilization rate of spectrum resources, and expand the coverage of the network. A D2D power control algorithm with price control based on non-cooperative game theory is proposed to solve the interference problem in the hybrid network system of cellular users and D2D users. Firstly, the utility function is established by analyzing the benefit and utility of D2D users and the interference to the cell base station, and by considering the service quality of the cell users and constraint of the interference tolerance of the base station to the transmission power. Then, by using the interference size of D2D users to cellular users, base station updates the interference price of the cellular system. Accordingly, the D2D users adjust the transmission power and maximize the profit and maximizes the overall profit of the system by using the Lagrange multiplier method to find the price. Next, the existence and uniqueness of Nash equilibrium solution are proved by taking Nash equilibrium as the result of non-cooperative game. Finally, numerical simulation results show that the algorithm can effectively control the transmission power of D2D users and effectively improve the total rate of D2D networks.

Keywords: Non-cooperative game · D2D · Interference price · Throughput

1 Introduction

With the rapid development of the next generation mobile network, the number of terminal devices and applications presents an explosive growth. How to solve the problem of large-scale user terminal access and the shortage of spectrum resources, and further expand the coverage of network is an important problem that must be solved in the development of network technology [1]. D2D (device-to-device) communication technology has developed rapidly because it can effectively improve spectrum utilization, expand network coverage and improve system throughput [2–5]. D2D communication refers to the technology of direct communication between adjacent devices

© Springer Nature Singapore Pte Ltd. 2020
J. Qian et al. (Eds.): ICRRI 2020, CCIS 1336, pp. 20–33, 2020.
https://doi.org/10.1007/978-981-33-4932-2_2

in the communication network, which is one of the key technologies of The5th Generation mobile communication (The5th Generation, 5G) [6]. Because D2D users share the spectrum resources of cellular users, the system will produce the same frequency interference, which will seriously affect the performance of cellular network communication. Therefore, reasonable methods should be adopted to reduce the interference caused by the introduction of D2D communication [7].

Power control is an effective method to limit interference directly. In the literature [8], the author presents an energy-saving algorithm to minimize the transmission power of D2D users, which can ensure that the cellular users can achieve the minimum signal interference noise ratio, so that D2D users can carry out signal transmission at the lowest transmission rate, and achieve a significant reduction in energy consumption. In literature [9], the power distribution problem is modeled as a reverse iterative combined auction game, and joint wireless resources and power distribution scheme are used to improve the performance of the uplink system. Literature [10] also considers the resource allocation scheme of system uplink, and proposes a greedy heuristic resource allocation algorithm based on interference degree to maximize the access quantity of D2D users and satisfy the service quality of both cellular users and D2D users, which greatly increases the throughput of cellular network. Literature [11] studied the power control strategy based on game theory and proved its excellent performance in energy efficiency. Literature [12] proposed joint scheduling and resource allocation algorithms for interference management and network throughput optimization. Literature [13] USES centralized resource management scheme in D2D communication network. Although the system energy efficiency is improved, the signaling cost and complexity are high. Literature [3] also adopts the centralized resource allocation method, which consider the improvement of spectrum utilization in the case of a pair of D2D users, but ignores the interference between D2D users. In order to reduce the complexity and signaling overhead, the optimal switching power control strategy is proposed in literature [14] according to the distributed power control method. Based on the density and path loss index of D2D link, the coverage of D2D link and the rate expression of D2D link are derived to maximize the transmission rate of D2D link. Literature [15] also put forward a distributed power control scheme. In the case of limiting the total transmission rate, the overall power consumption of the system is minimized. In literature [16], the author proposed a resource management method of joint channel allocation and power control by using Stackberg game. After resource allocation is stable, one D2D user USES one channel, and there are only at most one D2D user in one channel, which cannot guarantee the utilization of spectrum resources. Literature [17] studied the resource allocation scheme under the multi-user scenario, and used the knowledge of graph theory to solve the resource allocation problem. This scheme could achieve sub-optimal performance, but with little feedback information, the access rate and throughput could not reach the ideal level.

Aiming at the interference problem of D2D users reusing cellular users' spectrum resources for communication, considering the interference tolerance requirements of cellular users' service quality and base station, a D2D upstream power control scheme with interference price control is proposed based on the non-cooperative game theory. Firstly, the power control problem of D2D inter-user interference in the system is transformed into a game problem by the interference price. Considering the cellular customer service quality and the interference of base station tolerance of transmission power constraints, according to the interference of D2D users update price of reuse, adjust the transmission power, get the optimal response of the distributed power control scheme, using the Lagrange multiplier method to find prices to maximize the benefits, ensure the overall revenue maximization system. Simulation results show that the transmission power of D2D users can be effectively controlled and the system throughput can be effectively improved.

2 System Model

In this paper, the power control of uplink is studied under the single cell application scenario. The cell consists of a central base station and multiple users. Base stations and user terminals are configured with omni-directional antennas. The communication users in the system are divided into cellular users and D2D users. Each D2D user pair contains a transmitter and a receiver. D2D user multiplexes the spectrum resources of the cellular user, and D2D user terminal can realize direct communication without the forwarding of the base station. It can not only improve the utilization rate of spectrum resources in D2D network, but also expand the network coverage and system capacity. The system model is shown in Fig. 1 as the scenario when D2D users reuse cellular uplink resources.

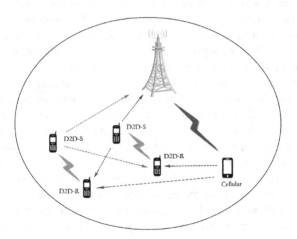

Fig. 1. System model

Let's say there are C orthogonal channels in the cell $C = \{C_i | i = 1, 2 ..., c\}$, D2D user pairs are denoted as sets $D = \{D_j | j = 1, 2 ..., n\}$. With the need of the D2D communication technology to communicate multiplexing cellular network and cellular network sharing spectrum, D2D communication link with the cellular communication link exists interfere with each other, including the base station signal by the cellular user when interference from the D2D launch, D2D user on the receiving end also received same cellular users and other D2D launch the interference to the user. If the interference is not managed effectively, it may even affect the normal communication of cellular users. Reasonable control of D2D users' transmitting power plays a crucial role in reducing interference, improving spectrum utilization and ensuring communication quality. P_c is the transmitting power of cellular users, P_j is transmitting power of D2D user j, $P_{j'}$ is transmitting power of D2D user j', G_{cj} is the channel gain of cellular user and D2D user j receiver, $G_{j'j}$ is the channel gain between adjacent D2D, G_{jj} is the channel gain of D2D transmitter and receiver, G_{cB} is the channel gain of cellular user c and base station, G_{jB} is the channel gain between the transmitter of D2D user j and the base station, σ^2 is the receiver noise power.

During the uplink communication, when multiple D2D users reuse the channel resources of the cellular users, the cellular users are interfered with from the D2D transmitter because the D2D users interfere with the signal reception of the base station. Therefore, the SINR of the cellular users is SINR_C:

$$\text{SINR}_C = \frac{P_c G_{cB}}{\sum\limits_{j \in D} P_j G_{jB} + \sigma^2} \tag{1}$$

The data transmission rate of cellular users can be expressed as

$$R_c = \log_2(1 + \text{SINR}_c) = \log_2\left(1 + \frac{P_c G_{cB}}{\sum\limits_{j \in D} P_j G_{jB} + \sigma^2}\right) \tag{2}$$

Similarly, cellular users will interfere with the receiving end of D2D users when they communicate. The corresponding SINR of D2D users is SINR_j:

$$\text{SINR}_j = \frac{P_j G_{jj}}{P_c G_{cj} + \sum\limits_{j' \in D/j} P_{j'} G_{j'j} + \sigma^2} \tag{3}$$

The data transmission rate of D2D users on the sub-channel is:

$$R_D = \log_2(1 + \text{SINR}_j) = \log_2\left(1 + \frac{P_j G_{jj}}{P_c G_{cj} + \sum\limits_{j' \in D/j} P_{j'} G_{j'j} + \sigma^2}\right) \tag{4}$$

In order to better manage the interference of D2D communication, a pricing mechanism is considered for D2D users to manage the interference of their transmitter. The base station control interference price factor $\lambda = [\lambda_1, \lambda_2, \ldots \lambda_n]^T$ represents the price required for each D2D user to reuse the cellular user channel resources. With the price given by the base station, D2D users adjust the transmission power to maximize their own benefits. All D2D users who reuse the same cellular users want to maximize their own interests and show the nature of non-cooperation. For the competition between D2D users, the non-cooperative game model can be used to solve the resource allocation scheme.

3 Non-cooperative Power Control Game Model

3.1 Non-cooperative Game Model

According to the system model, the non-cooperative game model between D2D users is defined as: $G = [N, \{P_j\}, \{u_j(p_j, p_{-j})\}]$, Where N represents the terminal set of D2D users, $\{P_j\}$ is the set of power policies for all D2D pairs, $P_j = [0, P_{max}]$ is the policy space of the terminal $j \in N$, P_{max} is the maximum transmission power of the user terminal, $u_j(p_j, p_{-j})$ is the set of utility functions for all D2D pairs.

Because the base station controls the price of D2D users' communication, and its goal is to get the maximum benefit from the interference quota sold to D2D users. Meanwhile, according to the interference size, the interference size of D2D users to the base station can be known. The utility function of the base station is

$$U_{BS}(\lambda, P) = \sum_{j=1}^{N} \lambda_j I(P_j) = \sum_{j=1}^{N} \lambda_j P_j G_{jB} \tag{5}$$

Where $I(P_j)$ interference power from D2D transmitter. The maximum interference that can be borne by the base station is limited. It is assumed that the maximum interference tolerance that the base station can bear is the threshold value Q, that is to say, the interference of all D2D users to the base station terminal shall not be greater than the maximum interference tolerance. Can be expressed as:

$$\sum_{j=1}^{N} P_j G_{jB} \leq Q \tag{6}$$

$$I = \sum_{j=1}^{N} P_j G_{jB} \tag{7}$$

there is $I \leq Q$. In general, because D2D communication is considered as a supplement to cellular systems, cellular users tend to have higher priority. Therefore, in order to guarantee the communication quality of cellular users, the base station must coordinate the interference caused by D2D transmission. For communication quality, a cellular user must ensure that the data transmission rate R_c is greater than the minimum data rate $R_{c,min}$, namely $R_c \geq R_{c,min}$. If the cellular user updates the transmission power without ensuring the formula base station will increase the interference price to force the D2D user to reduce the transmission power.

In order to maximize its total utility, the base station needs to find the optimal interference price to maximize its income, as shown below

$$\max U_{BS}(\lambda, P) \quad s.t \sum_{j=1}^{N} P_j G_{jB} \leq Q \tag{8}$$

The utility function of D2D users can be defined as:

$$u_j\left(P_j, P_{-j}\right) = \log_2\left(1 + \frac{P_j G_{jj}}{P_c G_{cj} + \sum\limits_{j' \in D/j} P_{j'} G_{j'j} + \sigma^2}\right) - \lambda_j P_j G_{jB} \qquad (9)$$

It is observed from formula (9) that the utility function of each D2D consists of two parts: benefits and costs. If the transmission power of D2D users increases, the transmission rate will be accelerated and the benefits will also increase. However, with the increase of transmitting power, D2D users cause more interference to the base station. The base station adjusts the price according to the interference, D2D users will pay more cost to the base station for this communication, so we need to find an optimal response and find the optimal state between the cost and the total revenue. The D2D user optimization model can be expressed as

$$\begin{aligned} \max u_j\left(P_j, P_{-j}\right) \forall j \in N \\ s.t\, P_{\min} \le P_j \le P_{\max} \end{aligned} \qquad (10)$$

The ideal result of the game is to enter the equilibrium state and obtain the Nash equilibrium point, which means that any D2D user cannot improve his utility by changing his power while maintaining the current strategy. The game strategy of D2D user terminal j is to determine the appropriate transmission power and maximize the utility value of the terminal in the Nash equilibrium state, that is:

$$(\text{NPG}) : \max_{p_j \in P_j} u_j\left(p_j, p_{-j}\right), \forall j \in N \qquad (11)$$

For j, $p_j^* \in P_j$ is the optimal strategy when other terminals decide to choose $p_{-j}^* = \left\{p_1^*, \cdots, p_{j-1}^*, p_{j+1}^*, \cdots, p_n^*\right\}$ that is

$$p_j^* = \arg \max_{p_j \in P_j} u_j\left(p_j, p_{-j}^*\right), \forall j \in N \qquad (12)$$

Then strategy portfolio $p^* = \left\{p_1^*, \cdots, p^*, \cdots, p_n^*\right\}$ is a Nash equilibrium. The optimal policy of each terminal is the response function of other terminal policies, defined as:

$$r\left(p_{-j}\right) = \min\left(p_{\max}, p_j^*\right) \qquad (13)$$

The following is an analysis of the existence and uniqueness of two key properties of Nash equilibrium solutions.

3.2   Existence of Nash Equilibrium Solution

Theorem 1 (existence): Nash equilibrium exists in power control mode G.
Proof: since the pure policy space of terminal j is defined as $\{0 \le P_j \le P_{max}\}$, So it's a convex set on Euclidean space that satisfies the conditions of nonnullity, closure,

boundedness, and the utility function $u_j(p_j,\ p_{-j})$ is a continuous function of P_j, Take the first and second derivatives of the effect function $u_j(p_j,\ p_{-j})$ with respect to P_j.

$$\frac{\partial u_j}{\partial P_j} = \frac{1}{\ln 2} \frac{G_{jj}}{P_c G_{cj} + \sum_{j' \in D/j} P_{j'} G_{j'j} + \sigma^2 + P_j G_{jj}} - \lambda_j G_{jB} \tag{14}$$

$$\frac{\partial^2 u_j}{\partial P_j^2} = -\frac{1}{\ln 2} \left(\frac{G_{jj}}{P_c G_{cj} + \sum_{j' \in D/j} P_{j'} G_{j'j} + \sigma^2 + P_j G_{jj}} \right)^2 < 0 \tag{15}$$

The second derivative of the effect function $u_j(p_j,\ p_{-j})$ to P_j is less than 0, so $u_j(p_j,\ p_{-j})$is a concave function of P_j, which means that the non-cooperative game model in this chapter has Nash equilibrium.

3.3   Uniqueness of Nash Equilibrium Solution

Theorem 2 (uniqueness): Power control model G has a unique Nash equilibrium.
Proof: As stated in theorem 1, there is a Nash equilibrium point in the non-cooperative power control game, which is represented by $r(P)$.

Where $r_j(P_j)$ represents the optimal response set of gambler j, where $r_j(P_{-j})$ is the response set of other gamblers. The key to proving that the game model has only a unique Nash equilibrium point is to prove that the optimal response $r(P)$ is a standard function because there is a unique solution to the standard function.

1. positive nature

Take the first derivative of $u_j(p_j,\ p_{-j})$ versus P_j

$$\frac{\partial u_j}{\partial P_j} = \frac{1}{\ln 2} \frac{G_{jj}}{P_c G_{cj} + \sum_{j' \in D/j} P_{j'} G_{j'j} + \sigma^2 + P_j G_{jj}} - \lambda_j G_{jB} = 0 \tag{16}$$

We have:

$$P_j^* = \frac{1}{\ln 2\lambda_j G_{jB}} - \frac{P_c G_{cj} + \sum_{j' \in D/j} P_{j'} G_{j'j} + \sigma^2}{G_{jj}} \tag{17}$$

Because $r(P) > 0$ can get the constraint condition that interferes with the price factor is

$$\lambda_j \leq \frac{G_{jj}}{\left(P_c G_{cj} + \sum_{j' \in D/j} P_{j'} G_{j'j} + \sigma^2 \right) \ln 2G_{jB}} \tag{18}$$

At the same time, it also indicates that if the interference price is too high, D2D user j cannot complete the communication. At this time, the j that cannot complete the communication will be deleted from this communication.

2. monotony

Proof: when $P \geq P'$, there is

$$r(p) - r(p') = \frac{1}{\ln 2\lambda_j G_{jB}} - \frac{P_c G_{cj} + \sum\limits_{j' \in D/j} P_{j'} G_{j'j} + \sigma^2}{G_{jj}}$$
$$- \left(\frac{1}{\ln 2\lambda_j G_{jB}} - \frac{P_c G_{cj} + \sum\limits_{j' \in D/j} P'_{j'} G_{j'j} + \sigma^2}{G_{jj}} \right) \qquad (19)$$

We get $r(P) - r(P') < 0$, which is $r(P) < r(P')$, so the function is a minus function of the variable, the optimal response strategy vector that satisfies the monotonicity.

3. scalability.

Scalability requires that any $a > 1$ has $ar(P) > r(aP)$.
So we assume $\forall a > 1$, we can get

$$ar(p) - r(ap) = (a - 1) \left(\frac{1}{\ln 2\lambda_j G_{jB}} - \frac{P_c G_{cj} + \sigma^2}{G_{jj}} \right) \qquad (20)$$

So $ar(P) - r(aP) > 0$, we can get $ar(P) > r(aP)$, so the scalability is proved.

In summary, since the optimal response $r(P)$ meets the requirements of positivity, monotonicity and scalability, the optimal response function $r(P)$ is the standard function.

It also proves that there is a unique Nash equilibrium in the non-cooperative game model.

4   Power Distribution

According to the results of the game analysis above, It is known that each D2D user is communicating in a "non-cooperative" manner. All D2D users will choose the power distribution scheme of the value of the maximum utility function, and each D2D user's choice is restricted by the following conditions.

$$\max u_j \left(P_j, P_{-j} \right) \forall j \in N$$
$$s.t P_{\min} \leq P_j \leq P_{\max} \qquad (21)$$

The objective of this optimization model is to maximize the value of the utility function, and the variable p of this optimization problem is a convex set, and the utility function u is a concave function, so the $-u$ in this model is a convex function. Therefore, the optimization problem is a convex optimization problem. The following Lagrange function is constructed by the constraint condition:

$$J_j = -u_j(P_j, P_{-j}) + a_j P_j - \beta_j(P_j - P_{min}) - \gamma_j(P_j - P_{max}) \tag{22}$$

$$J_j = \left[-R_D + \lambda_j^* P_j G_{jB}\right] + a_j P_j - \beta_j(P_j - P_{min}) - \gamma_j(P_j - P_{max}) \tag{23}$$

Where a_j, β_j, γ_j, are Lagrange multipliers (non-negative real Numbers), which can be obtained by using K.K.T condition [18].

$$P_j \geq 0 \,\forall j \in D$$
$$a_j P_j = 0$$
$$P_j - P_{min} \geq 0 \tag{24}$$
$$\beta_j(P_j - P_{min}) = 0$$
$$P_j - P_{max} \leq 0$$

$$\gamma_j(P_j - P_{max}) = 0$$
$$\frac{\partial J_j}{\partial P_j} = -\frac{\partial R_D}{\partial P_j} + \lambda_j^* G_{jB} + a_j - \beta_j - \gamma_j = 0 \tag{25}$$

So $$\lambda_j^* = \frac{1}{\ln 2 G_{jB}} \frac{G_{jj}}{P_c G_{cj} + \sum\limits_{j' \in D/j} P_{j'} G_{j'j} + \sigma^2 + P_j G_{jj}} \tag{26}$$

After adjusting the interference price, we can use the iterative method to reach the Nash equilibrium state, and the best response of the user is as follows

$$p_j(p_{-j}) = \left[\frac{1}{\ln 2\lambda_j G_{jB}} - \frac{P_c G_{cj} + \sum\limits_{j' \in D/j} P_{j'} G_{j'j} + \sigma^2}{G_{jj}}\right]_0^{P_{max}} \tag{27}$$

In the formula $[x]_a^b = \max\{\min(x, b), a\}$.

Combined with the optimal price and the maximum interference tolerance of the base station, the following power control algorithm can be obtained:

1. Initialization $P_j = 0$, $\forall j \in D$ base station initialization price is λ^*.
2. Number of iterations: $t = 0$
3. for $j = 1$ to N users do
4. Each D2D user finds the optimal response that meets the conditions and works with the optimal solution P_j^*
5. The base station measures the total interference received $\sum P_j G_{jB}$
6. Calculate the total interference $P_c G_{cj} + \sum P_{j'} G_{j'j} + \sigma^2$ received by user j
7. Update prices based on interruptions
   if $\sum P_j G_{jB} < Q$ Base stations reduce interference price
8. Update the best response function P_j^*
9. $t = t + 1$
10. Loop5
11. until $\left(\left\|P_j^t - P_j^{t-1}\right\| / \left\|P_j^{t-1}\right\|\right) \leq \xi$ for all $\forall j \in N$
12. The base station detects the total disturbance value and calculates the benefit.
13. Get the optimal transmission power P_j^t

5 Simulation

In order to evaluate the influence of D2D resource allocation algorithm on the system performance, simulation was used to verify and analyze. Users are randomly distributed in the cell. The base station is located in the center of the cellular system. Cellular users and D2D users are randomly distributed in the area with the base station as the center of the circle. Simulation parameter Settings are shown in Table 1.

Table 1. Main parameter Settings for simulation

Name of parameter	Parameter values
Cell system radius	500 m
D2D user versus distance	15–30 m
Channel bandwidth	5 MHz
Number of sub-channels	5
Maximum transmitting power of base station	43 dBm
D2D user transmission power	[0, 24] dBm
White Gaussian Noise	−174 dBm/Hz

Figure 2 shows the curve of the relationship between the transmission power of D2D users and the number of iterations. and define the transmitted power change state at the 0.01% level for the stable convergence, in the process of simulation, with the increase of the number of iterations, the D2D to the transmission power is increasing, under the algorithm to adjust the user power eventually stabilised, convergence speed in step 10 or so, has the characteristics of rapid convergence, and at the same time if the initial power setting values closer to convergence, will further improve the convergence speed of the algorithm.

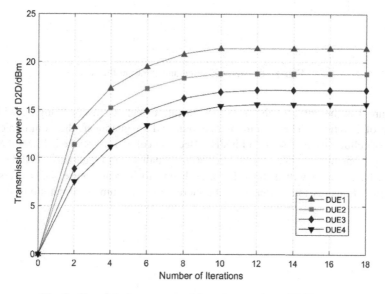

Fig. 2. Transmission power and iteration number of D2D users

Figure 3 shows the curve of the change of interference price with the number of iterations. Showed with the increase of the number of iterations, interference price convergence leveled off gradually, show that the iterative has better convergence. In addition, interference tolerance Q is larger, the faster convergence rate, the less the number of iterations required. With the decrease of the tolerance interference, Q convergence properties gradually become worse. Also you can see that the interference tolerance Q big system, interfere with the price will be relatively small in balance. This is because when the interference tolerance is large, the interference quota is sufficient, and the competition between D2D users is small, the interference price set by the base station will be reduced. In a system with small interference tolerance, when the interference quota is relatively rare, the competition between D2D users is large, and the interference price of the base station is relatively high in order to maximize the utility.

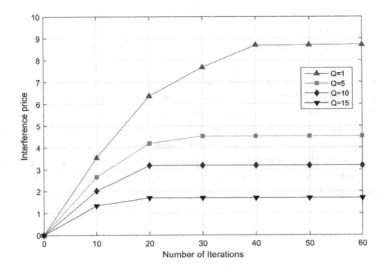

Fig. 3. Interference price and iteration times

Figure 4 is the curve of the transmission rate of cellular users changing with the number of iterations. The minimum rate required to guarantee the communication quality of cellular users is set to 5 bit/s/Hz. Because cellular users rate is lower than the minimum transmission rate, minimum service quality can't guarantee cellular subscribers, base station will disturb the price began to increase reuse, D2D users will reduce their transmission power, with the increase of the number of iterations, cellular user's transmission rate began to recover, the final stable near the minimum transmission rate, thanks to the D2D to communication, The rate of cellular users is significantly lower than it was at the beginning.

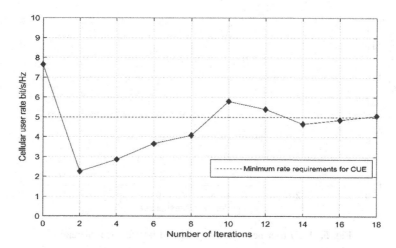

Fig. 4. Transmission rate and iteration number of cellular users

Figure 5 shows the curve of the total rate of D2D network with the limit of interference power. With the increase of the interference tolerance Q, the total speed of D2D network also increases. When the interference tolerance is small, the revenue of D2D network's total rate increases rapidly. When the interference tolerance is large, the total rate of D2D network increases slowly and even tends to be stable. And with the increase of the number of users, the total rate of D2D network is also increasing. This is because with the increase of the interference tolerance, more D2D users will be connected to the base station. Although the total rate of D2D network is increased, the interference amount is increased, and congestion may even occur, so the change trend of D2D network is slower. The total D2D network rate is positively correlated with the interference tolerance of the base station and the growth rate is gradually slow.

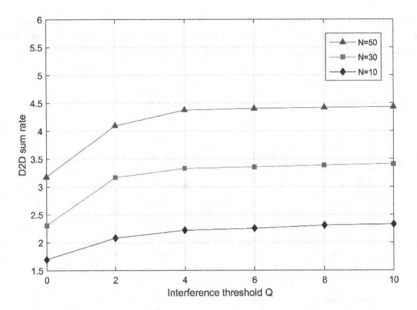

Fig. 5. D2D user network total rate and interference tolerance

6 Conclusion

A D2D communication resource allocation algorithm based on non-cooperative game theory is proposed to solve the problem of system interference when D2D users reuse cellular users' uplinked wireless resource communication. By pricing the interference of D2D users in the system, the uplinking power control problem in the hybrid network is transformed into a non-cooperative game problem, and the proposed power control algorithm not only maximizes the utility of D2D users, but also ensures that the system interference does not exceed the tolerance and has good convergence. Numerical simulation shows that the algorithm can effectively control the transmitting power of D2D users and effectively improve the total speed of D2D network.

References

1. Li, Y., et al.: SERS: social-aware energy-efficient relay selection in D2D communications. IEEE Trans. Veh. Technol. **PP**(99), 1 (2018)
2. Shi, L.: Wireless energy transfer enabled D2D in underlaying cellular networks. IEEE Trans. Veh. Technol. **PP**(99), 1 (2018)
3. Wang, J., Zhu, D., Zhao, C., et al.: Resource sharing of underlaying device-to-device and uplink cellular communications. IEEE Commun. Lett. **17**(6), 1148–1151 (2013)
4. Doppler, K., Rinne, M., et al.: Device-to-device communication as an underlay to LTE-advanced networks. Mod. Technol. Telecommun. **47**(12), 42–49 (2010)
5. Lei, L., Zhong, Z., et al.: Operator controlled device-to-device communications in LTE-advanced networks. Wirel. Commun. IEEE **19**(3), 96–104 (2012)

6. Jung, Y., Festijo, E., Peradilla, M.: Joint operation of routing control and group key management for 5G ad hoc D2D networks. In: 2014 International Conference on Privacy and Security in Mobile Systems (PRISMS). IEEE (2014)
7. Tehrani, M.N., Uysal, M., Yanikomeroglu, H.: Device-to-device communication in 5G cellular networks: challenges, solutions, and future directions. IEEE Commun. Mag. **52**(5), 86–92 (2014)
8. Hakola, S., Chen, T., et al.: Device-to-device (D2D) communication in cellular network - performance analysis of optimum and practical communication mode selection (2010)
9. Wang, F., et al.: Energy-efficient radio resource and power allocation for device-to-device communication underlaying cellular networks. In: 2012 International Conference on Wireless Communications and Signal Processing (WCSP). IEEE (2012)
10. Sun, H., et al.: Resource allocation for maximizing the device-to-device communications underlaying LTE-Advanced networks. In: IEEE/CIC International Conference on Communications in China-workshops. IEEE (2013)
11. Meshkati, F., et al.: A game-theoretic approach to energy-efficient power control in multi-carrier CDMA systems. IEEE J. Sel. Areas Commun. **24**(6), 1115–1129 (2006)
12. Wang, X., et al.: Joint scheduling and resource allocation for device-to-device underlay communication. In: Wireless Communications & Networking Conference. IEEE (2013)
13. Xu, C., Song, L., Han, Z., et al.: Efficiency resource allocation for device-to-device underlay communication systems: a reverse iterative combinatorial auction based approach. Sel. Areas Commun. (2013)
14. Lee, N., et al.: Power control for D2D underlaid cellular networks: modeling, algorithms and analysis. IEEE J. Sel. Areas Commun. **33**(1), 1–13 (2015)
15. Fodor, G., Reider, N.: A distributed power control scheme for cellular network assisted D2D communications. In: Proceedings of IEEE Global Telecommunications Conference, pp. 1–6 (2011)
16. Wang, F., Song, L., Han, Z., Zhao, Q., Wang, X.: Joint scheduling and resource allocation for device-to-device underlay communication. In: Proceedings of IEEE Wireless Communications and Networking Conference, Shanghai, China, pp. 732–737, April 2013
17. Guo, B., Sun, S., Gao, Q.: Graph-based resource allocation for D2D communications underlying cellular networks in multiuser scenario. Int. J. Antennas Propagat. **2014**, 1–6 (2014)
18. Zhang, R., et al.: Distributed resource allocation for device-to-device communications underlaying cellular networks. In: IEEE International Conference on Communications. IEEE (2013)

A Query Path Recommendation Method Supporting Exploratory Search Based on Search Goal Shift Graphs

Chao Ma and Yan Liang[✉]

Liaoning Shihua University, Funshun 113001, China
yanliang_chn@outlook.com

Abstract. Increasing new knowledge or learning about new skills is often the main factor that triggers search users to conduct exploratory searches. Due to the lack of knowledge of their problem domain, it is difficult to form a learning path about new knowledge or new functions in the minds of searchers in the early stage of performing exploratory search, often searching for unnecessary information, resulting in wasting search time, the search is inefficient. However, the current search system does not provide enough support to solve this problem. For this reason, this paper designs a query path recommendation method supporting exploratory search based on a search goal shift graph. In the initial stage of exploratory search, directly recommends a set of query paths for searchers based on the searcher's initial query, helping the searcher to find appropriate learning objects and build efficient learning paths, avoiding the search for irrelevant information, thereby shortening the search time and improving the search efficiency.

Keywords: Exploratory search · Query paths recommendation · Search goal shift graphs

1 Introduction

The main goal of exploratory search is learning and surveying [1], so the understanding of new knowledge or the learning of new skills is often the main motivation for search searchers to conduct exploratory search. Usually, such learning and investigation is not a simple question answer search, but through the continuous collection and analysis of relevant domain information, complete the construction and understanding of certain knowledge or concepts, and finally integrate it into one of their own knowledge systems. The process is not only about memorization of salient facts, but rather the development of higher-level intellectual capabilities [2]. However, before the searcher performs the exploratory search, the relevant domain knowledge is often few or even blank. Therefore, in the initial stage of the exploratory search, it is difficult to form a learning path about new knowledge or new skills, which often deviates from the task target. Some unnecessary information is searched and learned, resulting in a search time that is too long and inefficient. As in the following example:

© Springer Nature Singapore Pte Ltd. 2020
J. Qian et al. (Eds.): ICRRI 2020, CCIS 1336, pp. 34–45, 2020.
https://doi.org/10.1007/978-981-33-4932-2_3

A mother wants to take her children out of the country during the summer vacation to broaden their horizons. Since she has never been abroad before, this will be her first overseas trip. There is almost no plan in her mind, and even her destination is not determined. Therefore, she prepared to develop a reasonable travel plan by web search. This is a typical exploratory search task. In the early stage of the search, she opened her search process with the query "the best route to travel abroad". With the continuous browsing of the search results, she found that European culture is more intense, so she used the query "French Travel" to keep searching, and then she found that the Paris Orsay Museum may be very helpful for the expansion of the children's knowledge. She thus launched an in-depth search of the Musée d'Orsay with a series of queries such as "Paris Musée d'Orsay", "Orsay Gallery", "Matador Oil Painting". With the deep exploration of the Orsay Museum, she gradually felt that the current search content deviated from her mission goal – "Travel Plan", so she had to fall back to the initial query to re-find some new information related to the "Travel Plan".

From the above example, we can see that the "Orsay Gallery" and "Matador Oil Painting" queries are not very helpful to complete the search task – "Travel Plan". These queries are unnecessary. The reason for this problem is that the searcher did not know the process to achieve their goal, and did not generate a clear search idea in the mind to support future queries in the initial stage of the search. If the search system can provide searchers with a set of query paths related to search tasks in the early stage of search, help searchers to generate clear search ideas, such as "French Travel Guide" "Travel Visa Processing" "Ticket ordering". This problem will be solved very well.

However, the current query recommendation methods mainly focus on optimizing searchers' current query, and do not help searchers to directly build an efficient learning path which is far away from satisfying searchers' information needs of the whole search session. A recommendation system of exploratory search should propose a sequence of search goals in a well-defined order with a starting and ending point, rather than submit a sequence of unordered queries simply recommended because they are "similar" queries. We thus designed a query path recommendation method that learns from the behavior of many searchers to support new searchers engaged in exploratory search. Our recommendation method is based on a search goal shift graph. Firstly, we used the Dijkstra's shortest path algorithm to find all possible query paths in the graph. And then we used a greedy algorithm to select a set of paths from all the paths by maximizing a linear combination of similarity and diversity. Finally, we demonstrated the effectiveness of the method for exploratory search by comparing experiments with the other methods.

2 Related Works

Exploratory search is a type of information seeking and a type of sense-making focused on the gathering and use of information to foster intellectual development [3]. It is proposed by Marchionini to distinguish another well-known type of search activity:

lookup search. Lookup is the most basic kind of search [4] and refers to focused searches where the user has a specific goal in mind and also an idea of the expected result. A typical example would be a user wants to know who the current president of the United States is and looking for the information on the Web. In contrast to lookup search, the searcher may have a vague information need in an exploratory search system [5]. His goals are not necessarily well defined, neither are his the means to achieve them in the first place. For example, a user wants to know more about AI, he doesn't really know what kind of information he wants or what he will discover in this search session; he only knows he wants to learn more about that topic. Hence, the main goal in exploratory search is learning. The learning process is opportunistic, iterative, and multi-tactical [6].

Recently, exploratory search research focuses on the characteristics of the exploratory search process and the different types of support needed to help people make exploratory searches. Someone tries to construct an interactive Intent Modeling for Exploratory Search [7]. Some research efforts focus on traditional search techniques to support people engaged in exploratory search, such as query suggestions. For example, [8] employed a task graph to generate task recommendations for searchers to assist them in exploring different aspects of a topic, [9] proposed a topic-oriented query for exploratory search method, [10] designed a session-based concept suggestion model that supports information search by proposing concepts for query expansion. Other attempts have been made to design and research visual search interfaces to support exploratory search tasks. For example, [11] designed exploratory ranking interface; [12] used a negative feedback search intent radar interface to help users conduct exploratory search.

The above research work mainly focuses on refining user requirements, helping users to find the next search direction. But they do not pay attention to the confusion of users in the initial stages of exploratory search.

In contrast to these methods, our method mainly focuses on recommending a set of query paths that help searchers avoid searching for irrelevant information when they cannot construct a clear search path in the initial stage of exploratory search.

3 Paths Recommendation Framework

The basic framework of the query path recommendation method for exploratory search mainly consists of two parts, offline and online, as shown in Fig. 1:

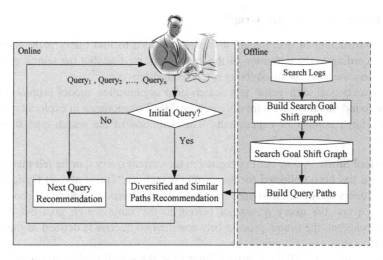

Fig. 1. Paths recommendation framework

1. Offline section: It mainly consists of two major steps: search goal shift graph building and query paths building. In the offline part, we construct a search goal graph. And then we use the Dijkstra's shortest path algorithm to find all possible query paths in the graph.
2. Online part also contains two steps, the initial query check and top-k recommend. In the online part, if the current query is the initial query, we will use a greedy algorithm to select a set of paths from all the paths that we found by maximizing a linear combination of similarity and diversity.

4 Query Paths Building

4.1 Related Definitions

1. **Search goal:** An atomic information need is reflected through a query or more queries. That is, two consecutive queries in the same session, and if they are similar to each other above a given thresh-old, they belong to the same search goal. The form is described as:

$$Same_{goal}(q, q') = \begin{cases} 1 & Sim(q, q') > threshold \; \theta \\ 0 & Sim(q, q') < threshold \; \theta \end{cases} \tag{1}$$

2. **Query path:** Suppose q1 is the initial query of the user, $U = (u_1, u_2, \ldots, u_n)$ is the search goal set. If there is a query sequence $\exists S = (q_1, q_2, \ldots, q_n)$, and each query represents a different search goal $q_i \in u_i \cup q_j \in u_j \cup u_i \neq u_j$, and there is a certain logical relationship between two adjacent search targets $u_i \overset{R}{\leftrightarrow} u_j$, the query sequence S is a query path related to the initial query q1.

4.2   The Search Goal Shift Graph

For the construction of query path, it is a simple idea to find query paths through a graph. According to the definition of the query path, we find that the search goal shift graph is very suitable for discovering query path discovery.

The search goal shift graph is a search goal organization model proposed by us [13]. We found that there are a lot of search goal shift phenomena in exploratory search through a large number of experiments. The definition of the search goal shift is as follows:

Search Goal Shift: A user isn't interested in the current query q or he fell that enough information has been collected for the current q such as "Travel abroad France", then submits a query q' for the next step search, such as "Passport application process". If the query q and the query q' do not belong to the same search goal but they are topically-coherent, the search process between the two queries is defined as the search goal shift.

Thus, we extracted all the queries submitted in the search goal shift processes from search engine logs and used the queries to build a search goal shift graph following the method designed in [13].

The search goal shift graph is a directed graph Ggs = (V, E, w) where:

1. The set of nodes is V = Q, Q is a distinct set of search goals;
2. $E \subseteq V \times V$ is the set of directed edges;
3. $w : E \to (0, 1]$ is a weighting function that assigns to every pair of search goals $\{g, g'\} \in E$ a weight $W(g, g')$.

The building process of the search goal shift graph is shown in Fig. 2 and 3.

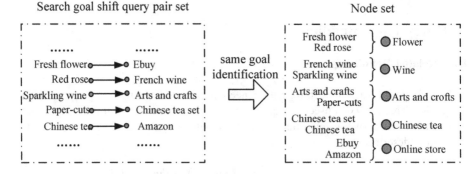

Fig. 2. Construction of node set.

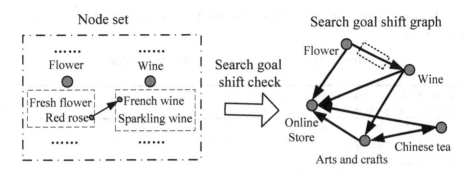

Fig. 3. Construction of edge set.

4.3 Query Paths Identification Algorithm

According to the definition of the query path and the search goal shift graph, if a node v as the initial goal of an exploratory search, all other nodes in the search goal shift graph may be the final goals for this exploratory search. Therefore, we define the shortest paths between node v and all other node are query paths for the node v.

Based on the above ideas, we adopt the Dijkstra algorithm [14] to find all the shortest paths between nodes in a graph as candidate recommendations.

Let the node at which we are starting be called the initial node. Let the distance of node Y is the distance from the initial node to Y. We assign some initial distance values and will try to improve them step by step.

1. Mark all nodes unvisited. Create a set of all the unvisited nodes called the unvisited set.
2. Assign to every node a tentative distance value: set it to zero for our initial node and to infinity for all other nodes. Set the initial node as current.
3. For the current node, consider all of its unvisited neighbors and calculate their tentative distances through the current node. Compare the newly calculated tentative distance to the current assigned value and assign the smaller one. For example, if the current node A is marked with a distance of 6, and the edge connecting it with a neighbor B has length 2, then the distance to B through A will be 6 + 2 = 8. If B was previously marked with a distance greater than 8 then change it to 8. Otherwise, keep the current value.
4. When we are done considering all of the unvisited neighbors of the current node, mark the current node as visited and remove it from the unvisited set. A visited node will never be checked again.
5. If the destination node has been marked visited (when planning a route between two specific nodes) or if the smallest tentative distance among the nodes in the unvisited set is infinity (when planning a complete traversal; occurs when there is no connection between the initial node and remaining unvisited nodes), then stop. The algorithm has finished.
6. Otherwise, select the unvisited node that is marked with the smallest tentative distance, set it as the new "current node", and go back to step 3.

The query paths identification algorithm based on the search goal shift graph as follows (Table 1):

Table 1. Algorithm for exploratory paths identification.

Input: the distance matrix between nodes A
Output: query paths Path[]

1.	**For** *i<n* **do**
2.	*Dis[i]=A[0][i];*
3.	*Path[i].add(v[0]);*
4.	**End for**
5.	*T={v[0]}*
6.	**While** *size(T)<n* **do**
7.	**Foreach** *Dis[i]* in *Dis* **do**
8.	*v[k]=find_min(v[0], Dis[i],Dis[]);*
9.	**For** *j<n* **do**
10.	**If** *Dis[k]+1≤Dis[j]* **then**
11.	*Dis[j]= Dis[k]+1*
12.	*Path[j].add(v[k]);*
13.	**End if**
14.	**End for**
15.	*T.add(v[k]);*
16.	*Dis.delete(v[k])*
17.	**End for**
18.	**End while**

5 Query Paths Recommendation

Our objective is to recommend a set of query paths related to the initial query that help the searcher generate a clear search idea in the early stage of exploratory search. So our recommendation paths should be both relevant and diverse.

Given a query paths set D and an initial query Q, our goal is to select top-k query paths ∂ that exhibits (i) high coverage of the result set of Q, and (ii) high diversity among the result set of ∂.

We refer to RQ as the results or the result set of Q. the result set of every query path p is always a subset of the result set of the initial query Q, i.e. $R_p \subseteq R_Q$ And we use coverage to measure the relevance between the recommended top-k paths and the initial query. The coverage of the query Q is defined as the number:

$$\text{cov}(\partial) = \left| \cup_{p \in Q} R_p \right| \qquad (2)$$

While the diversity between two queries p and p' is defined as:

$$dis(p, p') = \left| R_p \cup R_{p'} \right| - \left| R_p \cap R_{p'} \right|$$
$$= \left| R_p \right| + \left| R_{p'} \right| - 2 \left| R_p \cap R_{p'} \right| \tag{3}$$

Overall, the function we aim at maximizing is:

$$f(\partial) = \lambda \text{cov}(\partial) + (1 - \lambda) \sum_{P, P' \in \partial} dis(P, P') \tag{4}$$

where λ [0; 1] is a parameter that trades off between coverage and diversity. Therefore, the recommendation problem of the query paths is transformed into an optimal combination problem under a given constraint condition. That is, given a recommended number of conditions, how to select a set of exploration paths using the function $f(\partial)$ maximum.

$$\partial* = \arg \max f(\partial) \quad subject\ to\ |\partial| = k \tag{5}$$

For $\lambda = 0$, This problem corresponds to the well-known MaxiMumCoverge problem [15], which is known to be NP-hard. The problem can be solved by a greedy strategy, specific greedy algorithm as follows (Table 2):

Table 2. Algorithm for exploratory paths recommendation

Input: a query paths set D
Output: the top-k query paths
1. $\partial = \Phi$
2.   **While** $|\partial| < k$ **do**
3.   *Find* $P \in D - \partial$ max $w(\partial \cup \{p\})$
4. $\partial = \partial + P$
5.   **End While**
6.   **Return** ∂

6 Experiment

Our goal is to help search users generate a relatively clear search idea in the early stages of an exploratory search, rather than recommending the next search direction. Therefore, the traditional accuracy, error rate and recall rate methods could not meet our testing needs. In order to ensure the accuracy and objectivity of the evaluation results, we evaluated our recommendation method based on the user experiences.

We designed five exploratory search tasks similar to the "overseas travel planning" search task mentioned earlier, and then recruited 30 students from our university to participate in this experiment. All participants were divided into three groups on average, and performed the four exploratory search tasks designed in Sect. 3. The first group adopted the TOES method to support the user's exploratory search, the second group adopted the STES method and the third group adopted the GSES method. In addition, in order to ensure that the participants' own knowledge does not affect the exploratory nature of the tasks, we conducted a background survey of the participants to ensure that the participants were not experts or researchers in the topics of the search tasks.

6.1   Baseline Method

We compared our design recommendation method (GSPR) with the following two methods that also support exploratory search based on query recommendation.

1. Topic-Oriented Exploratory Search (TOES): TOES is a query recommendation method based on the topic semantic association graph. The topic semantic association graph has been built by hyperlinks on the Internet [8].
2. Exploratory Search Based on Search Task (STES). STES is a query recommendation method based on the task graph. The task graph is built by different search tasks that have been identified based on entities and syntactic structure patterns of queries from user logs [9].

6.2   Evaluation Metrics

1. **Recommendation utilization rate:** It indicates the proportion of recommended queries adopted by a user in a task.

$$Usage = \frac{Q_{recommend}}{Q} \times 100\% \tag{6}$$

where $Q_{recommend}$ represents the number of recommended queries adopted by the user. Q represents the number of all queries submitted by a user in a task.

2. **Search time:** It is the time spent by a user during the whole task.
3. **Exploring effect:** we used a concept map to evaluate how users change their knowledge structure as a result of exploratory search [16].

6.3   Experimental Results and Analysis

According to the comparison result in Fig. 4, the utilization rate of the method we designed is higher than the other two methods about 50%, while the comparison of Fig. 5 shows that the user's search time are less than the other two methods using our method, and the comparison results of Figs. 6, 7 and 8 show that the exploring effect produced by three different recommended methods is almost the same. This means that our method can better satisfy the user needs relative to the other two methods during exploratory search, and users can effectively shorten the search time to improve search efficiency using our method.

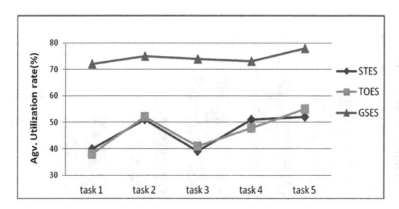

Fig. 4. Comparison result of utilization rate.

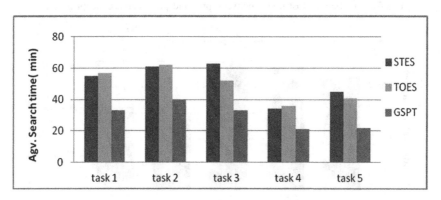

Fig. 5. Comparison result of search time.

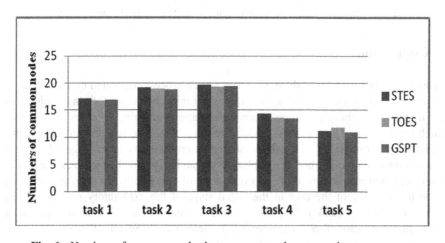

Fig. 6. Numbers of common nodes between pre- and post-search concept maps.

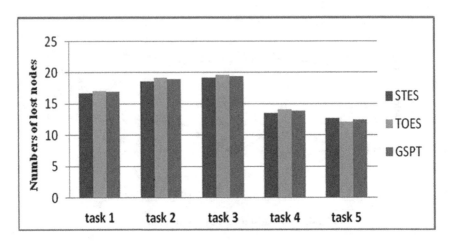

Fig. 7. Numbers of lost nodes between pre- and post-search concept maps.

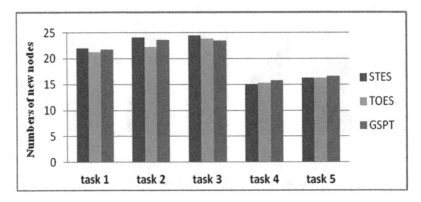

Fig. 8. Numbers of new nodes between pre- and post-search concept maps.

7 Conclusion

Through in-depth analysis of the exploratory search process, we found that it is difficult to form a relatively clear search idea in the user's mind before performing exploratory search, often searching for unnecessary information, making the entire exploratory search process full. The blindness leads to the user's long search time and the search efficiency is extremely low. In order to solve this problem, we have designed a search path recommendation method for exploratory search. The method utilizes the search target migration map to directly recommend a set of query paths for the user according to the initial query of the user in the initial stage of the exploratory search, thereby helping the user to quickly plan the learning content and avoid searching for those irrelevant information, thereby shortening the search time. Search efficiency.

References

1. Marchionini, G.: Exploratory search: from finding to understanding. Commun. ACM **49**(4), 41–46 (2006)
2. White, R.W.: Interactions with Search Systems. Cambridge University Press, Cambridge (2016)
3. White, R.W., Roth, R.A.: Exploratory search: Beyond the query-response paradigm. Synthesis Lect. Inf. Concepts Retrieval Serv. **1**(1), 1–98 (2009)
4. Athukorala, K., Głowacka, D., Jacucci, G., Oulasvirta, A., Vreeken, J.: Is exploratory search different? A comparison of information search behavior for exploratory and lookup tasks. J. Assoc. Inf. Sci. Technol. **67**(11), 2635–2651 (2016)
5. Marchionini, G., White, R.W.: Information-seeking support systems [Guest Editors' Introduction]. Computer **42**(3), 30–32 (2009)
6. White, R.W., Marchionini, G., Muresan, G.: Evaluating exploratory search systems. Inf. Process. Manage. **44**(2), 433–436 (2009)
7. Ruotsalo, T., Peltonen, J., Eugster, M.J., et al.: Interactive intent modeling for exploratory search. ACM Trans. Inf. Syst. **36**(4), 44.1–44.46 (2018)
8. Hassan Awadallah, A., White, R.W., Pantel, P., et al.: Supporting complex search tasks. In: ACM International Conference on Information and Knowledge Management, pp. 829–838. ACM, European (2015)
9. Sun, H., Jiang, C., Ding, Z., Wang, P., Zhou, M.: Topic-oriented exploratory search based on an indexing network. IEEE Trans. Syst. Man Cybern.: Syst. **46**(2), 234–247 (2016)
10. Mauro N., Ardissono L.: Session-based suggestion of topics for geographic exploratory search. In: Intelligent User Interfaces, pp. 341–352. ACM (2018)
11. Bespinyowong, R., Chen, W., Jagadish, H.V., et al.: ExRank: an exploratory ranking interface. Proc. VLDB Endow. **9**(13), 1529–1532 (2016)
12. Peltonen J., Strahl J., Floréen P.: Negative relevance feedback for exploratory search with visual interactive intent modeling. In: Intelligent User Interfaces, pp. 149–159. ACM (2017)
13. Ma, C., Zhang, B.: A new query recommendation method supporting exploratory search based on search goal shift graphs. IEEE Trans. Knowl. Data Eng. **11**(30), 2024–2036 (2018)
14. Mis, T.J., Frana, P.L.: An interview with edsger w. dijkstra. Commun. ACM **53**(8), 41–47 (2010)
15. Hochba, D.S.: Approximation algorithms for np-hard problems. ACM Sigact News **28**(2), 40–52 (1997)
16. Egusa, Y., Saito, H., Takaku, M., Terai, H., Miwa, M., Kando, N.: Using a concept map to evaluate exploratory search. In: 3th Symposium on Information interaction in context, pp. 175–184. ACM (2010)

Research on Image Super-Resolution Reconstruction Based on Generative Countermeasure Network

Rongzhao Jia and Xiaohong Wang[(✉)]

Liaoning Shihua University, Fushun 113001, China
wangxiaohong3815@sina.com

Abstract. Image super-resolution reconstruction technology is an important image processing technology used to improve image quality and video resolution in the field of computer vision. Nowadays, the main methods of super-resolution reconstruction based on convolutional neural network (CNN) are mostly to get higher peak signal-to-noise ratio (PSNR), which will produce many other problems, such as, the visual sensory effect of the image is not ideal, the gradient back propagation is difficult, the training effect of long training time is not ideal, the image lacks some high-frequency information and texture details, etc. In this paper, an improved image super-resolution reconstruction method based on generative multi anti network (GAN) is proposed. For the traditional generative adversary network, a VGG19 network assistant generator and discriminator model training are added, and the loss function is improved. Combined with the GAN adversary loss, the feature loss of the auxiliary VGG19 network is added. Experimental results show that compared with the existing methods, this method improves the overall visual effect of the reconstructed image, restores more realistic texture effect and higher MOS score under the condition of ensuring PSNR.

Keywords: Computer vision · Super-resolution reconstruction · Generative countermeasure network

1 Introduction

1.1 Research Background

Human beings obtain information mainly through vision. Image plays an important role in the process of information transmission. Compared with the ordinary definition image, the detail features of high-resolution image will be better, and it is more conducive for people to extract the key information to obtain the corresponding knowledge. In general, super-resolution [1] is defined as a process of obtaining a high-resolution image from a low-resolution image, which improves the resolution of the original image by means of software or hardware. Super-resolution has a relatively simple classification method, according to the number of input low-resolution images, the image super-resolution method can be classified into single image super-resolution and multi image super-resolution. Single image super-resolution is more challenging

© Springer Nature Singapore Pte Ltd. 2020
J. Qian et al. (Eds.): ICRRI 2020, CCIS 1336, pp. 46–62, 2020.
https://doi.org/10.1007/978-981-33-4932-2_4

than multi image super-resolution [2], because it needs to learn the relationship between them to visualize the low-resolution image of self-training set The details that are lost compared to high-resolution images. Image super-resolution reconstruction is widely used in military, medical [3] and public security fields.

1.2 Related Work

With the development of deep learning technology [4], more work will be applied to convolutional neural network [5] and generate confrontation network [6] or other deep learning models to perform single image super-resolution reconstruction. However, these methods are usually affected by many other factors. Compared with the vision of real high-resolution image and the high-resolution image generated by various depth learning technologies, many problems will be found. Most depth learning methods cannot avoid the visual effect of the deviation between the reconstructed image and the real image, because the feature images they generate are difficult to get from the context information Obtain the unique features of the input low resolution image. Moreover, the deep learning model is sensitive to the change of super parameters, and the training difficulty caused by the instability of the network during the training will bring certain difficulty to the super-resolution reconstruction task.

The methods of image super-resolution reconstruction can be divided into three categories, interpolation based, reconstruction based and learning based [7]. The method based on interpolation is simple and fast, but it is easy to produce fuzzy jagged edges, and the image perception effect is poor; the edge of the method based on reconstruction is relatively smooth, but the algorithm is complex and inefficient; the method based on learning adds prior information, which is greatly affected by the samples, and the efficiency is also low. This paper mainly uses the method based on learning. The basic idea of learning based method is to represent the image super-resolution reconstruction as a nonlinear mapping from low-resolution image to high-resolution image, and then get an approximate mapping function through machine learning algorithm under supervised learning. Many domestic research institutes and universities have studied the theory and algorithm of image super-resolution technology. Although the whole domestic research in this field started late, there is still good progress. Among them, the most prominent ones are the method based on deep convolution neural network proposed by Dong Chao, Tang Xiaoou, he Kaiming and others in the Department of information engineering of the Chinese University of Hong Kong [8] in 2014. SRCNN (2015 TPAMI) [9] proposed by Dong Chao and others is the first neural network model that applies deep learning technology to image super-resolution reconstruction. It can be divided into three steps,image block extraction and feature representation, feature nonlinear mapping and final reconstruction. Compared with the traditional methods, SRCNN has a very good reconstruction effect, especially in the aspect of double magnification, and inspired other methods based on convolutional neural network, mainly including pixel recursive super-resolution (PRSR) and perceptual loss [10]. However, although SRCNN has a high PSNR (peak signal-to-noise ratio), the resulting images usually lack high-frequency details, and the sensory effect is relatively general, which can not meet the expected fidelity in high-resolution.

At present, deep neural network has shown great potential in the field of image super-resolution reconstruction, but there are still many problems to be solved. Compared with the shallow model, the deep neural network can automatically learn the structure of the feature layer, and its powerful learning ability can more accurately restore more rich and detailed high-resolution content. However, it is still a very challenging task to train deep network for image super-resolution reconstruction. The existing methods based on the depth neural network have some shortcomings. For the image super-resolution reconstruction, the depth learning method may not use the pooling layer sometimes, so when the depth of the network increases, more parameters will be introduced, so it is easy to have the problems of over fitting, the model occupies too much space, and it is difficult to store and reproduce. Therefore, most of the existing methods are inspired by the residual prediction method and use relatively shallow networks. Kim et al. Trained the deep convolution neural network to learn the residual between the high-resolution image and the low-resolution image magnified by the bicubic [11] interpolation method, and combined the magnified low-resolution image with the predicted residual to obtain the reconstructed image. Therefore, we can explore the method of combining deep network and shallow network training, in which the shallow network can train stably, while the deep network can ensure accurate high-resolution image reconstruction. Another disadvantage of the existing depth neural network based methods is that the up sampling operation is performed in the image domain, which does not make full use of the layered features of the original low-resolution image, so the performance is relatively low. In view of this, Zhang et al. Of Northeastern University of the United States proposed a residual dense network RDN (2018 CVPR) [12] to make full use of the hierarchical characteristics of all convolutions.

There are two directions of super-resolution reconstruction, one is to recover the real and reliable details, and the application scenes such as super-resolution reconstruction on medical images, low-resolution camera face or shape recovery and other scenes that require strict details. The other is to pursue the overall visual effect, with low requirements for details. For example, the restoration of low resolution video TV, the restoration of camera blurred image and so on. This paper pursues the second direction. In the generation model of the combination of current deep learning technology and image super-resolution reconstruction algorithm, the visual effect of the image generated by the super-resolution model based on the generative countermeasure network is better. In order to obtain a higher PSNR, the traditional model based on the loss function of mean square error (MSE) [13] is solved. However, due to the ill posed problem of image super-resolution reconstruction and the defects in the structure design of the generated model, how to present the detailed features of the generated image more truly and clearly has become an important part of the current generated countermeasure network model to be improved. Compared with the existing methods, this paper changes the loss function to replace the traditional convolution neural network loss of mean square error with the generated network loss of confrontation and content, and transforms the content of the traditional pixel space into the similarity of confrontation property; introduces the depth residual network [14] to extract the detail texture in the image. In addition, because the current image quality evaluation standards are mainly aimed at the traditional methods and the super-resolution model reconstruction results based on convolutional neural network, the reconstruction results of the generation model such as the generation countermeasure network are not fully applicable, which will lead to the inconsistency between the subjective and objective quality evaluation results of the generated image. The reconstruction effect of SRGAN [15] model is far better than other methods, but it is the worst in the traditional quality evaluation standard. Therefore, whether other indicators should be considered for evaluation has become a question to be explored. Based on these problems, this paper studies image super-resolution reconstruction technology.

2   Network Model Design

2.1   Generative Countermeasure Network

The idea of generative adversary network originates from the idea of game theory [16], which is an unsupervised learning method. It is a machine learning technology by making two neural networks play games with each other. This method was proposed by Ian goodsell et al. in 2014. These two networks are called "generators" and "discriminators". The training process is shown in the Fig. 1.

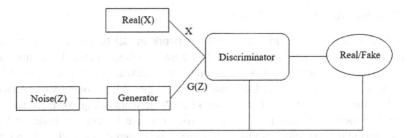

Fig. 1. The flow chart of alternating training between generating network and discriminating network for general generative countermeasure neural network.

The goal of the generator is to generate data that is indistinguishable from the training data set. The goal of the discriminator is to correctly determine whether a particular example is real (that is, from a training dataset) or fake (that is, created by the generator). Usually starting from the vector of random numbers, it indirectly learns to produce realistic examples by receiving feedback from the decision of the discriminator. When the discriminator is deceived to classify the false image as the real image, the generator knows that it does well. When the discriminator rejects the false image generated by the generator, the generator knows that it needs to be improved. The discriminator is also improving, for each category it makes, it gives feedback whether the guess is correct or not. Therefore, as the generator generates more realistic data, the discriminator can better improve its identification ability. Both networks improve at the same time. The goal of training is that when the false examples generated by the generator cannot be distinguished from the real data, the generator and the discriminator reach their Nash equilibrium, and the discriminator can at most randomly guess whether the examples are real or false.

Thers is the objective function of GAN.
Discriminator

$$\max_{D} V(D, G) = E_{x-P_{data}(x)}[\log(D(x))] + E_{z\sim P_z(x)}[\log(1 - D(G(z)))] \qquad (1)$$

Generator

$$\min_{G} V(D, G) = E_{z\sim p_z(z)}[\log(1 - D(G(z)))] \qquad (2)$$

Where X is the real sample, D (x) is the probability that x is judged as the real sample by the discriminator, Z is the sample of the input generator, G (z) is the sample generated by the generator, and D (g (z)) is the probability that the sample generated by the generator is judged as the real sample after the discriminator model. Goodsell theoretically proves the convergence of the algorithm, and when the model converges, the generated data has the same distribution as the real data (ensuring the model effect). Therefore, the role of generative adversary network is to add an additional discriminator network and two losses (discrimination loss and generation discrimination loss), and train two networks in an alternative training way.

2.2   Network Model of This Paper

The method of this paper is to optimize the network model based on the generative counterwork network, using the deep residual network (Resnet) and skip connection, and optimizing the objective function, adding a new auxiliary network VGG19 [17] network and a new feature loss. Combined with the generative counterwork network loss and feature loss, compared with the SRCNN network using MSE as the loss function, the method of this paper The recovered image is more realistic, and the generated image is more similar to the target image in semantics and style. The process of up sampling is put in the residual network, which greatly reduces the parameters and improves the training efficiency. The network model is shown in Fig. 2.

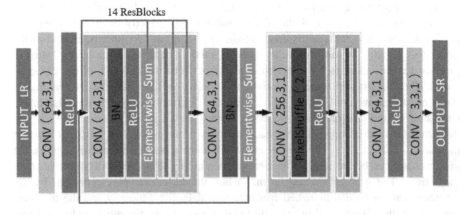

Fig. 2. Generator network. Network with corresponding kernel size, number of feature maps and stride indicated for each convolutional.

As shown in Fig. 2, the Generator network has 14 residual blocks of the same layout. We use three rollup layers, a small 3 × 3 kernel and 64 feature maps, then batch standardization layer and parametricrelu [18] as activation functions. Two sub-pixel convolution layers are trained to increase the resolution of the input image. Specifically, the generated network consists of three parts. After the low resolution image enters, it will pass through a convolution + relu function; then it will pass through 14 residual network structures. Each residual network is activated by convolution layer, BN layer and relu, then connected by a residual edge jump, then connected by a residual edge jump after being activated by convolution layer and relu; then connected with a residual edge jump after being connected by a convolution layer and a residual edge jump; finally, it will enter the upper sampling part, and connect After two times of up sampling, the image becomes 4 times of the original, and the resolution is improved. The first two parts are used for feature extraction, and the third part is used to improve the resolution.

In order to distinguish the real HR image from the generated SR sample, we train a discriminator network. The architecture is shown in Fig. 3. We use leakyrelu to activate (α = 0.2) and avoid maximum pooling across the network. It consists of eight convolution layers, among which the number of 3 × 3 filter cores increases. With the deepening of network layers, the number of features increases and the size of features decreases. The activation function is leakyrelu. Finally, the probability of natural image prediction is obtained by two full connection layers and the final sigmoid activation function. Discrimination network is composed of convolution and leakyrelu and standardization.

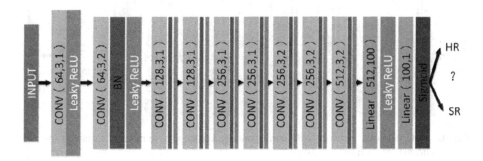

Fig. 3. Discriminator network. Network with corresponding kernel size, number of feature maps and stride indicated for each convolutional.

The residual structure mainly solves the problem of gradient disappearance. The multiplication of error back propagation [19] is the root of gradient disappearance. Adding one path is equivalent to adding another addition term, which can be returned through the added path, equivalent to adding 1, and the signal is strengthened to a certain extent. The residual network contains the possibility of numerous shallow networks, which can be transformed into shallow networks through training. So when there are many stacked networks, the result will be much better if the residual module is introduced. The introduction of the residual module represents a deeper network layer, which is more helpful to extract the detail texture of the image. BN [20] layer is batch standardization, which can translate and scale the distribution. The standardization of each layer can resist the disappearance of gradient and make parameter learning easier. Batch standardization is equivalent to data enhancement, because each random sampling is different, and the data becomes more and more to prevent over fitting. BN can greatly improve the speed and accuracy.

VGG19 network is to take the same feature map on a certain convolution layer, compare the differences on the feature map, so that the image details can be reconstructed better. Its network structure is shown in the Fig. 4. The traditional training of GAN discriminator network is to compare the real high-resolution image with 1, and generate the image with 0. We have improved it, the real high-resolution image and the false high-resolution image are introduced into the discrimination model through the feature map extracted by VGG19 network, and the result of the real high-resolution image is compared with the probability of determining that it is the generated image to get loss. Comparing the result of false high-resolution image feature map with the probability of true image, the loss is obtained. Use the obtained loss to train; train the generation model, transfer the low-resolution image into the generation model, get the high-resolution image, use the high-resolution image to get the discrimination result and compare with 1 to get the loss. The real high-resolution image and the false high-resolution image are introduced into the VGG network to obtain the features of the two images, and the loss is obtained by comparing the features of the two images.

Fig. 4. VGG network. Network with corresponding kernel size, number of feature maps and stride indicated for each convolutional.

2.3   Loss Function

The most important thing of a network is to specify an optimization target. The traditional task of image super-resolution reconstruction using convolutional neural network is based on the mean square error of pixel level as the optimization objective. The objective function used in this paper increases the loss of feature content of VGG19 network and the loss of Countermeasures of generative countermeasures network. That is, G loss = VGG loss + GAN loss; D loss = GAN loss. MSE loss is the minimum mean square deviation of image pixel space; GAN loss is the loss of generator/discriminator part defined by GAN model; VGG loss is the VGG loss based on the relu activation layer of pre-trained 19 layer VGG network, and the Euclidean distance between the generated image and the feature representation of reference image is calculated. A feature map of a certain layer is proposed on the trained VGG, and the feature map of the generated image is compared with that of the real image. The definition of loss function LSR is very important to the performance of our generator network.

MSE is the most widely used optimization target of image super-resolution reconstruction algorithm, and many methods rely on it. However, the solution of MSE optimization problem usually lacks high-frequency content while realizing extremely high PSNR. Our solution is not only to rely on pixel loss, but also to use a loss function closer to perceptual similarity. We define the VGG loss according to the relu activation layer of the pre trained 19 layer VGG network in Simonyan and Zisserman. As mentioned above, unlike the traditional GAN model discriminator D, we estimate the possibility that an input image is real and a generated image, and that the discriminator attempts to predict the probability that the real image x_r is more real than the false image x_f. We replace the traditional discriminator with the improved discriminator, which is expressed as D_{R_a}. The discriminator in GAN can be expressed as,

$$D(x) = \sigma(C(x)) \tag{3}$$

σ is sigmoid function and $C(x)$ is output of non transform discriminator. The improved discriminator can be expressed as follows,

$$D_{R_a}(x_r, x_f) = \sigma\left(C(x_r) - IE_{X_f}\left[C(x_f)\right]\right) \tag{4}$$

Where IE_{X_f} represents the operation of averaging all false data in mini batch processing. The discriminator loss can be defined as,

$$L_D^{Ra} = -E_{x_r}\left[\log\left(D_{Ra}(x_r, x_f)\right)\right] - E_{x_f}\left[\log\left(1 - D_{Ra}(x_f, x_r)\right)\right] \tag{5}$$

The counter loss of generator is in a symmetrical form,

$$L_G^{Ra} = -E_{x_r}\left[\log\left(1 - D_{Ra}(x_r, x_f)\right)\right] - E_{x_f}\left[\log\left(D_{Ra}(x_f, x_r)\right)\right] \tag{6}$$

Where $X_f = G(x_i)$ and x_i represent LR image input. The counter loss of generator includes x_r and x_f. The advantage of our generator is that it is suitable for the gradual change of the data generated in the confrontation training and the actual data, while in the traditional GAN, only the generated part takes effect. In the next part of the experiment, we can see that the improved loss function can help to learn clearer edges and more delicate textures.

3   Experimental Analysis

3.1   Evaluating Indicator

The evaluation indexes used in this paper are PSNR, SSIM and MOS. The peak signal to noise ratio (PSNR) unit is dB, which is the most common and widely used index to evaluate image quality. It is generally defined by mean square error (MSE). The mean square error of X and y of two m × n images is defined as,

$$MSE = \frac{1}{MN} \sum_{i=1}^{M} \sum_{j=1}^{N} [X(i, j) - \bar{X}(i, j)]^2 \qquad (7)$$

For RGB color image with three channels, each pixel has three RGB values, and its mean square error is defined as the sum of variance of all values divided by the image size and then divided by 3. Peak signal to noise ratio (PSNR) is defined as,

$$PSNR = 10\log_{10}\left[\frac{(2^n-1)^2}{MSE}\right] \qquad (8)$$

Each sampling point is represented by n-bit linear pulse code modulation, that is, the possible maximum pixel value in the image is 2N − 1. At present, peak signal-to-noise ratio (PSNR) is the most commonly used evaluation index for SR quality evaluation. The higher PSNR is, the higher image reconstruction quality is. However, it is also mentioned in the previous article that PSNR has its limitations. A high PSNR does not mean that the visual effect of the image is better. As shown in Fig. 5, the PSNR value of figure a is higher than that of figure B, but figure a is obviously more fuzzy and its visual effect is worse than that of figure B. The reason for this phenomenon is that the training target of a-map is MSE, which also shows that PSNR has great limitations in evaluating the visual effect of super-resolution reconstruction.

Fig. 5. Comparison of different super resolution reconstruction algorithms and Peak signal to noise ratio (PSNR) values.

In addition to peak signal noise, another commonly used evaluation index is structural similarity (SSIM). SSIM is a measure of similarity between two images. From the perspective of image composition, structure information is defined as independent of brightness and contrast, reflecting the attributes of object structure in the scene, and distortion is modeled as a combination of brightness, contrast and structure. The mean value is used as the estimation of brightness, the standard deviation as the estimation of contrast, and the covariance as the measurement of structural similarity.

The range of structural similarity is 0 to 1. When two images are as like as two peas, the value of SSIM is equal to 1. Similar to the peak signal-to-noise ratio (PSNR), SSIM can reflect people's subjective feelings and the visual effect of images, but it still has limitations in reflecting the visual effect of images.

The third commonly used evaluation index is subjective quality score (MOS), which is the most representative subjective evaluation method of image quality, and can directly reflect the evaluation of image quality by human eyes. The subjective quality evaluation uses the absolute classification scoring standard to map the poor and good scores to the number between 1–5. In the subjective quality evaluation test, the final score is calculated as the arithmetic mean value of the single score of the given image by the n-digit tester. As shown in Table 1.

Table 1. Subjective quality scoring table

Score	Quality
5	Best
4	Good
3	Average
2	Poor
1	Very bad

3.2 Experimental Environment and Data Set

The experimental environment of this paper is NVIDIA geforce GTX 1050, and the language is python 3.6.5, frame is pytorch1.2.0.

The data set used in the experiment is DIV2k, which is a newly released high-quality image data set for image restoration task, including 800 training pictures, 100 verification pictures and 100 test pictures. In addition, set5 and set14 standard datum data sets (including 5 and 14 images respectively) are added to further compare the model performance. All the experiments were carried out between the low resolution image and the high-resolution image with a scale factor of 4 times. The mean value of the weight of the network model and the discrimination model was 0, and the variance was 0.02. The Gaussian distribution is initialized randomly, and the error back propagation adopts the random gradient descent algorithm Adam, which is set as beta 0.9, the initial learning rate is 0.0001, the initial decay rate is 0.1, minimum batch size 16. In the training process, the generation model and the discrimination model are updated alternately. The experimental results are compared with VDSR, LapSRN, SRCNN, bicubic, ESPCN, SRGAN, ESRGAN models.

3.3   Experimental Process and Result Analysis

The whole network model and training process have been described in detail above. Generally speaking, the anti network architecture is the continuous improvement of the generator to cheat the discriminator, so as to improve the perception quality of SR image. This is not the same as the image generated using MSE optimization. We use two convolution layers with a small 3×3 kernel and 64 feature maps, and then use BN layer and ParametricReLU as activation functions. The discriminator network uses the leakyrelu activation function to avoid using the max pooling layer. It contains 10 volume layers and increases its kernel from 64 to 512. The 512 feature graphs are composed of two fully connected layers and the last sigmoid function to get the probability of final sample classification. The VGG19 network first transfers the image into the VGG19 network, which has been pre trained. First, the image subtraction operation (each color channel subtracting its own mean value) is carried out, otherwise the result of the image will be different. Then combine the three channels to form a new input layer image after conversion, and take the feature map of the last layer, in order to pay more attention to the main part of the image. The main parameters of each convolution layer of the discrimination network are shown in Table 2.

Table 2. Discriminator network structure

Convolution layer	kernel_size	in_channels	out_channels	Stride
Conv0	3×3	512	64	1
Conv1	3×3	64	64	2
Conv2	3×3	64	128	1
Conv3	3×3	128	128	1
Conv4	3×3	128	256	1
Conv5	3×3	256	256	2
Conv6	3×3	256	512	1
Conv7	3×3	512	512	1
Conv8	3×3	512	512	1
Conv9	3×3	512	512	2

In the above experiments Three point two Under environment and dataset. During the training, the training set image is sampled four times as the input low resolution image of the network, the convolution kernel is used 3×3, and the number of iterations is 500000. The Fig. 6 shows the changes of PSNR value in 200000 items before the training process using DIV2K data set. It can be seen that the PSNR value of this improved model has little change, but the effect of image reconstruction is obviously getting better and better, which proves that the traditional quality evaluation standard is not fully applicable to image super-resolution reconstruction. As shown in Fig. 7.

Next, choose "baboon", "face" and "Lenna" pictures respectively, and compare them with SRCNN, VDSR, bicubic and LapSRN models respectively. The effect is shown in Fig. 8, Fig. 9 and Fig. 10.

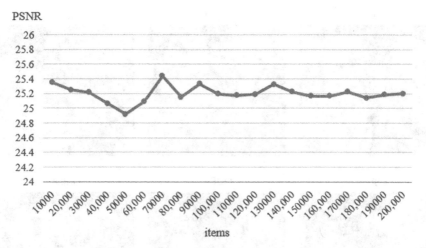

PSNR

items

Fig. 6. Using DIV2K data set to train the network, the change chart of PSNR score of the first 200000 items.

Items = 50000   Items = 100000

Items = 150000   Items = 200000

Fig. 7. Using DIV2K data set to train the network, comparison of effect pictures of some items in the process of network training.

SRCNN   VDSR   Bicubic

LapSRN   OURS   Origin

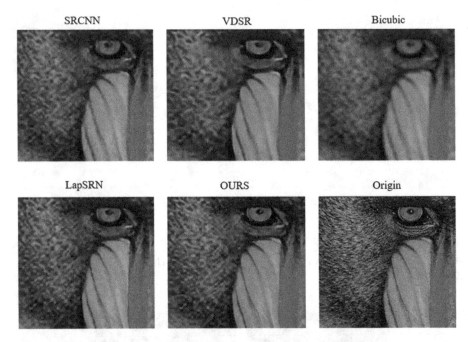

Fig. 8. The picture of the "baboon" compared with SRCNN, VDSR, bicubic and LapSRN models respectively.

SRCNN   VDSR   Bicubic

LapSRN   OURS   Origin

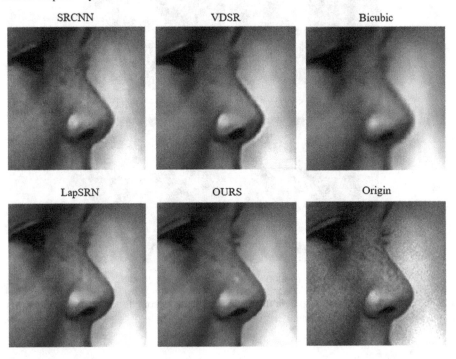

Fig. 9. The picture of the "face" compared with SRCNN, VDSR, bicubic and LapSRN models respectively.

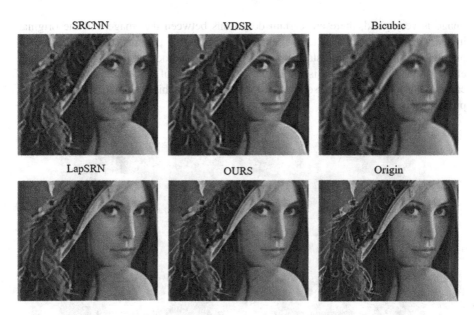

Fig. 10. The picture of the "Lenna" compared with SRCNN, VDSR, bicubic and LapSRN models respectively.

We use the subjective quality score (MOS) to evaluate the above images respectively. The experimental results of taking the average value of 50 test samples of each reconstruction model are shown in Table 3. As for 50 samples, the reason to choose the right number of samples is to ensure that the number of samples is less.

Table 3. Subjective quality scoring table.

	SRCNN	VDSR	LapSRN	Bicubic	OURS
Baboon	3.72	3.85	3.87	3.84	3.89
Face	3.69	3.86	3.91	3.89	3.91
Lenna	3.75	3.9	3.92	3.87	3.93
Average	3.72	3.87	3.9	3.87	3.91

It can be seen from the subjective quality score that the method used in this paper has better visual effect, which also proves that higher PSNR does not mean better visual sensory effect of image. As shown in Fig. 11, Here are some examples of other test images. The left image of each group is the first resolution image, the middle image is the high-resolution image, and the right image is the original image.

It can be seen that the reconstructed image has more high-frequency information and texture details, and the visual effect of the whole image has been very good, but some shortcomings have also been found, if the resolution of the original image is too low, the image is too fuzzy, although the visual sensory effect of the reconstructed image is very good, there are certain deviations between the image and the original image, such as the scarf pattern of the second group of pictures, the third group of pictures Picture shelf Book stacking. However, this method pursues the overall visual effect of the image, which is applied to the recovery of low-resolution video TV, camera blurred image and so on. It does not require high details, so these small problems are acceptable.

Fig. 11. Effect test drawing. The left image is the resolution image, the middle image is the high-resolution image, and the right image is the original image.

4   Conclusion

In order to make the image reconstruction model based on deep learning get more high-frequency information and texture details, and improve the visual effect of super-resolution reconstruction image, this paper proposes an image super-resolution reconstruction method based on generative countermeasure network based on the basic ResNet and SRCNN model, using Rrenet, optimizing the MSE loss in the past. In this paper, a new loss function is optimized based on the generative antagonism neural network. Here, we use the loss calculation of feature mapping based on VGG network to replace the content loss based on MSE, which is more sensitive to the change of pixel space and improves the efficiency of operation. Through the extensive MOS test of the image from the common benchmark data set, it is confirmed that the reconstructed image in this method has better visual effect and faster test time under the condition of ensuring good PSNR and SSIM values. Through a series of experiments, the effectiveness of this method is proved. The future research work will mainly focus on how to reduce the training parameters and network structure while ensuring the reconstruction quality; in addition, although the subjective quality score (MOS) is more intuitive to reflect the image reconstruction quality, it will cost a lot of manpower, while the traditional peak signal-to-noise ratio (PSNR) and structural similarity (SSIM) evaluation indicators can only be used as reference factors, which are not completely suitable It is necessary to develop a more practical evaluation index for super-resolution reconstruction. The future goal is to find a method that can not only ensure that the image has a higher PSNR value, but also has a better visual effect.

References

1. He, Y., Chen, S., Wei, J.: Assessment method of image super resolution reconstruction based on local similarity. Internet Multimed. Comput. Serv. (2013)
2. Damkat, C.: Single image super-resolution using self-examples and texture synthesis. Signal Image Video Process. **5**, 343–352 (2011)
3. Umehara, K., Ota, J., Ishida, T.: Application of super-resolution convolutional neural network for enhancing image resolution in chest CT. J. Digit. Imaging **31**, 441–450 (2018)
4. Shang, C., Yang, F., Huang, D., Lyu, W.: Data-driven soft sensor development based on deep learning technique. J. Process Control **24**, 223–233 (2014)
5. Malon, C.D., Cosatto, E.: Classification of mitotic figures with convolutional neural networks and seeded blob features. J. Pathol. Inf. (2013)
6. Leinonen, J., Guillaume, A., Yuan, T.: Reconstruction of Cloud Vertical Structure With a Generative Adversarial Network. Wiley, Hoboken (2019)
7. Bose, N.K., Ng, M.K., Yau, A.C.: A fast algorithm for image super-resolution from blurred observations. EURASIP J. Adv. Signal Process. **2006**(1), 1–14 (2006). https://doi.org/10.1155/ASP/2006/35726
8. Dong, C., Loy, C.C., He, K., Tang, X.: Learning a deep convolutional network for image super-resolution. In: Fleet, D., Pajdla, T., Schiele, B., Tuytelaars, T. (eds.) ECCV 2014. LNCS, vol. 8692, pp. 184–199. Springer, Cham (2014). https://doi.org/10.1007/978-3-319-10593-2_13
9. Dong, C., Loy, C.C., He, K., Tang, X.: Image super-resolution using deep convolutional networks. IEEE Trans. Pattern Anal. Mach. Intell. (TPAMI) **38**, 295–307 (2015). Preprint
10. Bi, S., Rancher, C., Johnson, E., McDonald, R., Jouriles, E.N.: Perceived loss of social contact and trauma symptoms among adolescents who have experienced sexual abuse. J. Child Sexual Abuse **28**, 333–344 (2018)
11. Xia, P., et al.: Performance comparison of bilinear interpolation, bicubic interpolation, and B-spline interpolation in parallel phase-shifting digital holography. Opt. Rev. **20**, 193–197 (2013)

12. Zhang, Y., Tian, Y., Kong, Y., Zhong, B., Fu, Y.: Residual dense network for image super-resolution. To appear in CVPR 2018 as spotlight. arXiv:1802.08797 (2018)
13. Chen, Y., Zhang, Z., Liu, H.: Study of the seismic performance of hybrid a-frame micropile/MSE wall. Earthq. Eng. Eng. Vibr. **16**, 275–295 (2017)
14. Mao, J., Zhong, D., Hu, Y., Sheng, W., Xiao, G., Qu, Z.: An image authentication technology based on depth residual network. Syst. Sci. **6**, 57–70 (2018)
15. Ledig, C., Theis, L., Huszar, F., et al.: Photo-realistic single image super-resolution using a generative adversarial. arXiv:1609.04802 (2015)
16. Shashibala, T., Gawali, B.W.: Research on unsupervised coloring method of Chinese painting based on an improved generative adversarial network. Veterinary Nurse (2014)
17. Ha, I., Kim, H., Park, S., Kim, H.: Image retrieval using BIM and features from pretrained VGG network for indoor localization. Build. Environ. **140**, 23–31 (2018)
18. Liu, H., Xu, J., Wu, Y., Guo, Q., Ibragimov, B., Xing, L.: Learning Deconvolutional Deep Neural Network for High Resolution Medical Image Reconstruction. Inf. Sci. **468**, 142–154 (2018)
19. Oh, S.-H.: Error back-propagation algorithm for classification of imbalanced data. Neurocomputing **74**, 1058–1061 (2010)
20. Wang, S.-H., Tang, C., Sun, J., Yang, J., Phillips, P., Zhang, Y.-D.: Multiple sclerosis identification by 14-layer convolutional neural network with batch normalization, dropout, and stochastic pooling. Front. Neurosci. (2018)

Fatigue Detection System Based on Facial Information and Data Fusion

Jinyi Ma[✉], Jing Wang, Jiahao Wang, Fuchao Yu, Yuqing Tian,
and Bini Pan

Department of Automation, Shenyang Aerospace University,
Shenyang 110135, China
dnmjrmjy@163.com

Abstract. With the continuous development of the automobile industry, the subsequent social problems have become increasingly serious. Not only traffic accidents cause huge harm to people, but also impose a heavy burden and impact on society. Fatigue driving is a major reason of traffic accidents, therefore it is essential to develop a reasonable and effective non-contact vehicle-mounted device that can accurately detect whether the driver is fatigued at the current time for cutting down traffic accidents and ensuring road safety. This paper proposes a new type of fatigue driving detection system. Firstly, it collects five kinds of face information such as PERCLOS value, blink frequency, and closed eye duration through Openmv. Secondly, the Technique for Order Preference by Similarity to Ideal Solution algorithm is used to fuse the standardized fatigue index with data. Finally, the quantification of the driver's fatigue value is realized. Experimental results show that the system can control the alarm more accurately to remind the driver to improve the driving state in time.

Keywords: PERCLOS · Face recognition · Fatigue detection · TOPSIS algorithm

1 Introduction

With the increasing number of motor vehicles in the country, safe driving has become a problem of concern in today's society. According to statistics from the transportation department, fatigue accidents account for a large proportion of traffic accidents every year. Therefore, it is of great practical significance to study a system that detects the fatigue state of drivers in real time. All the major domestic and foreign automobile manufacturers have been trying to develop in-vehicle equipment that can effectively avoid fatigue driving, and a variety of fatigue driving monitors are also emerging [1–3].

According to statistics, about 90% of the information in the driving of a driver is obtained visually. At present, driver fatigue detection methods can be roughly divided into detection based on driver physiological signals, detection based on driver operation behavior and vehicle state, and detection based on driver facial expressions. Because the first two methods are interfered by external factors, there are many disadvantages, and the measurement results are not accurate enough. The detection

© Springer Nature Singapore Pte Ltd. 2020
J. Qian et al. (Eds.): ICRRI 2020, CCIS 1336, pp. 63–75, 2020.
https://doi.org/10.1007/978-981-33-4932-2_5

method based on the driver's facial expression uses the driver's eyes and mouth movement characteristics to infer the driver's fatigue state. This type of method usually uses a camera to obtain the driver's facial information, and uses digital image processing technology to perform real-time eye state Identifying and judging the fatigue state of the driver based on this has a high detection accuracy, and has become a hotspot of research at home and abroad [4–6]. The PERCLOS method is currently the most commonly used detection method based on the driver's eye state. It estimates the driver's fatigue degree through the ratio of the eye closure time over a period of time. At present, the fatigue detection system usually uses P80 as the driver's fatigue recognition, in which P80 refers to 80% of the eye closure time as a specific time as the fatigue state evaluation index. However, the fatigue state of the driver is judged only based on the state of the eyes, which is prone to misjudgment and misjudgment [7, 8]. Therefore, this article uses the mouth information as one of the fatigue indicators, and proposes and develops a new vehicle fatigue detection system, as shown in Fig. 1. As shown, this system can more accurately determine the driver's fatigue state.

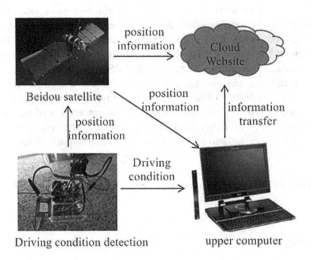

Fig. 1. Driving condition monitoring system

The rest of the article is summarized as follows: the second part introduces the hardware composition of the device; the third part analyzes the selection of various fatigue indicators; the fourth part introduces the data fusion method in detail; the fifth part is the experiment and conclusion.

2 Hardware Design

The hardware design of the system adopts modular thinking. The device is based on the Arduino single-chip microcomputer and is connected with the OPENMV vision module, shell Internet of Things module, alarm module, positioning module, and communication module. Realize the determination of human eye fatigue state based on computer vision. The block diagram of the system hardware is shown in Fig. 2.

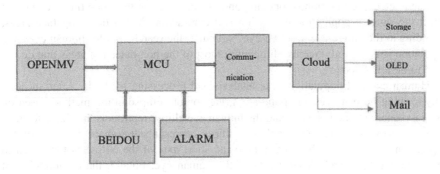

Fig. 2. Hardware design diagram

2.1 OPENMV Vision Module

The system uses the machine vision development component OpenMV vision module, as shown in Fig. 3. OpenMV is an open source, low-cost, and powerful machine vision module. It uses STM32F427CPU as the core and can communicate with other hardware through UART, I2C, SPI and GPIO interfaces. The OpenMV vision module provides a rich machine vision development module that integrates the OV7725 camera chip. On the small hardware module, the core machine vision algorithm is efficiently implemented in C language, and the API is provided. When developing, you can directly use the Python language to programmatically call image processing related algorithms and Python libraries on the IDE of the OpenMV vision module. And the highest pixel of OpenMV can reach 300,000.

Fig. 3. OpenMV vision module device

The machine vision algorithms on OpenMV include finding color patches, face detection, eye tracking, edge detection and landmark tracking. The OpenMV vision module can distinguish between two different states of human eyes opening and closing eyes and detect the duration of opening and closing eyes; at the same time, it can also detect the state of population expansion and contraction. When detecting face areas, according to the "three-eye and five-eye" structure, the search area for human eyes and population is narrowed, thereby eliminating most of the non-detection areas that do not need to be identified.

Human eye and population state analysis and processing methods: After successfully locating human eyes and population, the contour ellipse fitting method based on the least square method is used, and the human eye ellipse represents the shape features of the human eye, and the population ellipse represents the shape features of the population. The ratio of the long axis to the short axis of the ellipse is then used to determine the open and closed states of the human eye. Finally, the combination of PERCLOS and driving is used to judge the driving state, so that the driving state monitoring system is more accurate and reliable.

Fig. 4. OPENMV vision module detection result chart (COlor figure online)

As shown in the figure above, the device lights red when the human eye is not detected or when the eyes are closed or the population opening and closing size exceeds the threshold; when the human eye is detected and the closed eyes or the population opening and closing size does not reach the threshold, the device lights green. The detection results of the OPENMV vision module are shown in the Fig. 4 above.

The detection result of OPENMV vision module is displayed on the host computer as shown in Fig. 5:

Fig. 5. OPENMV vision module detection results displayed on the host computer

2.2   Positioning Module

Measuring the distance between a user's receiver and a satellite with a known location is the basic principle of GPS navigation. To realize this principle, we must first find out the corresponding position of the known satellite through the existing satellite ephemeris, and secondly, record the transmission time, that is, the time that the signal from the known satellite propagates to the user, and this time Multiply by the speed of light to achieve GPS navigation. Single-point positioning and differential positioning are two basic positioning methods in GPS navigation systems. According to the instantaneous position of the satellite moving at high speed as the known starting data, the position of the point to be measured is determined by the method of spatial distance resection. Assuming that the GPS receiving device is installed at the point to be measured on the ground at time t1, the time t2 when the GPS signal arrives at the receiving device can be measured, plus the satellite ephemeris and other parameter settings received by the receiving device to achieve positioning. The following figure shows the GPS positioning system (Fig. 6).

Fig. 6. Schematic diagram of GPS positioning system

Using the navigation and positioning module, you can locate the driver's location information in real time. When you make multiple alarms, you can send the geographic location letter to family members or the emergency contact set by the driver yourself.

The system locates the vehicle based on the navigation module, which can monitor the specific location of the vehicle in real time. When the driver is in a poor driving state or a traffic accident occurs, the system can directly alarm through the communication module. Because the system needs to be installed on the car and requires accurate location of the driver in real time, and because the device needs to be placed on the car, we need a positioning module that is small in size, easy to install, low power consumption, and strong endurance. After consulting the data, it was decided to use the ATGM336H-5N dual-mode navigation module as the Beidou navigation module of this system.

Figure 7 is the physical device diagram of Beidou navigation module; Fig. 8 is the display result of the upper computer of Beidou navigation module:

Fig. 7. Beidou navigation module device diagram

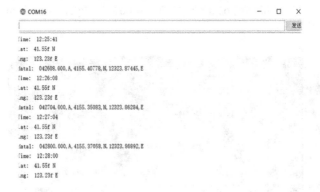

Fig. 8. Results of Beidou navigation module

3   Fatigue Characteristic Parameters

The eyes and mouth are the two organs that can best reflect the state of the human facial organs. The blinking of the eyes is frequent or the eyes are closed. The driver may blink or doze frequently; the mouth movements are large or open frequently, the driver may Yawning or talking with fellow travellers. Therefore, we extract information from these two organs to judge whether the driver is in a state of fatigue or inattention.

On the basis of the determination of the state parameters of the eyes, mouth and head, the fatigue degree of the driver is determined by the five fatigue characteristic parameters of PERCLOS value, blink frequency, duration of closing eyes, whether to yawn or open mouth frequency.

3.1   PERCLOS Principle

PERCLOS is the abbreviation of Percent Eye Closure, which refers to the proportion of time when the eyes are closed for a certain period of time. In the specific test, there are three measuring methods: P70 and P80. Among them, P80 is considered to be the most reflective of human fatigue. This paper uses the P80 method. Depending on the driver's eye state, we divide it into these situations:

1) Closed state: The eye state parameter is lower than 20% of its initial state parameter.
2) Fully open state: The eye state parameter is higher than 80% of its initial state parameter.
3) Half-open and half-closed state: the eye state parameter is 20% to 80% of its initial state parameter.
4) Sleepy state: The eye state parameter is lower than 80% of its initial state parameter.

According to the above, we define the meaning of different data as follows: The curve in the figure is the curve of the degree of opening with time during the process of eye closing and opening. According to this curve, a certain degree of closing or opening of the eye to be measured can be obtained Open the duration, thereby calculating the PERCLOS value. In the figure, t1 is the time from when the eyes are fully opened to 20% closed; t2 is the time from when the eyes are fully opened to 80% closed; t3 is the time from when the eyes are fully opened to 20% next time; t4 is the eyes are fully opened Open 80% of the time next time. The value f of PERCLOS can be calculated by measuring the value of t1 to t4.

$$f = \frac{t_3 - t_2}{t_4 - t_1} \tag{1}$$

Figure 9 shows the measurement principle of PERCLOS value. In the formula, f is the percentage of eye closure time in a specific time. For the P80 measurement method, we believe that when the PERCLOS value f > 0.15 within 1 min, the driver is considered to be in a fatigue state.

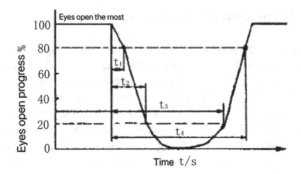

Fig. 9. Schematic diagram of measurement of PERCLOS value

3.2 Mouth Information Extraction

There may be two situations when the driver's mouth is open. One is yawning and is in a fatigue state; the other is that the driver is constantly talking and is in a distracted state. Both of these states may cause traffic accidents, so you can put your mouth. The action frequency of the part is used as the basis for fatigue judgment. The algorithm steps are:

The first step is to calculate the mouth opening:

Since the mouth area has a lower gray value than the surrounding area, for the lower half of the face image, it is easy to determine the position of the mouth and mark the position (Fig. 10) with a rectangular frame. The ratio of the height (G_m) to width (K_m) of the rectangular frame is the mouth. The opening of the department is O_m.

$$O_m = \frac{G_m}{K_m} \tag{2}$$

After the position of the mouth is determined, the width and height can be obtained by binding the rectangular frame to the mouth, and then O_m can be obtained.

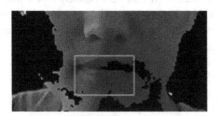

Fig. 10. Mouth recognition

The second step is to judge whether the mouth is open according to O_m:

When the driver does not speak, speak or yawn, the mouth closure is different:

When the mouth is closed, the opening of the mouth is kept within a certain range, between 0.0 and 0.35, with little change; when the ordinary mouth opens, the mouth suddenly opens and closes, and the duration is short. When the big mouth yawns, The opening of the mouth is large, generally greater than 0.5.

The third step is to count the duration of opening the mouth:

The duration of yawning and mouth opening when speaking is different. When yawning is greater than speaking, the duration of mouth opening can be used to distinguish whether the driver is speaking or yawning. The method to count the duration of the open mouth is: multiply the number of open frames of the mouth by the time of each frame to get the duration of the open mouth as follows:

The fourth step is to calculate the mouth movement frequency:

Define the frequency of mouth movements as the number of yawns per unit time:

$$F_y = \frac{N}{T} \tag{3}$$

In the formula, N represents the number of times the mouth is judged by O_m, T represents the unit time, and the frequency of mouth movement in the unit time is used as the basis for determining fatigue.

When the driver opens his mouth frequently, but the duration of opening the mouth is not long, and O_m is between 0.0 and 0.35, it can be judged that the driver is talking at this time; when the driver opens the mouth for a long time and O_m is greater than 0.5, then It can be judged that the driver is yawning. Speaking and yawning. The more the driver's mouth is opened, the higher the driver's fatigue or the number of times he speaks. The frequency of mouth movements when speaking must be greater than the frequency of mouth movements when yawning.

3.3   Blink Rate

When the human body enters a state of fatigue, there will be a short period of doze, and accompanied by obvious eyes closed completely, when the head is raised, the eyes are opened and the eyes blink frequently. In response to this phenomenon, statistics of the state parameters of the driver's head within a certain period of time, combined with blinking frequency and other information parameters, can determine the degree of driver fatigue.

Here we take 60 s as a cycle and count once in a blink of an eye. The following formula can be listed:

$$F_e = \frac{N}{T} \tag{4}$$

If the test blink frequency is greater than 0.3 within a week, the driver is considered fatigued.

3.4 Eyes Closed

The driver only completely closes his eyes in one of the following situations, that is, drowsiness and drowsiness cause the eyes to be completely closed for a long time. Record the length of eyes closed and combine with other fatigue information to judge the driver's fatigue.

Here we use 6 s as a cycle to record the duration of all closed eyes in a period:

$$F_t = x_1 + x_{2 + ...} x_n / T \tag{5}$$

After many tests and calculations, the length of closed eyes was standardized and transformed. If the duration of closed eyes is greater than 0.2, the driver is considered to be in a state of fatigue driving.

4 Face Data Fusion Based on TOPSIS Algorithm

4.1 Basic Principles and Evaluation Steps of TOPSIS

The TOPSIS method (Technique for Order Preference by Similarity to Ideal Solution) ranks the relative merits of each evaluation unit by measuring how close the evaluation unit is to the "ideal solution" and "negative ideal solution" and is one of the multi-objective decision analysis A common method. The "ideal solution" is a solution that assumes that each attribute value is the optimal value; the "negative ideal solution" is a solution that assumes that each attribute value is the worst value. If the evaluation object is closest to the "ideal solution" and far from the "negative ideal solution", it is the best; otherwise it is the worst. In this paper, when applying TOPSIS method, the expert experience method is used to determine the weights, which can not only combine multiple indicators of various research levels into a single indicator, but also avoid subjectively specifying the weight of each indicator. It is easy to operate and is conducive to the fusion of face information data. It is popularized in the evaluation of the obtained fatigue value. The evaluation steps are as follows:

Construct Evaluation Matrix. For the m indicators selected by a certain element layer of n evaluation objects in the fatigue evaluation system, the original data matrix $X = \{X_{ij}\}_{n*m}$ is obtained. In the formula, x_{ij} is the first. The value of the jth evaluation index of the i evaluation objects, $i = 1, 2, \ldots, n, \ j = 1, 2, \ldots, m$.

Matrix Standardization. Find the maximum value x_{jmax} and minimum value x_{jmin} of each column x_j in X separately, and then normalize x_{ij} to get the normalized value $r_{ij}(i = 1, 2, \ldots, n; \ j = 1, 2, \ldots, m)$, and get the normalization matrix $R = \{r_{ij}\}_{n*m}$.

Therefore, the value of "ideal solution" r_{jmax} is $r_j^+ = 1$; "negative ideal solution" $r_j^- = 0$.

Calculate the Euclidean Distance Between the Actual Value of Each Evaluation Object and the Ideal Solution r_j^+ and the Negative Ideal Solution r_j^-:

$$d_i^+ = \sqrt{\sum_{j=1}^{n} w_j \times \left(r_{ij} - r_j^+ \right)^2}, \quad d_i^- = \sqrt{\sum_{j=1}^{n} w_j \times \left(r_{ij} - r_j^- \right)^2}, \tag{6}$$

Calculate the Fatigue Value C_i After the Fusion of Facial Information of This Element.

$$C_i = d_i^- / \left(d_i^- + d_i^+ \right) (i = 1, 2, \ldots, n) \tag{7}$$

It can be seen that $0 \leq C_i \leq 1$ and C_i indicate that the driver status of the element is better; otherwise. It means the driver's condition is poor.

4.2 Comprehensive Calculation of Driver's Fatigue Evaluation Value

From the low level to the high level, the TOPSIS method is used repeatedly, and finally the comprehensive benefit C of land use in various regions of the country is finally obtained. According to the measured value of driver fatigue in each state, the pros and cons of the driver's state are arranged and compared with the actual state to verify the accuracy of the algorithm.

5 Experiment

In this experiment, videos of three classmates driving simulation were collected, and 30 consecutive frames were randomly selected from each video as experimental samples. The experiment calculates the fatigue parameters according to the recognition results of the fatigue detection device, uses the TOPSIS method to fuse the face data, and then obtains the driver's fatigue state value, so as to determine whether the driver needs to be reminded to adjust the driving state in time.

The experimental results are shown in Table 1. It can be seen that the fatigue value sorting results are in line with the actual tester's status, indicating that the system's reasoning results have a high accuracy. When the fatigue value is less than 0.1, the driver is determined to be in a fatigue state, reminding the driver to promptly Adjust your own driving status, this device can meet the needs of driver fatigue warning. The analysis shows that the discriminant error is mainly due to the large movement of the subject's head that causes the eyes to be blocked, or the edge of the mouth is blurred due to light problems, which causes problems in segmentation.

Table 1. Experiment

	PERCLOS value	Blink rate	Eyes closed	Population opening and closing frequency
wide awake	0.00	0.10	0.00	0.10
	0.00	0.08	0.00	0.00
	0.01	0.12	0.00	0.12
	0.05	0.12	0.00	0.15
normal	0.04	0.18	0.00	0.13
	0.07	0.16	0.00	0.18
	0.10	0.31	0.03	0.49
general	0.12	0.24	0.01	0.52
	0.16	0.42	0.18	0.40
	0.20	0.66	0.37	0.66
sleepy	0.18	0.71	0.17	0.70
	0.50	0.69	0.91	0.54

	Population opening and closing range	Fatigue value	Rank
wide awake	0.20	0.1366	2
	0.00	0.1170	1
	0.30	0.1089	3
	0.28	0.1086	5
normal	0.27	0.1083	4
	0.34	0.1018	6
	0.64	0.0746	9
general	0.70	0.0680	7
	0.39	0.0665	8
	0.84	0.0499	11
sleepy	0.51	0.0368	10
	0.63	0.0233	12

6   In Conclusion

A real-time monitoring method for driver fatigue state based on eye state recognition is proposed. First, locate the eyes through face detection and heuristic rules, obtain the status information of the eyes through Openmv, and then calculate five fatigue indicators such as PERCLOS, average eye opening degree, and maximum eye closure time to infer driver fatigue status. The results show that the five fatigue indicators extracted have significant differences under different fatigue levels. The TOPSIS data fusion of different indicators can accurately and effectively detect the fatigue state of the driver. The research in this paper is mainly aimed at indoor environments with uniform background light. In the future, the research will be conducted under complex working conditions such as changes in outdoor lighting environment, occlusion and facial expression changes. On the basis of further improvement of the algorithm, the driver's operating behavior and Relevant information such as vehicle trajectory, through multi-source information fusion, further improves the reliability and robustness of the recognition system in the actual road environment.

References

1. Zhang, F., Su, J., Geng, L., et al.: Driver fatigue detection based on eye state recognition. In: 2017 International Conference on Machine Vision and Information Technology (CMVIT), pp. 105–110. IEEE (2017)
2. Fu, R., Wang, H., Zhao, W.: Dynamic driver fatigue detection using hidden Markov model in real driving condition. Expert Syst. Appl. **63**, 397–411 (2016)
3. Mandal, B., Li, L., Wang, G.S., et al.: Towards detection of bus driver fatigue based on robust visual analysis of eye state. IEEE Trans. Intell. Transp. Syst. **18**(3), 545–557 (2016)
4. Mu, Z., Hu, J., Min, J.: Driver fatigue detection system using electroencephalography signals based on combined entropy features. Appl. Sci. **7**(2), 150 (2017)
5. Li, K., Wang, S., Du, C., et al.: Accurate fatigue detection based on multiple facial morphological features. J. Sens. (2019)
6. Sikander, G., Anwar, S.: Driver fatigue detection systems: a review. IEEE Trans. Intell. Transp. Syst. **20**(6), 2339–2352 (2018)
7. Min, J., Wang, P., Hu, J.: Driver fatigue detection through multiple entropy fusion analysis in an EEG-based system. PLoS One **12**(12) (2017)
8. Zhou, L., Xu, E., Chen, D., et al.: Design of rolling ball control system based on OpenMV. J. Phys.: Conf. Ser. IOP Publishing **1303**(1), 012102 (2019)
9. Zyoud, S.H., Fuchs-Hanusch, D.: A bibliometric-based survey on AHP and TOPSIS techniques. Expert Syst. Appl. **78**, 158–181 (2017)
10. Biswas, P., Pramanik, S., Giri, B.C.: TOPSIS method for multi-attribute group decision-making under single-valued neutrosophic environment. Neural Comput. Appl. **27**(3), 727–737 (2016)
11. Kuo, T.: A modified TOPSIS with a different ranking index. Eur. J. Oper. Res. **260**(1), 152–160 (2017)
12. Liang, D., Xu, Z.: The new extension of TOPSIS method for multiple criteria decision making with hesitant Pythagorean fuzzy sets. Appl. Soft Comput. **60**, 167–179 (2017)

Human-Robot Interaction

Emotion Detection During Human-Robots Interaction in Visual Reality Environment

Bo Yang[✉], Baiqing Sun, Qiuhao Zhang, Yina Wang, Yong Li,
Junyou Yang, and Li Gu

Shenyang University of Technology, Shenyang 110870, China
1274670125@qq.com

Abstract. With the increasing of aging and disability population, care robots are going to be an assistant for elderly and disabled people. However, it is not satisfied any more with just ordering them to carry out some simple assignments mechanically. It is desirable that the robot is capable to avoid negative emotions of objects, and make them more secure and comfortable feelings psychologically during the work process. In this paper, in order to explore how objective environmental factors influence subjective feelings independently, an emotion detecting and recognition platform was developed including a virtual-reality-interact immersive environment for inducing object emotions and an implemented physiological signal detection device in which physiological signals including Electrocardiograph (ECG), Electromyogram(EMG), Electrodermal activity (EDA), functional near-infrared spectroscopy (fNIRS) were recorded synchronously. Verification experiments were executed and the results showed its availabilities and efficiencies of the proposed emotion detection system.

Keywords: Virtual environment · Physiological signal · Emotion detection

1 Introduction

With the acceleration of aging society in China, elderly caring issues are worsening in the coming decades. Service robots will have to be alternative in most of the situations. In the process of human-robot interaction, it is very important for human's physical and mental health to be human's positive emotions, to eliminate the alertness of the robot and the psychological state of safety and freedom [1]. If the service robot cannot perceive and respond to the user's emotions and psychology, the user will easily generate negative emotions [2]. Therefore, what is becoming more and more important is that service robots 'perception and feedback of users' emotions and psychology.

Studies have shown that the users may be affected emotionally by the robot's appearance and behavior [3]. Professor Morales of Nagoya university in Japan use robots as mobile vehicles to a variety of factors affecting human comfort (linear velocity, linear acceleration, angular velocity, angular acceleration, obstacle distance, etc.). Professor Anna G. developed a new robot training method [4]. The robot interacts naturally and reasonably with the subjects through an EMG acquisition device that captures the muscles of the human face. Different EMG signals were isolated from facial expressions to obtain positive and negative feedback [5]. Affective computing is

J. Qian et al. (Eds.): ICRRI 2020, CCIS 1336, pp. 79–92, 2020.
https://doi.org/10.1007/978-981-33-4932-2_6

an interdisciplinary field based on psychology, computer science and biomedical engineering proposed by Picard [6]. Picard considers that human physiological signals contain a lot of emotional information, and researchers can infer emotions by detecting human physiological signals [7]. The key for emotion research is the method of inducing and detecting emotions.

With regard to the method of detecting emotions, researchers can indirectly detect human emotions by detecting physiological signals related to emotions. Emotional physiological signals are controlled by the body's central nervous system and autonomic nervous system [8]. The central nervous system is derived primarily from the cerebral cortex. The main function is to produce human thinking and emotional changes. Electroencephalography (EEG) and functional near infrared spectroscopy (FNIRS) have been added in more and more studies as a beneficial means to study the lower emotional changes in the central nervous system [15, 16]. The autonomic nervous system is not dominated by human will, it is closely related to emotions. When human beings are emotionally stimulated, various organs and tissues of the organism will be extensively activated, resulting in an obvious physiological response beyond the normal physiological rhythm [9]. The rich sources of information on the physiological response of the human body include brain and heart activity, blood pressure and respiration and skin temperature changes, Muscle changes, and sweat secretion. When human beings are emotionally stimulated and enter a more intense state of stress, brain wave disorders will occur, such as rapid heartbeat, increased sweating, muscle stiffness and other physiological phenomena [17–19]. However, under the same stimulus, the difference in individual's psychology and personality makes the physiological changes they produce have differences, so a single physiological signal change is not enough to express emotions. Different types of physiological signals have different sensitivity to emotions, and fusion of multiple physiological signals can solve this problem [10].

With regard to the method of inducing emotions, with the development of science and technology, Virtual Reality (VR) technology has gradually become an important tool for psychological research [11]. Virtual reality technology can simulate the real world and provide users with a high degree of immersion [12]. In the study of emotions, virtual reality technology can not only effectively induce people's emotions [13], but also can more easily design the scenes needed for the experiment according to the needs of the experiment, and can avoid various external interference [14].

This article studies the impact of obstacles on human emotions during the interaction between humans and wheelchair robots. Aiming at the problem of emotion induction and emotion detection, this paper has designed a new 4D human-computer interaction experiment platform. The platform is composed of two parts, including an emotional induction platform and a physiological signal detection platform. The emotion-inducing platform uses virtual reality technology to design the scenes needed for the experiment. The platform is used to induce the subjects' emotions. The hardware equipment includes an intelligent wheelchair robot and a set of specialized VR equipment. The physiological signal detection platform integrates four emotion-related physiological signals. The hardware equipment includes an electrocardiogram module, electromyogram module, Electrodermal activity module, and a functional near-infrared spectroscopy instrument. The 4D human-computer interaction experiment platform is built to effectively induce people's real emotions and detect emotions comprehensively and accurately. This has laid a solid foundation for future research in the field of emotion recognition.

2   4D Human-Computer Interaction Experiment Platform

2.1   Mobile Device Module

The intelligent wheelchair uses four brushless dc motors to drive omni-directional wheels. The operation mode of intelligent wheelchair can be divided into manual control mode and PC control mode. Manual control mode manual control mode switch position in the command input module through the command input module. The command input module has two knobs, an emergency stop button and a joystick. Two knobs are used to control mode switching and speed level adjustment respectively. When the wheelchair is in PC control mode, it can be driven by receiving information from the external computer through the RS232 data communication interface. As an experimental platform, the use of this intelligent wheelchair helps with a variety of control needs (Fig. 1).

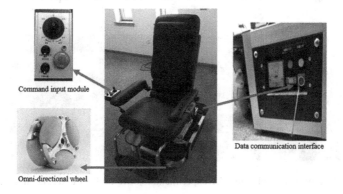

Fig. 1. The function and structure of the intelligent wheelchair

2.2   Environmental Equipment Module

Any subtle changes in space can lead to physiological responses that is difficult to be detected. In order to minimizing the psychological and emotional reaction of non-experimental factors in the experiment. We developed an environmental device for reasonably efficient emotional induction. Oculus rift cv1 professional VR virtual reality equipment was employed as a visual device. It can simulate an interactive and realistic 360° 3D scene. wo screens with resolution of 1080 × 1200 pixels are mounted simultaneously that ensures the smooth operation of 360° 3D scene without interruption. It helpful for users to fully immerse themselves in the 3d virtual world, and makes it possible to have a high degree of immersion and presence in the scene.

To ensure clear details of the virtual scene, we used Autodesk 3Dmax to design the scene model (room, floor, window, ceiling, human body, wheelchair, obstacles). After the drawing is completed, the 3D model is rendered with VRAY to simulate the real scene. Then, we used the UNITY 3D integrated game development tool as a platform for scene visualization and virtual scene construction. The 3D model drawn in advance was imported, and the software development kit based on Unity3D environment provided by Oculus was used for programming modification. Call the VR camera provided by Oculus for the main camera. Users can also control the perspective freely according to their will on the basis of ensuring continuous effects (Figs. 2 and 3).

Fig. 2. Oculus rift cv1 professional VR virtual reality equipment

Fig. 3. The human model and the experimental scene built in unity3D environment

2.3 Physiological Signal Detection Equipment Module

This research is mainly to build an emotional detection platform based on physiological signals. So that enrich the comprehensive study of human physiological changes under the emotional fluctuations. We collected the blood oxygen content of the prefrontal lobe under the control of the central nervous system and the electrical signals of the heart, muscle and skin under the control of the autonomic nervous system, respectively. From the active and passive two aspects to examine the physiological response of human body.

Cardiac, muscle and skin conductance signals are collected using MP160 multi-channel physiological signal acquisition instrument from BIOPAC. MP160 is a powerful computerized multi - guide physiological recorder. An integrated approach is provided for the collection and recording of physiological signals from a variety of different position. The signal of blood oxygen content in the frontal lobe of brain was selected from the functional near-infrared spectroscopy (fNIRS) of WOT-100 produced by Hitachi. It can monitor changes in cerebral blood flow in the frontal lobe of the brain. Functional near infrared spectroscopy (fNIRS) is a new non-invasive brain imaging technique. FNIRS performs functional brain imaging on a similar principle to functional magnetic resonance imaging (fMRI). When the brain receives an external stimulus, neural activity in the brain causes local hemodynamic changes.

3 Main Software Design

3.1 4D Effect Driver Function

It realizes the synchronization of virtual environment and real environment, achieves the consistency of physical sensation (wheelchair) and visual effect (virtual environment), and constructs a 4D human-computer interaction environment. To do this, you'll need a laptop, an Oculus rift cv1 virtual reality device, and an intelligent wheelchair and an RS232 USB cable. The specific connection mode is shown in Fig. 4.

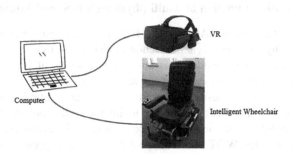

Fig. 4. Laboratory equipment connection diagram

We use Visual Studio 2012 programming platform and C++ language to write the program. The control procedure of the intelligent wheelchair follows the instruction manual and communication protocol of the intelligent wheelchair. Four brushless dc motors are driven by control instructions. The entire control instruction contains a total of 21 bytes. Every 3 bytes controls the operation of a motor, including enable, rotation direction, and motor speed. 1–3 bytes, 4–6 bytes, 7–9 bytes and 10–12 bytes are respectively responsible for controlling the left front wheel, right front wheel, left rear wheel and right rear wheel (Fig. 5).

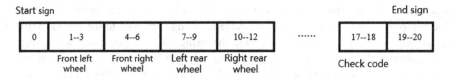

Fig. 5. Intelligent wheelchair control commands

Based on the Visual Studio 2012 programming platform, we created a Visual interface by adding MFC project resource dialog while writing the wheelchair driver. Firstly, manually enter a real speed (0 m/s–1.6 m/s). After clicking start, open the serial port and save the real speed into the set speed array. Brushless dc motor is driven by hexadecimal assignment (command speed = real speed *219.8). At a sampling rate of 50 ms, the speed array is written continuously to the serial port, and then the RS232 data line is used to pay the brushless dc motor under the four omnidirectional wheels for motion control.

On the other hand, the above speed array is written into the set virtual serial port through virtual serial port technology. Use the Visual Studio 2012 platform to program using the C# language, Load the script into the wheelchair model. When the startup button is opened, the virtual serial port is opened at the same time, and unity3d automatically executes the prepared script. Receive the realistic speed from the visualization window. The virtual wheelchair model will move in line with the world coordinate system in 3D unity (displacement = speed * time * direction of movement). After the stop button is pressed, each serial port is closed, the serial port handle is cleared, and the intelligent wheelchair and the virtual wheelchair are stopped.

3.2   Synchronization Function of Multi-physiological Signal Equipment

The Biopac MP160 and the HITACHI WOT-100 functional near-infrared spectroscopy (fNISR) are two physiological signal detection devices. To ensure the consistency of sensor detection data. We consider programming two kinds of physiological signal detection equipment to achieve synchronous acquisition.

Synchronization program using Visual Studio 2012 programming platform, using C++ language. it is convenient for people to interact with the program during its operation by using MFC. The formation and construction of the interface is done by the computer. Because of the WOT-100 functional near-infrared spectroscopy equipment, the equipment is equipped with software ktControlPC, which can receive the trigger signal and the termination signal through the serial port. Change the parameter set in ktControlPC to set the serial port trigger. Use the docking between two virtual serial port modules. Realize the trigger and termination of upper computer control.

The Biopac MP160 host device can receive signals from external triggering devices via the Biopac STP100 isolated digital interface. In order to realize the function of control trigger and stop of upper computer, C8051 single chip microcomputer is used to connect with STP100 isolated digital interface. The digital signal generated by the MCU will consist of two levels. The digital signal generated by the MCU will consist of two levels, +5 v and 0 v. Positive 5 V is interpreted as binary 1, 0 V is interpreted as binary 0. When the serial port is opened and the switch is off (the button is pressed), the rising edge signal changes from 0 V to +5 V. When the serial port is closed, the drop edge signal changes from +5 V to 0 V when the switch is on (the button is pressed). The Setuptrigger option in the software AcqKnowledge attached to the Biopac MP160 device was set as the External mode, and the rising edge trigger mode was selected. The equipment for recording physiological signals is a Biopac M160 basic unit (with corresponding ECG, EMG, and EDA modules) and a WOT-100 functional near-infrared spectroscopy device. Record the signal on the laptop using AcqKnowledge 5.0 and ktControlPC software.

Firstly, AcqKnowledge and ktControlPC are configured separately. Run the MFC program to initialize the serial port. After pressing the start collection button, the program will simultaneously send a trigger signal to the serial port communicating with the MCU. At the same time, the data collection and data storage of the sensor will also start to run. The data is stored in a file as a time series for easy analysis and processing. After pressing the end collection button, each serial port is closed, the serial port handle is cleared, and the sensor stops data collection (Fig. 6).

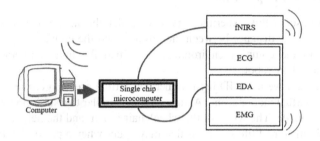

Fig. 6. Schematic diagram of physiological signal detection equipment connection

4   Emotional Induction Test

4.1   The Subjects

The consent of all participants has been obtained in this study. A total of 15 participants aged from 21 to 26 (M = 23.2, SD = 2.2). All the subjects were students and teachers from Shenyang university of technology. The subjects were assured that they had no history of chronic or psychiatric illness. And all subjects had no or minimal experience with virtual environments.

4.2   Wearing Mode

Before the experiment, the test equipment should be properly worn for the subjects. CATHAY CH3236TDY disposable physiological electrode was used as a medium to connect the BIOPAC wireless detection module with the human body. The skin conductance EDA module in BIPACMP160 connects the left index finger, middle finger and lower left metacarpal respectively to measure. EMG is connected to the brachioradialis muscle of the right arm. ECG is connected to the lower left chest. Head Hitachi functional near infrared spectroscopy and Oculus VR virtual reality equipment. As far as possible, subjects should be comfortable to wear and have clear vision before the experiment so that avoiding adverse effects on the experiment (Fig. 7).

Fig. 7. Wear mode of physiological signal acquisition equipment

4.3   Design of the Experiment

The main purpose of this experiment is to consider the factors that may affect the human emotions (the distance between the obstacle and the human body, the shape of the obstacle) in the complex environment in which the human interacts with the intelligent wheelchair robot.

For this reason, we used 4D high immersion virtual scene to create a 10 m*20 m full simulation virtual indoor scene. A virtual model similar to a subject in an intelligent wheelchair was made. The width of the wheelchair is 1 m and the height of the human eye is 1 m. Simulate real life scenarios that may occur when a person interacts with an intelligent wheelchair robot. Then two independent factor evaluation experiments were conducted.

The transverse distance between the subject and the obstacle was taken as an independent evaluation factor: during the experiment, the wheelchair would move in a straight line at a constant speed, with an obstacle on the left side of the road. obstacle plane parallel to the runway. In the experiment, the distance d between the wheelchair and the obstacle (a cube with side length r = 1 m) was 0.2 m, 0.3 m, 0.4 m, 0.5 m, 0.6 m, 0.7 m, 0.8 m (Figs. 8 and 9).

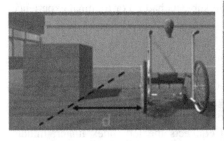

Fig. 8. Distance experiment   **Fig. 9.** Shape experiment

The obstacle shape was taken as an independent evaluation factor: during the experiment, the wheelchair would move in a straight line at a constant speed, with an obstacle on the left side of the road. Different from the distance experiment in the previous group, the obstacle in this group faces the runway vertically at the sharp Angle. The height of each obstacle is 1 m, and the radius of the outer circle is 0.5 m. The distance between the wheelchair and the obstacle is fixed as d = 0.2 m. The subjects performed each independent factor evaluation in an experiment in which intergroup comparisons were in random order. Before each experiment, the subjects were given a 3-min break to calm down and adjust their mood, reduce the possibility of interference between groups.

5   Data Analysis

In the distance experiment, we obtained 105 data samples from 15 subjects (each subject,7 experiments and 7 physiological data samples). In the shape experiment, we obtained 36 data samples from 12 subjects (each subject,3 experiments and 3 physiological data samples).

Each sample contains four physiological signals, including ECG, EMG, EDA and prefrontal blood oxygen content. Tables 1, 2 and Figs. 10, 11 and 12 are sample information and physiological signal data, where the emotion- inducted variable is the distance between the person and the obstacle. Tables 3 and 4 and Figs. 13, 14 and 15 are sample information and physiological signal data, in which the emotion-induced variable is the shape of the obstacle.

Table 1. Distance experimental sample information

Sample	Specific information
Number of subjects	15 persons (9 males and 6 females)
Number of samples	105
Sampling frequency	2000 Hz
The sample time	20 s

Table 2. Sample 1 2 3 of a subject in distance experiment

Distance variable	Obstacle shape	Wheelchair speed	The runway length
0.2 m	quadrangular	0.4 m/s	8 m
0.4 m	quadrangular	0.4 m/s	8 m
0.8 m	quadrangular	0.4 m/s	8 m

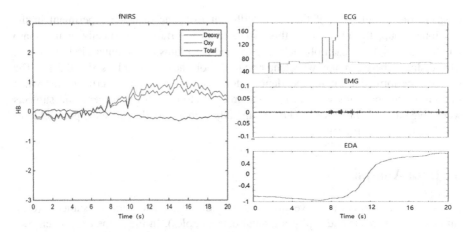

Fig. 10. Physiological data from sample 1

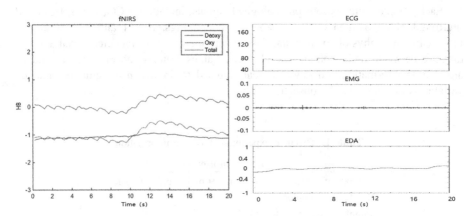

Fig. 11. Physiological data from sample 2

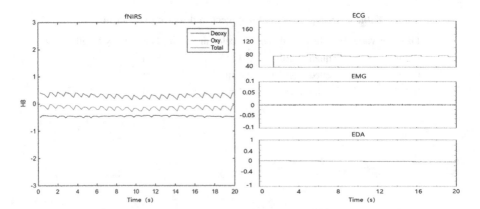

Fig. 12. Physiological data from sample 3

Table 3. Shape experimental sample information.

Sample	Specific information
Number of subjects	12 persons (8 males and 4 females)
Number of samples	36
Sampling frequency	2000 Hz
The sample time	15 s

Table 4. Sample 4 5 6 of a subject in shape experiment.

Shape variable	Distance	Wheelchair speed	The runway length
Triangular prism	0.2 m	0.4 m/s	6 m
Quadrangular	0.2 m	0.4 m/s	6 m
Cylinder	0.2 m	0.4 m/s	6 m

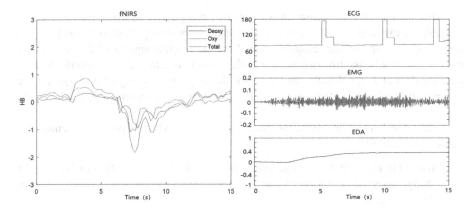

Fig. 13. Physiological data from sample 4

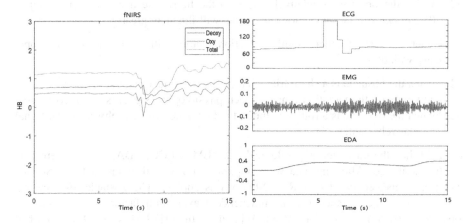

Fig. 14. Physiological data from sample 5

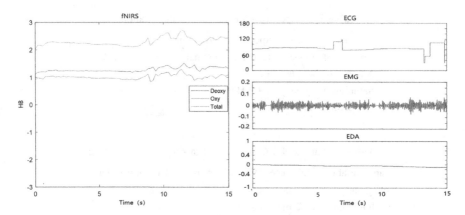

Fig. 15. Physiological data from sample 6

After the end of the experiment, we carried out a general analysis and research on the distance experiment and the shape experiment respectively. Both the blood oxygen content signal and the skin electrical signal are the original signals of the sensor. We carried out band-pass filtering on the EMG signal to eliminate the interference of baseline drift and signal noise, and detected the original ECG signal with QRS wave group so that calculating the corresponding RR interval to convert it into the heart rate signal.

Under the test of independent evaluation factors with shape as the emotional inducing variables:

(1) When the obstacle has a large lateral distance to the subject (0.6 m–0.8 m), the change range of each physiological signal is relatively flat, even without obvious fluctuation.
(2) When the lateral distance between the obstacle and the subject is relatively average (0.4 m–0.5 m), each physiological signal has a certain range of change.
(3) When the obstacle is relatively close to the transverse distance of the subject (0.2 m–0.3 m), the physiological signals will fluctuate greatly.

Under the test of independent evaluation factors with shape as the emotional inducing variables:

(1) when the obstacle is a sharp triangular prism and a quadrilateral prism with edges and corners, the variation range of each physiological signal is relatively large.
(2) when the obstacle is a round cylinder, the fluctuation of each physiological signal is relatively gentle.

And then the measured physiological data (EMG, ECG, EDA, fNIRS) were normally distributed. After the data were imported into Statistical Product and Service Solutions, repeated two-factor RMANOVA was applied. The obstacle distance test results, where the distance was an independent evaluation factor, showed that the obstacle distance [F(3,51) = 44, P = 0.0+] and the time [F(2,34) = 3.7, P = 0.03] had significant influence on the experiment. The distance (0.2 m–0.3 m) was significantly different from other groups in the distance experiment (P = 0.02). Shape test results showed that shape of obstacle [F(1,16) = 4.56, P = 0.05] and time [F(2,34) = 10, P = 0.004] had significant influence on the experiment. When the obstacle is a rounded object, there is a significant difference between the groups of other sharp objects in the shape experiment (P = 0.01).

To this end, we can come up with a theory: With the proximity of obstacles, the pressure on people's emotional and physiological reactions showed an increasing trend in the distance experiment. When the obstacle is an object with edges and corners, the stimulation to human body is more obvious than that of the round object in the shape laboratory.

According to the above analysis, it is not difficult to find that changes inmood lead to fluctuations in the physiological signal spectrum. There is a strong correlation between emotions and physiological signals.

6 Conclusion

This paper studies the influence of obstacles on human emotions during the interaction between human and wheelchair robot, and builds a 4D human-computer interaction experiment platform. The platform consists of two parts, including an emotional induction platform and a physiological signal detection platform.

The emotion-inducing platform uses virtual reality technology to design an immersive virtual experiment scene for human and wheelchair robot interaction, and can simultaneously run the wheelchair robot in the real environment and the wheelchair in the virtual environment, further increasing the immersion. The platform can effectively induce the emotions of the subjects.

The physiological signal detection platform integrates four emotion-related physiological signals controlled by the autonomic nervous system and the central nervous system, and the platform realizes the synchronous acquisition of the four physiological signals.

This paper carefully designed the experimental process, and studied the influence of a single factor (the distance between people and obstacles and the shape of obstacles) on people's emotions. According to the later report of the subjects, the majority of subjects subjectively indicated that the distance between the obstacles and the shape of the obstacles did have the ability to affect their own emotional changes. With the increase of the distance of the obstacle and the sharpness of its shape, negative emotions such as tension, anxiety and fear appeared at the psychological level. By statistical analyzing the physiological data obtained in the experiment, This paper finds that the distance between the person and the obstacle and the shape of the obstacle can indeed affect the emotional change of the person, verifies the correlation between the emotion and the physiological signal, proves the feasibility of the experiment design, and proves the effectiveness of the human-computer interaction experiment platform.

References

1. Zhou, Q.: Multi-layer affective computing model based on emotional psychology. Electronic Commerce Res. **18**(1), 109–124 (2017). https://doi.org/10.1007/s10660-017-9265-8
2. Reeves, B., Nass, C.: The media equation: how people treat computers, television, and new media like real people and places. Philos. Soc. Sci. **30**(1), 120–124 (1996)
3. Negi, S., Arai, T.: Psychological assessment of humanoid robot appearance and performance using virtual reality, pp. 718–719. Robot and Human Interactive Communication (2008)
4. Morales, Y., Miyashita, T.: Social robotic wheelchair centered on passenger and pedestrian comfort. Robot. Autonomous Syst. **87**(1), 355–362 (2017)
5. Gruebler, A., Berenz, V.: Coaching robot behavior using continuous physiological affective feedback. In: International Conference on Humanoid Robots, pp. 26–28 (2011)
6. Sidney, M.: The affective computing approach to affect measurement. Emotion Rev. **10**(2), 174–183 (2017)
7. Deaudelin, C., Dussault, M.: Human-computer interaction: a review of the research on its affective and social aspects. Can. J. Learn. Technol. **29**(1), 89–110 (2003)
8. Schacter, D., Wegner, D.: Emotion and motivation. Psychology (2016)
9. Hagemann, D., Waldstein, S.: Central and autonomic nervous system integration in emotion. Brain Cogn. **52**(1), 79–87 (2003)
10. Verma, G., Tiwary, U.: Multimodal fusion framework: a multiresolution approach for emotion classification and recognition from physiological signals. Neuroimage **10**(2), 162–172 (2014)
11. Stone, R., Small, C.: Virtual natural environments for restoration and rehabilitation in healthcare. Intell. Syst. Ref. Libr. **24**(1), 497–521 (2014)
12. Chirico, A., Cipresso, P.: Effectiveness of immersive videos in inducing awe: an experimental study. Sci. Rep. **7**(1), 12–18 (2017)
13. Ulrich, R.: Natural versus urban scenes: some psychophysiological effects. Environ. Behav. **13**(5), 523–556 (1981)
14. Felnhofer, A., Kothgassner, O.: Is virtual reality emotionally arousing? Investigating five emotion inducing virtual park scenarios. Int. J. Hum.-Comput. Stud. **8**(2), 48–56 (2015)
15. Mohammadpour, M., Hashemi, S.M.R., Houshmand, N.: Classification of EEG-based emotion for BCI applications (2017)
16. Michiko, O., Yoshiki, K., et al.: E6(S)-2 measurement of fNIRS to estimate emotion in a VR system. Japanese J. Ergon. (2017)
17. Wexler, B.E., Warrenburg, S., Schwartz, G.E., et al.: EEG and EMG responses to emotion-evoking stimuli processed without conscious awareness. Neuropsychologia **30**(12), 1065–1079 (1992)
18. Gouizi, K., Maaoui, C., Reguig, F.B.: Negative emotion detection using EMG signal. In: International Conference on Control. IEEE (2014)
19. Canento, F., Fred, A., Silva, H., et al.: Multimodal biosignal sensor data handling for emotion recognition, pp. 647–650. IEEE (2011)

A Novel Auxiliary Strategy with Compliance and Safety for Walking Rehabilitation Training Robot

Donghui Zhao[1(✉)], Junyou Yang[1], Yizhen Sun[1], Jiawei Feng[1], and Yokoi Hiroshi[2]

[1] Shenyang University of Technology, Shenyang 110870, China
zhaodonghui@sut.edu.cn
[2] The University of Electro-Communications, Tokyo 1820021, Japan

Abstract. The walking rehabilitation training and comfortable auxiliary walking in daily life are crucial in improving the life quality for the elderly and the disabled. A walking rehabilitation training robot is developed in this paper, which integrates passive walking, active walking and hybrid walking mode. To achieve a non-contact detection approach of motion intention, we develop a multi-channels proximity sensor, and establish the detection platform. A motion control method with compliance and safety based on adaptive neural fuzzy inference system is proposed to investigate the user's motion intention and generate the control target. Finally, comparative experiments are implemented to verify the superiority and effectiveness of the proposed method. The results showed that the rehabilitation training robot with the proposed controller has the capability to precisely control motion state after calculating the user's walking intents. Our proposed method can effectively provide the assist strategy with compliance and safety for the elderly and the disabled. The walking assisted-robot with similar structure, which integrates the proposed approach, has a good universal property for the elderly and the disabled with different motion capability in the hospitals, pension centers, and families.

Keywords: Walking rehabilitation training robot · Multi-channels proximity sensor · Safety and compliance · Adaptive neural fuzzy inference system

1 Introduction

Owing to the growth of aging population and the disabled people suffering from lower limb disorders, there exist great demands for assisted walking tools under the condition of relative lack of professional nursing [1, 2]. Meanwhile, walking rehabilitation training and comfortable auxiliary walking by assisted walking tools are crucial in behavioral assistance of daily life. Assisted walking robot generally can be categorized into two types: exoskeleton robots [3, 4] and walking vehicle robots [5, 6]. Although different kinds of auxiliary devices effectively provide multiple auxiliary functions in public places, they would prefer to stay in their home where they feel more confident than moving to any expensive adult care or healthcare facilities. Therefore, our laboratory developed a welfare robot: Walking Rehabilitation Training Robot (WRR),

© Springer Nature Singapore Pte Ltd. 2020
J. Qian et al. (Eds.): ICRRI 2020, CCIS 1336, pp. 93–104, 2020.
https://doi.org/10.1007/978-981-33-4932-2_7

which can be applied to interior space for the elderly and the disabled who have lost their walking balance. This WRR integrates three auxiliary modes, passive walking, active walking and hybrid walking mode, which provides the appropriate assisted walking according to the user's motion capability.

It's crucial to accurately identify the user's motion intention during the process of operating the robot [7]. In recent years, some interaction strategies, such as handle manipulation, voice interaction and bioelectric signal interaction, have been applied to intention recognition [8–10]. Using the joystick is often adopt to identify the user's direction intention, however, it's difficult and dangerous for the users to focus on controlling the joystick during rehabilitation training walking [8]. Speech recognition can accurately recognize human language through deep neural network technology [9], but it is difficult to accurately express the ambiguous position and direction. Surface electromyography has been applied to recognize the motion intention, because of its rich information and mature non-invasive acquisition technology [10]. However, it is troublesome for the patients to wear the sEMG devices repeatedly in the application process. Moreover, due to the high time-varying of sEMG signal, the lack of lower extremity function leads to the large individual difference, which affects the stability of the system. To improve the safety and convenience of patients, this paper proposes a non-contact motion intention recognition approach. Through the application of multi-channel proximity sensors and pressure sensors, the motion intention is investigated, which includes the walking speed intention and abnormal gait. Finally, the motion controller is proposed based on the motion intention. Various experiments and comparative analysis are performed to verify the correctness and effectiveness of the method.

The remaining part of this paper is organized as follows. Section 2 introduces the walking rehabilitation training robot and multi-channels proximity sensor. Section 3 illustrates details about the motion method based on the adaptive neural fuzzy inference system. In Sect. 4, the experiments and comparative analysis are conducted. Finally, the conclusions are given in Sect. 5.

2 Walking Rehabilitation Training Robot

Aiming to resolve convenient assistance of daily behavior based on the user's appeal, our laboratory developed WAR in Fig. 1 for people with walking disabilities. WAR provides functions of auxiliary standing and auxiliary walking. WAR equipped with the omnidirectional wheel could swerve in narrow spaces with no turning radius. It has three gait training modes: (i) Passive training (based on the preset path); (ii) Active training (based on recognition of users' intention); (iii) Hybrid training (the combination of Passive training and Active training). These directions and velocities of the two robots can be controlled by telecontroller, joystick, and the body center of gravity, which can achieve active motion, passive motion and hybrid interactive strategy combined with the actual situation of the patient.

To detect the walking direction intent and gait information, the test platform was implemented, as shown in Fig. 2. Proximity sensors (Fig. 2(c)) were adopted to measure a user's walking while using the walking assistive robot. The proximity sensor can detect linear distance from the robot plate to both legs, and the communication frequency of whole system is 100 ms. The proximity sensor measures the distance and sends data to a microcomputer (Fig. 2(b)). The microcomputer converts CAN communication data to Serial communication data, and sends them on a PC (Personal Computer). The application on the PC has two functions, recording the distance data and controlling the walking assistive robot.

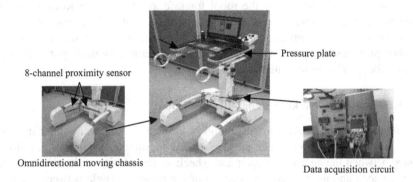

Fig. 1. Walking rehabilitation training robot

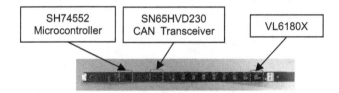

Fig. 2. 8-channels proximity sensor

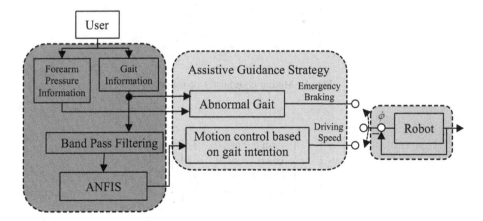

Fig. 3. Overall block diagram of the proposed assistive strategy with safety and compliance

3 Motion Control Method

The Fig. 3 shows the block diagram of the proposed assistive strategy with safety and compliance, which serves as the controller for the robot. In Fig. 3, the ANFIS (adaptive neural fuzzy inference system) receives the relative distance data of both legs from the proximity sensors, and then generates an output driving speed. This output value, combined with the emergency braking force determined by the abnormal gait recognition system, is sent to control the robot for safe assistance. The forearm pressure information and gait information of user are applied to identify the abnormal gait [11]. It is crucial to provide emergency braking force with the aim of walking safety. Finally, the robot provides the functions of assisted-walking and abnormal gait recognition based on the user's motion intention.

3.1 Kinematics of Walking Rehabilitation Robot

According to the robot's structure, the motion coordinates of the walking rehabilitation robot is established in Fig. 4. The parameters and coordinate system are as follow: l: distance between the geometric center and wheel; v: the velocity of the robot; α: angle between x-axis and the movement direction of the robot; θ_i: angle between the wheel i ($i = 1,2,3,4$) and the x-axis. The kinematics equation of robot is:

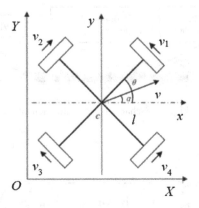

Fig. 4. Motion coordinates of robot

$$\begin{bmatrix} v_1 \\ v_2 \\ v_3 \\ v_4 \end{bmatrix} = \begin{bmatrix} -\sin\theta & \cos\theta & l \\ \cos\theta & \sin\theta & -l \\ -\sin\theta & \cos\theta & -l \\ \cos\theta & \sin\theta & l \end{bmatrix} \begin{bmatrix} v_x \\ v_y \\ \dot{\theta} \end{bmatrix} \qquad (1)$$

where v_x, v_y are the x and y component of velocity of robot. The relationship between the four wheels speed is:

$$v_1 + v_2 = v_3 + v_4 \tag{2}$$

Then we obtain the following equations:

$$\begin{aligned} v_x &= \gamma_x G_v \cdot \cos \theta \\ v_y &= \gamma_y G_v \cdot \sin \theta \end{aligned} \tag{3}$$

where γ_x, γ_y are the velocity parameters along x-axis and y-axis, and G_v is walking intention speed.

3.2 Adaptive Neural Fuzzy Inference System

Considering the fuzziness properties in this computational process, especially the mapping between the user's intention speed and measured gait information, the fuzzy approach is applied in this paper. To be applicable for users with various motion capability, the adaptive neural fuzzy inference system (ANFIS) is adapted for its excellence at adaptation and low computational cost. The designed architecture of ANFIS is shown in Fig. 5, which includes five layers. The calculation process is shown in Fig. 6.

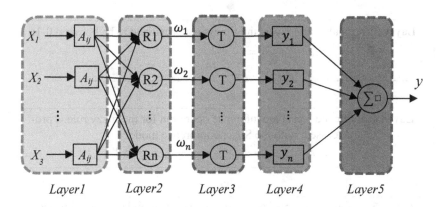

Fig. 5. The ANFIS architecture

Where c_{ij}, σ_{ij} are the premise parameters to be adjusted by using the back propagation gradient method in Fig. 6. ω_i expresses the fitness of the fuzzy rule. $p_{i0}, p_{i1}, \cdots p_{i10}$ are the parameters of the inference part. For the training of the learning scheme, the input data of the 32 distance data and intention velocities, and output data of the driving speed are collected.

4 Experiments and Discussion

4.1 Experiment of Auxiliary Walking

The experiment of auxiliary walking is conducted based on the proposed method. Four subjects walk 10 times (20 m/time) assisted by the rehabilitation robots, as shown in Fig. 7. Two walking states are included, normal gait and restraint gait. To simulate the restraint gait of the elderly and the disabled, the knee joint of subject is fixed with the holder.

Layer 1:(input layer) each code in this paper represents the degree of the membership function for the input variables with Gaussian shaped membership function:

$$\mu_{A_{ij}}(x_j) = \exp[-\frac{1}{2}(\frac{x_j - c_{ij}}{\sigma_{ij}})^2], \ i = 1,2,..10, \ j = 1, 2,....10$$

Layer 2:(rule layer) the fitness of each fuzzy rule is determined by using the t-norms algebraic product operation:

$$\omega_i = \mu_{A_{i1}}(x) \times \mu_{A_{i2}}(x) \times \cdots \times \mu_{A_{i10}}(x), i = 1, \cdots, 10$$

Layer 3:(normalization layer) the fitness is normalized to be between 0-1:

$$\bar{\omega}_i = \frac{\omega_i}{\sum_{i=1}^{10} \omega_i}$$

Layer 4:(inference layer) the inference operation for the fuzzy rule is processed, with the Sugeno inference model:

$$y_i(x_1, x_2, ..., x_{10}) = p_{i0} + p_{i1}x_1 + \cdots + p_{i0}x_{10}, i = 1, \cdots, 10$$

Layer 5:(output layer) this layer is the defuzzification layer. It infers each fuzzy rule for the final output:

$$y = \sum_{i=1}^{10} \bar{\omega}_i y_i$$

Fig. 6. The five layers of ANFIS

Figure 8 shows relative distance data between both legs and the rehabilitation robot under the conditions of normal gait and restraint gait. For the purpose of exact goal, data of start-up time and end time are excluded during the whole calculation. The subject 1 is the normal gait assisted by the rehabilitation training robot in Fig. 8(a) and the subject 2 is the restraint gait assisted by the rehabilitation training robot in Fig. 8(b). The normal gait of subject 1 is alternate swing in Fig. 8(a), which implies that the rehabilitation training robot effectively keep up with the user's intentional walking speed. The right leg of subject 2 is nearer to the robot in restraint gait than left leg, which is an effective and ease method to support the body weight according to the subject's walking habit. The experiments show that the proposed approach accurately recognize the gait rhythm, which provides the theoretical basis for calculating the walking speed based on user's subjective intention. The rehabilitation training robot based on the proposed approach effectively keep up with the users' walking speed (Fig. 9).

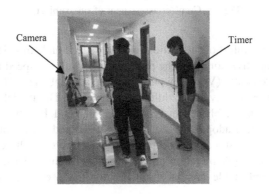

Fig. 7. Experiment of auxiliary walking

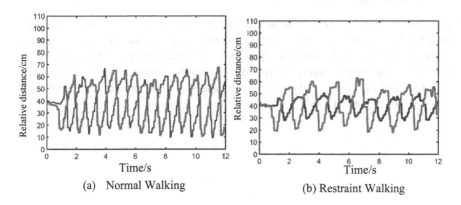

(a) Normal Walking

(b) Restraint Walking

Fig. 8. Gait information

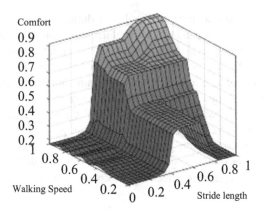

Fig. 9. Comfort assessment of one subject

To quantify the comfort of interaction, walking speed and stride length are applied to evaluate the subjective comfort. The parameters of walking speed and stride length are divided into five grades (Very Low, Low, Common, High, Very High). It represents the user's five perception levels for different parameters. The model of comfort evaluation for subject 1 is established in Fig. 7. The subject 1 has the most comfortable feeling when the system adopts the fast walking speed and the medium stride length. The proposed auxiliary strategy close to active gait improves the comfort of users. Therefore, the walking rehabilitation training robot with the proposed controller provides the most comfortable auxiliary strategy for the users with different motion capability.

4.2 Method Comparison

To examine robot's movement accuracy, the following experiment and method comparison are implemented. When one subject performs 100 forward steps of uniform length (approximately 30 cm), we check how accurately the rehabilitation robot followed the stride length. Figure 10(a) and (b) show the results conducted by previous control method in [12] and [13], respectively, and Fig. 10(c) presents the results based on the proposed control method.

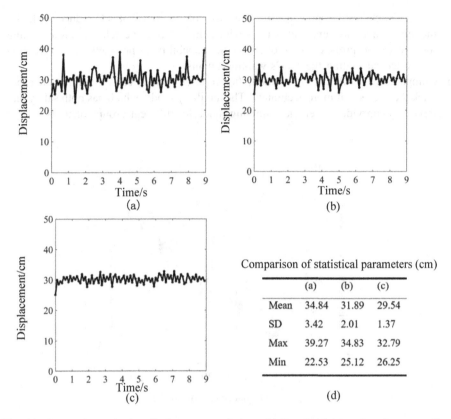

Comparison of statistical parameters (cm)

	(a)	(b)	(c)
Mean	34.84	31.89	29.54
SD	3.42	2.01	1.37
Max	39.27	34.83	32.79
Min	22.53	25.12	26.25

(d)

Fig. 10. Comparison of robot displacement variations. (a) Result of the prelearning controller. (b) Result of the PID controller. (c) Result of the proposed controller. (d) Comparison of statistical parameters

Compared to Fig. 10(a) and (b), the signals in Fig. 10(c) have the minimum fluctuation. The mean value, standard deviation, maximum value, and minimum value with different control methods are shown in Fig. 4(d), which includes the statistical results of subjects with various motion capability. The experiment shows that the proposed control method effectively generates straight-line motions that closely corresponded to the stride lengths and stride rates of the subject. It provides a solid theoretical basis for active-assisted walking training and semi-assisted walking training in walking rehabilitation training.

Moreover, experiment with time-varying parameters is performed where the forward movements of a subject are retested by varying the stride lengths in the following order: 10, 15, 20, 25, and 30 cm in Fig. 11. The experiment shows that the proposed method has more accurate displacements compared to the results with the PID controller. Specifically, in Fig. 11, the widest fluctuation range appears in the 15 cm-stride, and the strides exceeding 20 cm showed relatively fewer fluctuations. We can verify the advantages of the proposed method through experiments:

a. The statistical results show that the proposed control contributes to reduce the faulty measurements and generations in previous time steps. The rehabilitation training robot with the proposed controller has the capability to precisely control motion state after calculating the user's walking intents.

b. Compared with PID control method, the proposed method can ensure that the robot track the user's gait more accurately. The auxiliary mode, which has both safety and flexibility, provides an efficient training mode for different groups of users.

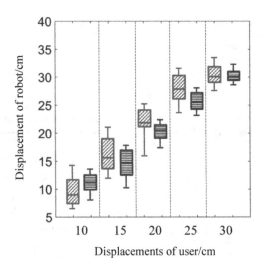

Fig. 11. Comparison of robot displacements based on PID controller and the ANFIS controller, respectively, for different stride lengths

To verify the economy and reliability of this method, we compare the proposed method with traditional method, as shown in Table 1. Most of the abnormal gait recognition methods need to wear special wearable equipment. Although the recognition rate is almost the same as other recognition methods, the proposed method increases the recognition of dragging abnormal gait, and it's more comfortable, economic and convenient. For the purpose of better service for the elderly and people with walking disabilities, our laboratory established smart house based on multiple welfare-robots: GRR, WAR, Intelligent Wheelchair Robot (IWR), Excretory Support Robot (ESR), and Transportation Robot (TR). These robots integrate the proposed method, as shown in Fig. 12. It includes 3 areas: recreation area, living area, and rehabilitation area. The method proposed in this paper can be applied to similar walking aids, which has a better promotion value for the scene of assisting the elderly and the disabled in daily life.

Table 1. Comparison of recognition system of abnormal behavior

	COP-FD	SVM	Our method
Wear equipment	Proprietary wearable devices	Proprietary wearable devices	No
Wearing position	Shoes	Waist	No
Recognition in multiple directions	Yes	Yes	Yes
Dragging gait	Yes	No	Yes
Accuracy rate	90%	95.83%	91.22%
False recognition rate	18.18%	0.89%	6.27%

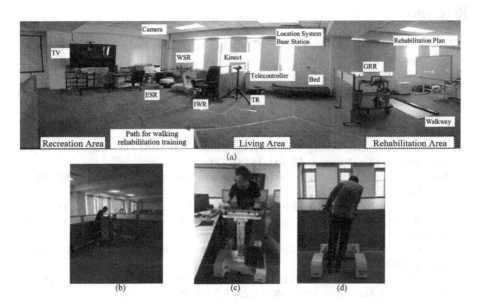

Fig. 12. Experimental scenes in daily life. (a) Application scene: the smart house based on five welfare robots (b) Walking rehabilitation training based on preset path (c) Auxiliary walking and take things on the table (d) The recognition of abnormal gait

5 Conclusions

This study proposes a non-contact auxiliary strategy with compliance and safety for walking rehabilitation training robot. Firstly, we develop the walking rehabilitation training robot for the disabled and the elderly. To achieve a non-contact motion intention detection, we develop the 8-channel proximity sensor, and establish the detection platform of gait information. Then, we propose the motion controller with compliance and safety. Especially, the ANFIS is proposed to investigate the user's motion intention and generate the control target. The experiments and comparative analysis are implemented to verify the superiority and effectiveness of the proposed method. The results show that our method can effectively provide the assist strategy with compliance and safety for the elderly and the disabled with different motion capability. Compared with the previous methods, our method has improved the system economy, which is more suitable for an intelligent walker with a similar structure. The assist-robots integrated with the proposed method have a good universal property for the elderly and the disabled with weak motion capability in the hospitals, pension centers, and families. However, the proposed algorithm is verified on the simulated scenarios in our lab. It may be different from the real scene of assisting the elderly and the disabled. Therefore, we plan to conduct detailed research and continue to optimize the algorithm in some specific scenarios in future work.

References

1. Wilson, G.: Robot-enabled support of daily activities in smart home environments. Cogn. Syst. Res. **54**, 258–272 (2019)
2. Shishehgar, M., Kerr, D., Blake, J.: A systematic review of research into how robotic technology can help older people. Smart Heath **17**, 1–18 (2018)
3. Ma, Q., Ji, L., Wang, R.: The development and preliminary test of a powered alternately walking exoskeleton with the wheeled foot for paraplegic patients. IEEE Trans. Neural Syst. Rehabil. Eng. **26**(2), 56–61 (2012)
4. Miura, K., et al.: Gait training using a hybrid assistive limb (HAL) attenuates head drop: a case report. J. Clin. Neurosci. **52**, 141–144 (2018)
5. Eck, D., Schilling, K.: Evaluation of a drive assistance function for older adults. IFAC Proc. Vol. **45**(4), 176–181 (2012)
6. Park, J., Kim, T., Kim, J.: Model-referenced pose estimation using monocular vision for autonomous intervention tasks. Auton. Robots **44**(2), 205–216 (2019). https://doi.org/10.1007/s10514-019-09886-9
7. Zhang, X., Yue, Z., Wang, J.: Robotics in lower-limb rehabilitation after stroke. Behav. Neurol. **2017**(4), 1–13 (2017)
8. Hu, J., Hou, Z.G., Chen, Y.X., et al.: Lower limb rehabilitation robots and interactive control methods. Acta Automatica Sinica **40**(11), 2377–2390 (2014)
9. He, K., Tian, Y.Q., Li, Q., et al.: Design of speech-recognition intelligent trash based on LD3320. Foreign Electron. Measur. Technol. **6**, 85–88 (2016)
10. Park, S., Chung, W.K., Kim, K.: Training-free Bayesian self-adaptive classification for sEMG pattern recognition including motion transition. IEEE Trans. Biomed. Eng. **67**(6), 1775–1786 (2020)
11. Zhao, D.H., Yang, J.Y., Bai, D.C., et al.: A non-contact recognition method of abnormal gait based on node-iteration type fuzzy petri net. Chinese J. Sci. Instrum. **40**(4), 227–236 (2019)
12. Akane, F., Naoya, M., Jiang, Y.L., et al.: Gait analysis based speed control of walking assistive robot. In: Proceedings of the 2018 IEEE International Conference on Intelligence and Safety for Robotics, pp. 88–92 (2018)
13. Lee, G., Jung, E.J., Ohnuma, T., et al.: JAIST robotic walker control based on a two-layered Kalman filter. In: Proceedings of IEEE International Robotic Automation, pp. 3682–2687 (2011)

Adaptive Collision-Free Reaching Skill Learning from Demonstration

Uchenna Emeoha Ogenyi, Dalin Zhou, Zhaojie Ju$^{(\boxtimes)}$, and Honghai Liu

University of Portsmouth, Portsmouth, UK
{uchenna.ogenyi,dalin.zhou,zhaojie.ju,honghai.liu}@port.ac.uk

Abstract. In this paper, we considered the task of the robot learning low-level trajectory task in a novel clustered constraint environment. We propose a novel adaptive trajectory algorithm used to generate the necessary trajectory which satisfies the constraint of avoiding collision with an obstacle. Our approach is based on Gaussian mixture model which decomposes the trajectory into several ellipses since the isoline of a single Gaussian model is also an ellipse. Moreover, we employed the principle of the artificial potential field to modify the direction of the motion in the presence of obstacles. Since our approach is based on the underlying reactive skill dynamics, it does not share the same disadvantages as approaches which assume both the model of the task trajectory and the response from the obstacle should be learned from the demonstrations.

1 Introduction

The most essential skill in humans is the reaching skill. Human beings apply reaching skill in most of our daily life activities to bring a human hand to the object of interest without collision. This strong environmental adaptive skill is required for collaborative robot arms that must coexist with a human operator at work places such as in part assembly tasks. For industrial manipulators, it is usually assumed that the motion of the end-effector is mostly disturbed by any obstacle especially when the task learning is a low-level trajectory task. In such a situation, the task execution must be interrupted and a control algorithm has to be used to move the end-effector around the obstacle. Addressing this challenging problem requires a robot to possess a set of skills which may be difficult to pre-program since the position of the obstacle is usually not known in advance.

Ignorance of such circumstance led to most prior approaches proposed in the past [1,2] assuming that demonstrations are performed in uncluttered constrained environments. But the presence of cluster in the environment can introduce additional constraints which if not accounted for can undermine the underlying human intent of reaching the desired goal. So, their approach is impractical in a real-world scenario since human environment keeps changing with time.

© Springer Nature Singapore Pte Ltd. 2020
J. Qian et al. (Eds.): ICRRI 2020, CCIS 1336, pp. 105–118, 2020.
https://doi.org/10.1007/978-981-33-4932-2_8

Further approaches employed the principles of motion planning algorithms which can compute collision-free path required for a robot to successfully complete a task [3]. Researchers have employed various motion planning algorithm, such as Dynamic roadmaps (DRM) [4,5], Rapidly Exploring Random Trees (RRT) [6,7], and elastic [8,9]. Although, significant improvements have been made using this approach, in general, motion planning algorithms is computational complex due to the high dimensionality of the configuration space that must be considered.

Another school of thought believe that Learning from Demonstration (LfD) could better be employed in learning skills from human demonstration in a clustered environment. Amongst other benefits of LfD approach is its flexibility in learning complex motions and simplicity/easy to implement [10]. Also, LfD has the advantage of allowing people without programming skills to transfer tasks skills to the robot just by demonstrating and allowing the robot to learn from those demonstrations [11]. This attribute of LfD makes the robot more adaptive to a new environment with only minor adjustment to the learning parameters.

Several works have been proposed to learn the dynamic movement primitive as well as coupling terms for obstacle avoidance from demonstrations. For example, [12] learn a DMP which encapsulates correlation information of the coupling motor control variables and employed reinforcement learning to modulate the optimal parameters of the dynamical system in a new an environment. Furthermore, Chi et al., [13] method integrated dynamic potential filed with the acceleration equation of the DMP to realise reactive action depending on the distance and velocity between the robot's end-effector and the obstacle. [14] also employed this method and further introduced cost function to estimate the deviation from the mean of demonstrations to the distance of the obstacle from the environment. A major drawback of this approach is the assumption that only the mean of the demonstration sufficiently expresses demonstrated skills. In contrast, we propose a control strategy which aims to in addition to the mean of the distributed demonstrations identify and incorporate how spread the distribution is in order to efficiently express the demonstrated skills. Moreover, DMPs allows only a single demonstration which limited its ability to potentially learn different ways of executing tasks skills, hence limiting its robustness in new scenarios.

On the other hand, the statistical model (GMM/GMR) permits variant demonstrations for task learning, hence promising robust generalisation of tasks in new scenarios. The probabilistic approach 'GMM/GMR'was used for modelling and predicting motion trajectory [15]. In another work, [16] presented a method that uses GMM/GMR to encode and compute an average trajectory from a set of sub-optimal task demonstrations. Some approaches [17] added the skill constraints at the demonstration stage such that the influence of the constraints is learned to achieve the desirable trajectory at the reproduction stage. However, learning the influence of the entire constraints from the demonstration stage is not feasible and therefore this approach is not suitable for an adaptive reaching tasks considered in our work. Our proposed approach estimates the parameter both from the demonstration state and the reproduction stage to more accurately acquire the desired trajectory.

In our work, we tackle the problem of learning skills from human demonstrations which can be influenced by the presence of obstacles. We use kinaesthetic teaching for data collection from human demonstrations. From these demonstrations, robot joint angles and its associated timesteps are recorded and used to learn the direction which will lead the end-effector through the desired motion. We adopt the GMM to compute the optimal path for the robot to reach the goal point and proposed a modified virtual repulsive potential function to avoid obstacles.

2   Motion Trajectory Learning

The most popular method for learning human motion trajectory is by using Gaussian Mixture Model (GMM). The GMM is a mixture of a sequence of Gaussian distributions that can allow an arbitrarily large number of Gaussian components and a small number of variances. Rather than having hard assignments into clusters, like in K-means, GMM supports soft assignments which implies that each distribution has some level of responsibility for producing a given data point.

2.1   Gaussian Mixture Model

To extract the characteristics of a human arm motion trajectory for any collection of time $t = (t_1, t_2, ..., t_N)^T$, the probability density function of the Gaussian distribution of a vector $x = (x_1, x_2, ..., x_N)^T$ with K-components of a D-dimension is defined as:

$$p(x_j|\theta) = \sum_{m=1}^{M} \pi_m \mathbb{N}(x_j|\mu_m, \Sigma_m) \tag{1}$$

Where $p(x_j|k) = (x_j; \mu_m, \Sigma_m)$ is the set of the model parameters; μ_m is the mean and Σ_m is the covariance matrix of the Gaussian, and (π_m) is the prior of the j-th component, and exp is the exponential function. The priors are adjusted to satisfy the constraints in (8) and (3) where the sum of all the mixture weights must be equal to 1 and each of the weight must lie between 1 and 0.

$$\sum_{m=1}^{M} \pi_m = 1 \, for \, m = 1, ..., M \tag{2}$$

$$0 < \pi_m < 1 \tag{3}$$

$$\mathbb{N}(x_m; \mu_m, \Sigma_m) = \frac{1}{(2\pi)^{\frac{D}{2}} \sqrt{|\Sigma_m|}} \exp(-\frac{1}{2}(x_i - \mu_m)^T \Sigma^{-1}(x_i - \mu_m)) \tag{4}$$

In most cases, the optimal number of K-component is unknown. One way to estimate it is by using the information comparing criteria. The popular ones are the Bayesian Information Criterion (BIC) and the Akaike Information Criterion (AIC). While both can estimate the required components, however, care must be taken on the choice of the information criteria, as the BIC tend to choose more complex models that might overfit whereas, the AIC tends to choose simple models that might underfit.

2.2 Gaussian Mixture Regression

In order to retrieve smooth trajectories of the estimated trajectories, the trajectories are considered as a regression problem and Gaussian Mixture Regression (GMR) is applied. Following the regression method, the conditional expectation of x_s given x_t is estimated as follows:

$$\mu_k = \begin{bmatrix} \mu_{t,m} \\ \mu_{s,m} \end{bmatrix}, \quad \Sigma_m = \begin{pmatrix} \Sigma_{tt,m} & \Sigma_{ts,m} \\ \Sigma_{st,m} & \Sigma_{ss,m} \end{pmatrix} \tag{5}$$

By applying the weighted mean and variance, the expected distribution of x_s given x_t can be computed as a block decomposition of the data-point x_j. For each component K, the conditional expectation of $x_{s,m}$ given the temporal value x_t and the estimated conditional covariance of x_s given x_t are given by the theorem of Gaussian conditioning based on combination property of Gaussian distribution as follows:

$$x_{s,m} = \mu_{s,m} + \Sigma_{st,m}(\Sigma_{t,m})^{-1}(x_t - \mu_{t,m}), \tag{6}$$

$$\Sigma_{s,m} = \Sigma_{s,m} - \Sigma_{st,m}(\Sigma_{t,m})^{-1}\Sigma_{ts,m} \tag{7}$$

2.3 Learning GMM Parameters

Learning requires estimating the model parameters (the mean, covariant and mixing coefficient) of the distribution to permit soft assignments of the K-components to the data points. It helps to find the best-fit parameters for the model. To find the maximum likelihood of a data point being fitted into the k-components, EM employs Bayes theorem to compute the probability that given observation belongs to each cluster. It works by choosing random values for the mixing data points and using those guesses to estimate the next set of the data [18]. The EM algorithm consists of two steps: E-step or Expectation step and the M-step or Maximisation step.

E-step: Estimates the distribution of the hidden variable given the data and the current values of the parameters. To achieve that, a latent variable Z_m is introduced to each data point. The latent variable Z_m indicates the probability that the ith data point is generated from the mth Gaussian components. In order to start the E-step, we need to initialise the values of the parameters. A good practice is to estimate them using k-means.

$$Z_m^i = \frac{\pi_m \mathbb{N}(x_i|\mu_m, \Sigma_m)}{\sum_{m=1}^{M} \pi_m \mathbb{N}(x_i|\mu_m, \Sigma_m)} \tag{8}$$

$$Z_m = \sum_{i=1}^{N} Z_m^i \tag{9}$$

M-step: After calculating the posterior, we then need to estimate the parameters of each Gaussian and evaluate the log-likelihood. To do this, we first compute the maximum likelihood of the parameters estimated in E-step. This set of iteration continues until the threshold defined in the log-likelihood is met.

$$\pi_m^{new} = \frac{1}{N} \sum_{j=1}^{N} Z_m^i \tag{10}$$

$$\mu_m^{new} = \frac{1}{Z_m} \sum_{i=1}^{N} Z_m^i x_i \tag{11}$$

$$\Sigma_m^{new} = \frac{1}{Z_m} \sum_{i=1}^{N} Z_m^i (x_j - \mu_m^{new})^T (x_j - \mu_m^{new}) \tag{12}$$

The EM is well known for its convenient and easily extensibility to incremental learning, however, the major limitation of the EM algorithm is that it requires to keep all the historical data in memory for accurate order update to be achieved. Moreover, traditional GMM cannot adjust the parameters of the distribution as new data points arrive or are acquired, and can hone in on a local maxima that is not close to the optimal global maxima. To overcome this problem, [19] proposed an algorithm which merges density components to improve the log-likelihood and reduce the number of clusters. This proposed algorithm failed to account for the novel data points that might arrive one-by-one as it universally assumed that new data comes in blocks. Addressing this challenge, [20] proposed an approach to tackle novel data points which arrives one-by-one by keeping only two GMM components in the memory and without historical data.

3   Theoretical Background of Artificial Potential Field

The traditional potential field method as proposed by [21] assumes that the robot is moving by the influence of abstract artificial force fields. The artificial field consists of two components: the repulsive potential field and attractive potential field. Considering a robot that needs to move from its current position "A" to a goal position "B" in the presence of an obstacle "O". Using the principle of APF, the goal point generates attractive potential fields which make the robot move towards it. Conversely, the obstacle generates a repulsive force which is inversely proportional to the distance between the robot and the obstacle with a defined distance threshold to push the robot away from the obstacle. In other words, attractive force (15) is applied for the robot to reach the goal point and repulsive force (17) to avoid obstacles. The attractive artificial filed is represented as:

$$U_{att}(q) = \frac{1}{2}\eta\rho^2 \tag{13}$$

Here, η is a positive scaling factor, ρ is the distance between the robot q and the goal. Assuming the goal is a point object in a 2-D plane. The position of the goal and that of the end-effector could be expressed as vectors of $[x_g, x_g]^T$ and $[x_r, x_r]^T$ respectively. So, the parameter ρ which is the Euclidean distance between the robot end-effector position and the goal position which is calculated as below.

$$\rho = \sqrt{(x_r - x_g)^2 + (y_r - y_g)^2} \tag{14}$$

The corresponding attractive force is given by the negative gradient of the attractive potential (15). Consequently, there is a move from higher to lower potential field along the negative of the attractive field in other to reach the goal position.

$$F(att)(q) = -\nabla U_{att}(q) = \eta\rho \tag{15}$$

To prevent collision between the robot and the obstacle, the traditional repulsive potential function is represented as follow:

$$U_{rep}(q) = \begin{cases} \frac{1}{2}k(\frac{1}{d(x)}) - \frac{1}{d_0})^2 & : d(x) \leq d_0 \\ 0, & : d(x) > d_0 \end{cases} \tag{16}$$

where k is the repulsive potential gain coefficient, $d(x)$ is the minimum distance between the robot's current position and the obstacle, and d_0 is the distance threshold which limits the range of the repulsive potential field. So, when the distance between the robot and the obstacle is greater than d_0, the robot will not be affected by the repulsive force. Assuming, the position of the obstacle is represented as $[x_o, yo]^T$. Then, the associated repulsive forces could be computed by finding the negative gradient of the repulsive potential function (16):

$$F_{rep}(q) = -\nabla U_{rep}(q) = \begin{cases} k(\frac{1}{d(x)}) - \frac{1}{d_0})\frac{1}{d(x)^2}\frac{\partial d(x)}{\partial q} & : d(x) \leq d_0 \\ 0 & : d(x) > d_0 \end{cases} \tag{17}$$

$$d_x = \sqrt{(x_r - x_o)^2 + (y_r - y_o)^2} \tag{18}$$

The total repulsive potential field can be obtained by summing up the potentials caused by all of the obstacles within the workspace.

$$U_{rept}(q) = \sum_{i=1}^{N} U_{repi}(q) \tag{19}$$

Using this function, moving the robot to the goal while avoiding obstacle is achieved by following the direction of the resultant forces obtained by summing all the negative gradients attractive/repulsive potentials of the target and obstacle obtained from a given point in the plane. Doing this, the total potential force field vectors will point away from the obstacle and towards the goal as influenced by the combined gradient of the attractive/repulsive potentials.

$$F_{net} = F_{att} + F_{rep} \tag{20}$$

4   Desired Direction Estimation

In this section, we presented an approach to fit ellipses on the generalised trajectory to permit elliptical potential field force around the segmented ellipse that formed the trajectory. In addition, we modelled the optimal closest point from the ellipse to the obstacle to enhance the robot's ability to avoid obstacle with minimal computational power.

4.1   Trajectory Segmentation with Fitted Ellipse

Considering that a Gaussian mixture model consists of several single Gaussian models (SGM) which can be represented by an ellipse since the isoline of probability density is also an ellipse [22]. Based on that, we hypothesized to fit ellipses on the generalised trajectory from the GMR to permit elliptical potential field force influence around the ellipse rather than on the generalized trajectory. To formed the ellipse and computed the orientation of the ellipse; firstly we computed the eigenvalues and corresponding eigenvectors and used then to find the axes of the strain ellipses which is used to calculate the ellipse orientation.

Using the general equation of an ellipse which is centred at $(0,0)$, aligned at the major and minor axes that are defined by σ_x and σ_y; where $\sigma_x > \sigma_y$ is given as:

$$\left(\frac{x}{\sigma_x}\right)^2 + \left(\frac{y}{\sigma_y}\right)^2 = 1 \tag{21}$$

Assuming the same single GMM is sampled from the underlying Gaussian distribution having the centre determined by the mean μ_i. Then, the ellipsoidal probabilistic contour of the eigenvectors $\vec{V}_i$ with a corresponding covariance matrix Σ_i for a given K-component is given as $[\vec{V}_i, D_i] = \text{Eig}(\Sigma_i)$. This is an indication that the shape and size of the ellipse are underpinned by its associated covariance matrix Σ_i.

The orientation of the component is constructed as the angle ϕ of largest eigenvector towards the x-axis:

$$\phi = arctan\frac{\vec{V}x_2}{\vec{V}x_1} \tag{22}$$

where $\vec{V}x_2$ and $\vec{V}x_1$ are the eigenvectors of the covariance matrix that corresponds to the largest eigenvalues. This is an indication that the magnitude of the component alignment to the desired direction can be tuned by modifying the eigenvalues of the confidence ellipse and computing the angle of the largest matrix eigenvalues.

4.2   Obstacle Closest Distance

Various distance threshold calculation approaches have been proposed in the literature [23]. The most common approach is to introduce a fixed threshold value which must be maintained at every situation. This approach is not adaptive since the threshold value is fixed and is independent of the inclined angle of the object to the position of the robot. The most adaptive approach to calculate the threshold distance by sampling different point locations on the surface of the obstacle in order to estimate the closest point from each of the K-component ellipses of the trajectory.

The case of finding the closet point from the obstacle to a single GMM component (the trajectory) is established by analyzing the steps captured by [24] and summarized here. Considering an ellipse that is centred along the x and y axes, and parameterised by (x_e, y_e) with radius a and b representing the semi-major and semi-minor axis of the ellipse at an angle ϕ. With the values of point $P_p(x_p, y_p)$ and point $P_e(x_e, y_e)$ given, the distance $D(P_p, \phi)$ between the two points can be calculated as in Eq. 23.

$$(D(P_p, \phi)) = \sqrt{(x_p - x_e)^2 + (y_p - y_e)^2} \tag{23}$$

From Eq. 23, we can deduce that the distance $D(P_p, \phi)$, is a function of ϕ, and any value of ϕ which drive the differential of $D(P_p, \phi)$ to zero will form the minimum distance which is D_{min}.

$$\frac{D(P_p, \phi)}{d\phi} = 0 \tag{24}$$

By substituting the values of $x_e = a.cos\phi$ and $y_e = a.sin\phi$, into the Eq. 23, we can compute the minimum distance as shown in Eq. 25.

$$D_{min} = \sqrt{(x_p - a.cos\phi)^2 + (y_p - b.sin\phi)^2} \tag{25}$$

4.3   Trajectory Component Adjustment

Haven computed the closest point from the obstacle point to the desired trajectory, and also the desired orientation of the fitted ellipse, this section presents the schematics for the component adjustment. The component adjustment will ensure that the robot maintains the threshold and adjust any component(s) that falls below it in the presence of an obstacle. By ensuring that only affected components are adjusted, the computational complexity is reduced and the motion

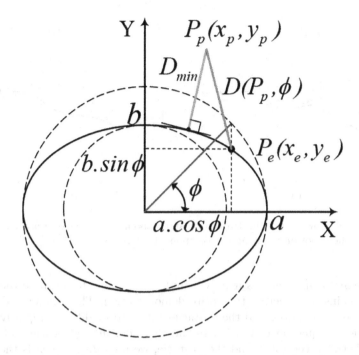

Fig. 1. Geometry for calculating the shortest distance between the robot and obstacle

of the robot will become similar to the way human being cloud have performed the same task.

The overview of component adjustment approach is illustrated in Fig. 2. Figure 2 (a) shows the presence of an obstacle within the workspace of the robot. The black dotted lines represent the optimal trajectory the robot would follow in the absence of an obstacle. Now that the environment has changed, the robot must avoid the obstacle while moving to the target point and it must maintain a close distance from the optimal trajectory. To apply our approach, we fitted the acquired trajectory with ellipses in the form of 6-GMM components and measures the distance of the obstacle to the components. It was computed that only 3-components are closer to the robot the obstacle. Thus, they are the only components that must be adjusted in order to satisfy the task and environmental constraints. As illustrated in Fig. 2 (b), those identified components were adjusted by applying the modified AFP presented in Subsect. 4.4.

4.4 Modified APF

We modified the traditional APF so we can apply it in our obstacle avoidance case. Here the focus is on the repulsive potential field as the attractive potential field has the form of the traditional attractive potential field. To find the repulsive points, the trajectory is segmented and interest is focused on the areas with potential of been influenced by the obstacle. This approach reduces the

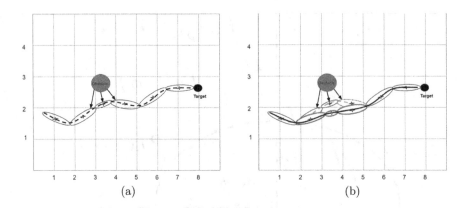

Fig. 2. Figure showing (a) the trajectory at the point of obstacle approach (b) the response of the robot on applying the component adjustment method.

computational complexity as it only samples few points to estimate the closet point as against considering the entire demonstration. The minimum distance between the fitted ellipse and the obstacle is used to modify the repulsive force computational equation in Eq. 16. In the equation, instead of using $d(x)$, the distance between the robot and the obstacle, we used d_{min} which is the closet point from the robot to the segmented GMM component as shown in Fig. 1 and presented in Eq. 26.

$$\hat{U}_{rep}(q) = \begin{cases} \frac{1}{2}k(\frac{1}{(d_{min})} - \frac{1}{d_{th}})^2 - e^{\sigma_i^2} & : d_{min} \leq d_{th} \\ 0, & : d_{min} > d_{th} \end{cases} \quad (26)$$

where σ_i represents the width of the distribution for each of the given K-component of the Gaussian.

$$\hat{F}_{net} = \hat{F}_a(q) + \hat{F}_r(q) \quad (27)$$

5   Experimental Setup

The operator is tasked to perform a realistic robot sweeping an object into a dustpan while avoiding multiple stationary obstacles within the scene. Using the robot, a user pushes a piece of rubbish denoted by T on 14 cm 14 cm table into a dustpan (G) while avoiding obstacles (o1, o2 and o3) as depicted in Fig. 3. The operator repeated this task for three time in order to collect a set of demonstrations required for the GMM/GMR.

5.1   Data Acquisition and Preproccessing

The experimental data is acquired from a widely used Sawyer robot platform. The research version of Sawyer robot is compatible with ROS and integrated

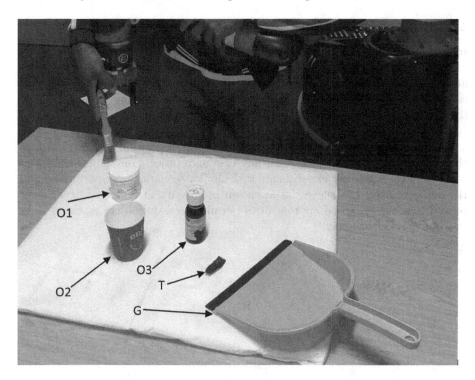

Fig. 3. In the experiment, a sample demonstration of a sweeping task is performed.

with a variety of sensors such as vision and force sensors. In the experiment, the Sawyer robot learns how to sweep an object into the dustpan while avoiding multiple objects (obstacles). A similar approach has been studied to allow a robot such as Roomba Vacuum Cleaning Robot to better adapt to an unseen environment [25]. The teaching process can be achieved in two ways: (i) using a remote joystick to teleoperation through the desired task (ii) using kinaesthetic teaching. In this work, we adopt the second method which allows the human demonstrator to move the robot manipulator through the desired motion while the robot records the trajectory. The key benefits of this method are that it allows for accuracy and direct recording of the control commands, and ensures that the demonstrations are constrained to actions that are within the robot's abilities, hence eliminating correspondence problem.

We assume that the important sensing information comes from (i) the state of the robot's end-effector which is obtained using the Sawyer robot built-in encoder, (ii) the distance between the robot and the target, and (iii) the shortest distance between the robot and the obstacles retrieved from the Kinect sensor since it can return distance information representing an absolute position of the obstacle in a specific range and we choose the shortest distance as our desired value as illustrated in Eq. 25.

5.2   Experimental Result

In order to validate the proposed learning by demonstration with obstacle avoidance approach proposed, we conducted several experiments for a robot to generate a trajectory needed to satisfy both the task constraints and the environmental constraints. The experiment is a sweeping task which involves using the robot to push a piece of rubbish denoted by T on 14 cm 14 cm table into a dustpan (G) while avoiding obstacles (o1, o2 and o3) as depicted in Fig.4 (a). From Fig. 4 (a), the robot successfully moved through the path position of the rubbish to the dustpan without colliding with any obstacles objects on the table as presented in Fig. 3. Further experiments with similar scenarios were performed and the outcome came out successful as shown in (b)–(d). Overall, the experiment proved that the proposed approach can permit successful object reaching tasks in a clustered environment.

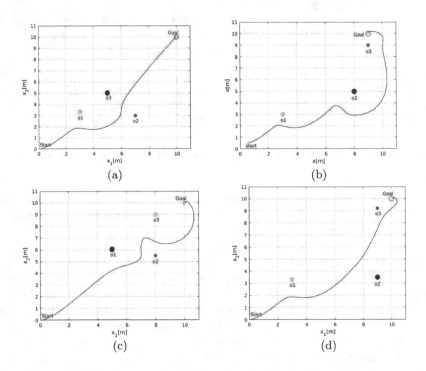

Fig. 4. Figure (a)–(d) shown samples of the reproduced trajectories. The circular points represent the position of obstacles in the environment while the start point is the origin and the goal point is defined ahead.

5.3   Conclusion

A learning by demonstration framework to enable a robot to move its manipulator from a start position to a goal position while avoiding obstacles is presented in this work. The kinesthetic teaching technique was employed to acquire motion demonstrations from the human teacher while DTW was used to align the demonstrated trajectories. The optimal trajectory is modelled using the Gaussian mixture model and Artificial potential difference. Our approach allows for the part of the demonstration prone to collision to be identified such that only a section of the demonstration will be adjusted to satisfy both the task and environmental constraints. Hence, reducing the amount to data to be left in the memory to achieve a satisfactory reproduction.

Although, obstacle avoidance is studied, there are still many problems to be further discussed. While our approach can thrive in an environment with static obstacles it is not guaranteed to perform greatly in an environment with moving obstacles. This area will be further investigated in future.

References

1. Khansari-Zadeh, S.M., Billard, A.: Learning stable nonlinear dynamical systems with Gaussian mixture models. IEEE Trans. Robot. **27**(5), 943–957 (2011)
2. Calinon, S., Guenter, F., Billard, A.: On learning, representing, and generalizing a task in a humanoid robot. IEEE Trans. Syst. Man Cybern. Part B (Cybern.) **37**(2), 286–298 (2007)
3. Shin, K., McKay, N.: A dynamic programming approach to trajectory planning of robotic manipulators. IEEE Trans. Automatic Control **31**(6), 491–500 (1986)
4. Kunz, T., Reiser, U., Stilman, M., Verl, A.: Real-time path planning for a robot arm in changing environments. In: 2010 IEEE/RSJ International Conference on Intelligent Robots and Systems, pp. 5906–5911, October 2010
5. Leven, P., Hutchinson, S.: A framework for real-time path planning in changing environments. Int. J. Robot. Res. **21**(12), 999–1030 (2002)
6. Bircher, A., Alexis, K., Schwesinger, U., Omari, S., Burri, M., Siegwart, R.: An incremental sampling-based approach to inspection planning: the rapidly exploring random tree of trees. Robotica **35**(6), 1327–1340 (2017)
7. Zamani, A., Galloway, J.D., Bhounsule, P.A.: Feedback motion planning of legged robots by composing orbital lyapunov functions using rapidly-exploring random trees. In: 2019 International Conference on Robotics and Automation (ICRA), pp. 1410–1416. IEEE (2019)
8. Brock, O., Khatib, O.: Elastic strips: a framework for motion generation in human environments. Int. J. Robot. Res. **21**(12), 1031–1052 (2002)
9. Yang, Y., Brock, O.: Elastic roadmaps-motion generation for autonomous mobile manipulation. Auton. Robots **28**(1), 113 (2010)
10. Akgun, B., Cakmak, M., Jiang, K., Thomaz, A.L.: Keyframe-based learning from demonstration. Int. J. Soc. Robot. **4**(4), 343–355 (2012)
11. Konidaris, G., Kuindersma, S., Grupen, R., Barto, A.: Robot learning from demonstration by constructing skill trees. Int. J. Robot. Res. **31**(3), 360–375 (2012)
12. Kormushev, P., Calinon, S., Caldwell, D.G.: Robot motor skill coordination with EM-based reinforcement learning. In: 2010 IEEE/RSJ International Conference on Intelligent Robots and Systems, pp. 3232–3237, October 2010
13. Chi, M., Yao, Y., Liu, Y., Zhong, M.: Learning, generalization, and obstacle avoidance with dynamic movement primitives and dynamic potential fields. Appl. Sci. **9**(8), 1535 (2019)

14. Paxton, C., Hager, G.D., Bascetta, L.: An incremental approach to learning generalizable robot tasks from human demonstration. In: IEEE International Conference on Robotics and Automation (ICRA), pp. 5616–5621. IEEE (2015)

15. Wiest, J., Höffken, M., Kreßel, U., Dietmayer, K.: Probabilistic trajectory prediction with gaussian mixture models. In: 2012 IEEE Intelligent Vehicles Symposium, pp. 141–146, June 2012

16. Billard, A., Calinon, S., Dillmann, R., Schaal, S.: Survey: robot programming by demonstration. In: Handbook of Robotics, vol. 59, no. BOOK_CHAP (2008)

17. Rana, M.A., Mukadam, M., Ahmadzadeh, S.R., Chernova, S., Boots, B.: Learning generalizable robot skills from demonstrations in cluttered environments (2018)

18. Dempster, A.P., Laird, N.M., Rubin, D.B.: Maximum likelihood from incomplete data via the EM algorithm. J. Royal Stat. Soc.: Ser. B (Methodol.) **39**(1), 1–22 (1977)

19. Song, M., Wang, H.: Highly efficient incremental estimation of gaussian mixture models for online data stream clustering. In: Intelligent Computing: Theory and Applications III, vol. 5803. International Society for Optics and Photonics, pp. 174–183 (2005)

20. Arandjelovic, O., Cipolla, R.: Incremental learning of temporally-coherent gaussian mixture models. Society of Manufacturing Engineers (SME) Technical Papers, p. 1 (2006)

21. Khatib, O.: Real-time obstacle avoidance for manipulators and mobile robots. In: Proceedings of 1985 IEEE International Conference on Robotics and Automation, vol. 2, pp. 500–505, March 1985

22. Lin, H., Zhang, T., Chen, Z., Song, H., Yang, C.: Adaptive fuzzy Gaussian mixture models for shape approximation in robot grasping. Int. J. Fuzzy Syst. **21**(4), 1026–1037 (2019)

23. Certad, N., Acuna, R., Terrones, A., Ralev, D., Cappelletto, J., Grieco, J.C.: Study and improvements in landmarks extraction in 2D range images based on an adaptive curvature estimation. In: VI Andean Region International Conference, pp. 95–98. IEEE (2012)

24. Weerakoon, T., Ishii, K., Nassiraei, A.A.F.: An artificial potential field based mobile robot navigation method to prevent from deadlock. J. Artif. Intell. Soft Comput. Res. **5**(3), 189–203 (2015)

25. Denning, T., Matuszek, C., Koscher, K., Smith, J.R., Kohno, T.: A spotlight on security and privacy risks with future household robots: attacks and lessons. In: Proceedings of the 11th International Conference on Ubiquitous Computing, pp. 105–114 (2009)

Multi-robot Systems and Control

Event-Based Cooperative Control Framework for Robot Teams

Igor Bychkov[ID], Sergey Ul'yanov[(✉)][ID], Nadezhda Nagul[ID], Artem Davydov, Maksim Kenzin, and Nikolay Maksimkin

Matrosov Institute for System Dynamics and Control Theory at Siberian Branch of Russian Academy of Sciences, 134, Lermontova Street, Irkutsk, Russia
{bychkov,sau,sapling,artem,mnn}@icc.ru

Abstract. Efficient and autonomous execution of large-scale missions using a group of robots implies the use of an advanced control system, usually consisting of multiple subsystems arranged in a hierarchy. For the purpose of unification of interaction of separate subsystems, an event-based cooperative control framework for robot teams is developed. In the paper, we demonstrate how this framework can be used in solving challenging problems in robotics: path-following problem, real-time path-planning problem, group routing problem, and action-planning problem. A novel approach to formalization and analysis of logic discrete event systems, which are the main component of the framework, based on logic calculus and automatic theorem proving is also briefly described.

Keywords: Cooperative control · Discrete event system · Path-following · Path-planning · Group routing · Action planning

1 Introduction

The rapid development of unmanned technologies in recent decades has led to a significant growth in commercial, environmental and military robotic applications. In situations where extensive research is required in an area of interest, but no resources or time is available to deploy a static surveillance network, it is justified to use a team of autonomous robots.

To ensure long-term autonomous operation in an unknown environment, robots must be equipped with an advanced control system that typically consists of many interrelated components. There are two basic approaches [23] to organize a control system. Horizontal organization assumes the equality of individual system components, while vertical organization establishes their hierarchy. In practice, as a rule, hybrid solutions are used. Regardless of the type of organization, the issues concerning interaction of system components and formation of their behavioral logic are important. Many control architectures have been designed to treat these issues. However, for the most part

The event-triggered control framework has been mostly designed under support of the RFBR (Projects No. 20-07-00397 and No. 19-08-00746). Results of Sects. 4, 7 have been obtained under support of the Russian Science Foundation (Project No. 16-11-00053).

ⓒ Springer Nature Singapore Pte Ltd. 2020
J. Qian et al. (Eds.): ICRRI 2020, CCIS 1336, pp. 121–137, 2020.
https://doi.org/10.1007/978-981-33-4932-2_9

they are aimed at solving specific problems, for example, task allocation and planning [30], formation control [28], and path-planning [8,24], and suffer from scaling and uniformity deficiencies.

Shortcomings of known architectures can be overcome by using an event-driven approach [7,13], within which the behavior of individual system components is described in terms of events. Indeed, while considering some system, it may be noted that in fact a designer is interested in system's discrete state changes only at certain points in time through instantaneous transitions. If a concept of event is associated with each such transition, the notion of discrete event system is obtained. A discrete event system (DES) is a discrete-state, event-driven system, that is, its state evolution depends entirely on the occurrence of asynchronous discrete events over time. In many cases consideration of a complex system with the event-driven point of view proves to be helpful. In addition, event-based control systems benefit from being able to change the pattern of behavior only when it is necessary, thus reducing the computational load.

In the paper, we adopt and enhance a hierarchical control system for autonomous underwater vehicles (AUV) from [1,2]. In contrast to the analogues [7,13], the designed event-based framework uses events as triggers at different levels of control. Another essential feature of the control architecture presented is the usage of logical inference machine at different levels of the control system, for example, to constrain DES behavior according to paradigm of supervisory control and modify DES at the middle control level, or to design action plans due to the new information obtained at the high level of the control system. The paper describes general concepts of the framework (Sect. 2) with providing a formal approach to the analysis of its main component (Sect. 3) and demonstrates how it can solve challenging problems in robotics: cooperative path-following problem (Sect. 4), path-planning problem (Sect. 5), group routing problem (Sect. 6), and action planning problem (Sect. 7).

2   Event-Based Control Framework

For efficient implementation of joint operations, it is assumed that each AUV in the team is provided with a hierarchical control system consisting of three levels (see Fig. 1). The function of the lower level is to execute control laws corresponding to the current operating mode determined by the intermediate level. The planner at the intermediate (tactical) level determines the way of solving the task activated as a result of strategic planning at the upper level. The high-level planner at the upper level performs spatio-temporal decomposition of the mission into a sequences of tasks (or actions). Although placed at the intermediate level, the multicomponent DES, as a subsystem that provides a response to events detected by analyzing data from sensors and means of communication, is used for tasks corresponding to different control levels. So switching the DES state can initiate updating of control actions at the lower level or replanning a path at the intermediate level.

There are several levels of abstraction which a designer may implement when constructing DES. Timed and stochastic ones allow taking into account moments of time at which events occur. When such information is not necessary and only the order of events matters, untimed DES are employed, describing so called logical behavior of the

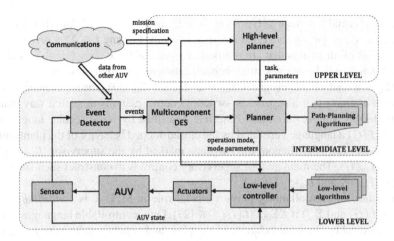

Fig. 1. Control system architecture.

system. The problems modeled with the help of logical DES include restricting state set which system may achieve, a proper ordering of events, finding unwanted events in the event sequences, and many others. A convenient way to formalize logical DES is automata model. Finite state automata explicitly represent system states and state transition structure what makes easier the system-theoretical properties study, such as, for example, safety and liveness of the system. Exactly automata-based logical DES are used in presented control system at the intermediate level.

A wide variety of problems traditional for the lower level can be efficiently solved with controllers synthesized using sublinear vector Lyapunov functions [15,33]. In Sects. 4 and 5, the control system design approach based on these functions are applied to build path-following controllers for both a single AUV and multi-AUV formation.

For the representation and processing of knowledge at the upper level, the logical calculus of positively-constructed formulas (PCFs) will be used [32]. A detailed discussion of the calculus, its characteristics and features can be found in [19]. In this paper, among other issues, we consider how the PCF calculus, primary used at the upper level of control system, may be applied to control DES.

3   PCF-Based Approach to des Control

In the late 1980s, for DES modeled as automata the supervisory control theory (SCT) was established. SCT supposes that some events may be forbidden from occurring by an acting device called a *supervisor*, in order to restrict system behavior according to a set of constraints. Comprehensive survey on the state of the art SCT may be found in [4,29,34]. In this section we consider how the logical inference in the PCF calculus may be employed to solve one of the basic problems of SCT.

Consider DES in the form of an automaton $G = (Q, \Sigma, \delta, q_0, Q_m)$, also called a *plant* in the automatic control theory [27]. Here Q is the set of states q; Σ is the set of events; $\delta: \Sigma \times Q \to Q$ is the transition function; $q_0 \in Q$ is the initial state; $Q_m \subset Q$ is the set of marked states. Let Σ^* denote a Kleene closure. δ is easily extended on strings from Σ^*.

G generates formal language $L(G) = \{w : w \in \Sigma^* \text{ and } \delta(w, q_0) \text{ is defined}\}$ and marked language $L_m(G) = \{w : w \in L(G) \text{ and } \delta(w, q_0) \in Q_m\}$. For any $L \subset \Sigma^*$ a *prefix-closure* of L is the set of all strings that are prefixes of words of L, i.e. $\overline{L} = \{s | s \in \Sigma^* \text{ and } \exists t \in \Sigma^* : s \cdot t \in L\}$. Symbol $\cdot$ denotes string concatenation and is often omitted.

Let Σ_c be a controllable event set, $\Sigma_{uc} = \Sigma \setminus \Sigma_c$, $\Sigma_c \cap \Sigma_{uc} = \emptyset$. The supervisory control framework assumes that a supervisor switches control patterns in such a way that the supervised DES achieves a control objective described by some regular language K. Denote $L(\mathcal{J}/G)$ a language generated by the closed-looped behavior of the plant and the supervisor. Let $L_m(\mathcal{J}/G)$ denote the language marked by the supervisor: $L_m(\mathcal{J}/G) = L(\mathcal{J}/G) \cap L_m(G)$. The main goal of supervisory control is to construct such supervisor that $L(\mathcal{J}/G) = \overline{K}$ and $L_m(\mathcal{J}/G) = K$.

The notion of controllability plays a crucial role in SCT. K is *controllable* (with respect to $L(G)$ and Σ_{uc}) if $\overline{K}\Sigma_{uc} \cap L(G) \subseteq \overline{K}$ [27]. Only controllable languages may be exactly achieved by the joint behavior of the plant and supervisor. To verify controllability condition, a product of automata for the system and the specification is built to check if the same uncontrollable transitions present in both specification and the plant. This idea is borrowed for controllability checking via PCFs.

The general form of PCF representing some automaton consists of the single base $\mathcal{B} = \{I(S), L(\varepsilon, S), L_m(\varepsilon, S), \delta(S_1^i, \sigma^i, S_2^i), \delta_m(S_1^i, \sigma^i, S_2^i), \Sigma_c(\sigma^j), \Sigma_{uc}(\sigma^j)\}, i \in \{1, \ldots, n\}$, $j \in \{1, \ldots, m\}$, n is the number of transitions, m is the number of events, and two questions shown in Fig. 2. Here the predicate $L(s, S)$ denotes "s is a current sequence of events in the state S" and $L_m(s, S)$ denotes "s is a current sequence of events in the state S, and s is a marked string". The first arguments of these atoms will accumulate the strings of languages generated and marked by the automaton. Predicate of the form $\delta(S_1, \sigma, S_2)$ is interpreted as the automaton transition from a state S_1 to a state S_2 with an event σ. Predicates $\delta_m(S_1, \sigma, S_2)$ are employed in case when S_2 is a marked state. The predicate $I(\_)$ denotes the initial state of the automaton. Controlled and uncontrolled events will be represented by the predicates $\Sigma_c(\_)$ and $\Sigma_{uc}(\_)$, respectively. As usual, the function symbol "$\cdot$" denotes strings concatenation, and the "ε" symbol corresponds to the empty string. Applying the inference rules to this PCF, the strings of the languages generated and marked by the automaton will be built as the first arguments of the atoms $L(s, S)$, $L_m(s, S)$ in the base.

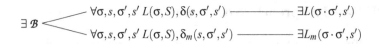

$$\exists \mathcal{B} \begin{cases} \dfrac{\forall \sigma, s, \sigma', s'\ L(\sigma, S), \delta(s, \sigma', s')}{} & \exists L(\sigma \cdot \sigma', s') \\[2ex] \dfrac{\forall \sigma, s, \sigma', s'\ L(\sigma, S), \delta_m(s, \sigma', s')}{} & \exists L_m(\sigma \cdot \sigma', s') \end{cases}$$

Fig. 2. General form of PCF representation of some automaton.

The PCF $\mathcal{F}_{G \times \mathcal{H}}$ constructing the product of automata G and $\mathcal{H}$, a recognizer for K, consists of one base subformula, which base conjunct is $B_{G \times \mathcal{H}} = \{I_1(S_0), I_2(P_0), \delta_1(S_1^i, \sigma^i, S_2^i), \delta_2(S_1^k, \sigma^k, S_2^k)\}$, containing atoms for transitions δ_1, $i = \overline{1, n_1}$, of the first automaton and transitions δ_2, $k = \overline{1, n_2}$, of the second one. The questions of $\mathcal{F}_{G \times \mathcal{H}}$ are as follows below:

$Q_{1,p} : \forall s, p \, I_1(s), I_2(p) - \exists I_3(s \cdot p), \delta_3(\varepsilon, \varepsilon, s \cdot p)$

$Q_{2,p} : \forall \sigma, s_1, p_1, s_2, p_2, \sigma', s' \, \delta_1(s_1, \sigma, s_2), \delta_2(p_1, \sigma, p_2), \delta_3(s', \sigma', s_1 \cdot p_1) -$
$$- \exists \delta_3(s_1 \cdot p_1, \sigma, s_2 \cdot p_2).$$

Index 1 in the subscripts of atoms corresponds to the atoms of the automaton G, 2 to the H, and 3 to the atoms corresponding to the product of G and H. Here, the functional symbol · is used to trace the pairs of states in the product automaton. $Q_{1,p}$ adds to the base the initial state of the product automaton and undetermined transition atom corresponding to the third automaton, thanks to which the connectivity of the product automaton is controlled by $Q_{2,p}$. The latter adds the transition atoms of the product automaton to the base.

According to the paradigm of the PCF calculus, to verify controllability of K we built an inference for the negation of the corresponding formula. To check if K is not controllable we employ the PCF $\mathcal{F}_{G \times H}$ with additional rules Q_1–Q_6 depicted in Fig. 3. In these rules, states which are simultaneously achieved from the states of G and $G \times H$ by the same events will be called the *neighbouring* states and stored as arguments of the predicate $N(\_, \_)$.

$Q_1 : \forall s_1, s_2, p_1, p_2, \sigma \, I_1(s_1), I_3(p_1), \delta_1(s_1, \sigma, s_2), \delta_3(p_1, \sigma, p_2) - \exists N(s_1, p_1)$

$Q_2 : \forall s_1, s_2, p_1, \sigma \, N(s_1, p_1), \delta_1(s_1, \sigma, s_2), E_{uc}(\sigma) - \exists Chk(p_1, \sigma, 0)$

$Q_3 : \forall p_1, p_2, \sigma \, Chk^*(p_1, \sigma, 0), \delta_3(p_1, \sigma, p_2) - \exists Chk(p_1, \sigma, 1)$

$Q_4 : \forall p_1, \sigma \, Chk(p_1, \sigma, 0) - \exists UC(p_1, \sigma)$

$Q_5 : \forall s_1, s_2, p_1, p_2, \sigma \, N(s_1, p_1), \delta_1(s_1, \sigma, s_2), \delta_3(p_1, \sigma, p_2) - \exists N(s_2, p_2)$

$Q_6 : \forall p_1, \sigma \, UC(p_1, \sigma)$

Fig. 3. Questions of the PCF checking the controllability.

Q_1 adds to the base the initial states of G and $G \times H$ as a first pair to be checked for uncontrollable transitions. Although such s_1, p_1 are not neighboring states in our sense, $N(s_1, p_1)$ is necessary for further inference.

Q_2 checks if there is an uncontrollable transition from some state of the automaton corresponding to the plant. Upon successful answer to this question, the atom $Chk(p_1, \sigma, 0)$ is added to the base. The first argument of this atom denotes a state p_1 of H, in which the transition on the uncontrollable event σ should exist for K to be controllable. By default we suppose that the transition does not exist so the third argument in $Chk()$ is 0 before checking by the next rule.

Q_3 is aimed to check if both neighboring states share the same uncontrollable event. A successful answer to this question means that the uncontrollable event is legal, i.e., allowed by the specification, since a transition with it exists in both G and H. We need to delete the atom $Chk(p_1, \sigma, 0)$ with the help of * operator from the base to ensure that there will not be additional answers to this question with the same pair of states.

Q_4 checks if the atom $Chk(p_1, \sigma, 0)$ is present in the base. Its presence in the base means that an uncontrollable event is not allowed by the specification, i.e., K is not controllable. So the atom $UC(p_1, \sigma)$ is added to the base upon a successful answer. It contains information about the state and the event that violate the controllability condition. This atom is further used in the goal question to complete the inference. It's worth noting that the inference search strategy must be configured in such a way that this question is checked for an answer immediately after the previous one.

Q_5 adds the next checked pair of states to continue the inference search. Q_6 is the goal question that ends the inference. If the inference has ended with the exhaustion of the search options for substitutions and the answer to the goal question has not been found then the specification language is controllable. Since the formalized automata are finite with finite sets of events, then the search space for the inference in this formalization is finite. Therefore, the inference ends in both cases – either the specification is controllable or not.

Further development of the PCFs-based approach to DES will allow solving important SCT problems such as centralized and decentralized supervisors construction, partially observed DES design and study.

4   Formation Path-Following Problem

This section extends the results from [31] by introducing event-triggering mechanisms for communication and control of AUVs. Event-based control and communication strategies are widely used in multi-robot systems [5,21,25] because they allow reducing the load on communication channels and onboard computing devices. Despite some progress in this area, the event-triggered formation control problem subject to parameter uncertainties, control saturation, and measurement errors still remain relevant.

We consider a leader-follower formation of underactuated AUVs whose movement in the horizontal plane is described by the following equations [17]:

$$\begin{cases} \dot{x}_i = u_i \cos(\psi_{Bi}) - v_i \sin(\psi_{Bi}), \\ \dot{y}_i = u_i \sin(\psi_{Bi}) + v_i \cos(\psi_{Bi}), \\ \dot{\psi}_{Bi} = r_i, \\ \mathcal{F}_i = m_{ui}\dot{u}_i + d_{ui}, \\ 0 = m_{vi}\dot{v}_i + m_{uri}u_i r_i + d_{vi}, \\ \mathcal{G}_i = m_{ri}\dot{r}_i + d_{ri}, \quad i = \overline{1, N} \end{cases} \tag{1}$$

where x_i, y_i are the coordinates of the i-th AUV in a global reference frame, ψ_{Bi} is the yaw angle; u_i v_i are the surge and sway speeds, r_i is the yaw rate; $m_{\{\cdot\}i}$ define the mass-inertial characteristics of the AUV, which depend on the nominal mass m_i and moment of inertial I_{zi}; $d_{\{\cdot\}i}$ are the disturbances; $\mathcal{F}_i$, $\mathcal{G}_i$ are the control force and torque.

We assume that the AUV with the number $l \in \overline{1, N}$ is selected to be the formation leader. The leader's task is to track the movement of a virtual target (VT), which as a point moves along a given path. Other members of the formation should maintain its relative position with respect to the leader.

Denote by (x_0, y_0) the global coordinates of the VT and by u_0, r_0, ψ_{B0} its forward speed, angular speed, and heading angle respectively. We assume that the follower, having either the direct measurements (distance and bearing angle) or its global coordinates and the global coordinates of the leader obtained through communications, can calculate its relative coordinates

$$\begin{bmatrix} s_{ei} \\ y_{ei} \end{bmatrix} = R \begin{bmatrix} x_i - x_l \\ y_i - y_l \end{bmatrix}, \quad R(\psi_{B0}) = \begin{bmatrix} \cos \psi_{B0} & \sin \psi_{B0} \\ -\sin \psi_{B0} & \cos \psi_{B0} \end{bmatrix}, \quad i \neq l. \tag{2}$$

For the leader $(i = l)$ in (2) we have $i = l$ and $l = 0$.

Using the relative coordinates, dynamic equations for the leader-follower pair can be written as

$$\begin{cases} \dot{s}_{ei} = -v_{tl} \cos \psi_l + r_0 y_{ei} + v_{ti} \cos \psi_i, \\ \dot{y}_{ei} = -v_{tl} \sin \psi_l - r_0 s_{ei} + v_{ti} \sin \psi_i, \\ \dot{\psi}_i = r_i + \dot{\beta}_i - r_0, \\ \dot{v}_{ti} = \frac{\mathcal{F}_i - d_{ui}}{m_{ui}} \cos \beta_i - \frac{m_{uri} + d_{yi}}{m_{vi}} \sin \beta_i, \end{cases} \tag{3}$$

where $v_{ti} = (u_i^2 + v_i^2)^{1/2}$ is the absolute speed, $\psi_i = \psi_{Bi} + \beta_i - \psi_{B0}$ is the angular deviation from the VT's heading direction; $\beta_i = \arctan(v_i/u_i)$ is the side-slip angle. Considering the VT as a leader for the formation leader, the equations for the VT-leader pair are easily obtained.

Let the desired position of the follower with respect to the leader in coordinates (s_{ei}, y_{ei}) be given by vector (s_{ei}^*, y_{ei}^*). For the leader, it is natural to take $s_{el}^* = 0$, $y_{el}^* = 0$. To improve formation accuracy, each AUV must move in such a way as to satisfy

$$\psi_i = \psi_i^* \triangleq \arctan \frac{r_0 s_{ei}^*}{u_0 - r_0 y_{ei}^*}, \quad v_{ti} = v_{ti}^* \triangleq \sqrt{(r_0 s_{ei}^*)^2 + (u_0 - r_0 y_{ei}^*)^2}.$$

The formation path-following problem consists in deriving control laws for $\mathcal{F}_i$ and $\mathcal{G}_i$ that ensure

$$\lim_{t \to \infty} |\Delta s_{ei}(t)| \leq s_{ei}^\infty, \quad \lim_{t \to \infty} |\Delta y_{ei}(t)| \leq y_{ei}^\infty, \quad \lim_{t \to \infty} |\Delta \psi_i(t)| \leq \psi_i^\infty, \quad \lim_{t \to \infty} |\Delta v_{ti}(t)| \leq v_{ti}^\infty,$$

where $\Delta s_{ei}(t) \triangleq s_{ei}(t) - s_{ei}^*$, $\Delta y_{ei}(t) \triangleq y_{ei}(t) - y_{ei}^*$, $\Delta \psi_i(t) \triangleq \psi_i(t) - \psi_i^*(t)$ and $\Delta v_{ti}(t) \triangleq v_{ti}(t) - v_{ti}^*(t)$; s_{ei}^∞, y_{ei}^∞, ψ_i^∞, v_{ti}^∞ characterize the steady state errors.

In order for followers to be able to maintain a desired formation, the leader changes the VT's speeds depending on the current value of the path curvature c_c as

$$\begin{cases} \dot{u}_0 = k_0(u_0(t) - u_0^*(t)), \quad u_0^*(t) = \bar{u}_0 + (\underline{u}_0 - \bar{u}_0)|c_c(t)|/\bar{c}_c, \\ r_0 = c_c(t)u_0, \end{cases}$$

where $\bar{u}_0$, $\underline{u}_0$ are upper and low bounds for VT's speed u_0, $\bar{c}_c$ is the curvature constraint $(|c_c(t)| \leq \bar{c}_c)$. The leader broadcasts information about the state of TV at moments $t_k = t_{k-1} + g_0 h$, $k = 1, 2, \ldots$ with minimal positive integer g_0 satisfying condition

$$|x_0(t_{k-1} + g_0 h) - x_0(t_{k-1})| > \sigma_0 |x_0(t_{k-1} + g_0 h)|, \quad \sigma_0 > 0, \tag{4}$$

where $x_0 \triangleq [u_0 \ r_0 \ \psi_{B0}]^T$, h is the sampling interval, which is common for all AUVs.

The formation path-following control laws are designed as

$$\mathcal{F}_i(t) = \mathcal{F}_{ci}(t_k) + \mathcal{F}_{si}(t_k), \quad \mathcal{G}_i(t) = \mathcal{G}_{ci}(t_k) + \mathcal{G}_{si}(t_k), \quad t \in [t_k, t_{k+1}),$$
$$\mathcal{F}_{si}(t_k) = \operatorname{sat}(k_{1i}\triangle\hat{s}_{ei,k} + k_{2i}\triangle\hat{v}_{ti,k}, \overline{\mathcal{F}}_{si}), \qquad (5)$$
$$\mathcal{G}_{si} = \operatorname{sat}(k_{3i}\triangle\hat{y}_{ei,k} + k_{4i}\triangle\hat{\psi}_{i,k} + k_{5i}\triangle\hat{r}_{i,k}, \overline{\mathcal{G}}_{si}),$$

where $\mathcal{F}_{ci}$, $\mathcal{G}_{ci}$ are the feedforward controllers aimed at canceling disturbances and tracking the heading direction of the the VT (see [31]); $\mathcal{F}_{si}$, $\mathcal{G}_{si}$ are the feedback controllers, $\overline{\mathcal{F}}_{si}$, $\overline{\mathcal{G}}_{si}$ are control resources allocated for stabilization, $\triangle\hat{s}_{ei,k}$, $\triangle\hat{y}_{ei,k}$, $\triangle\hat{\psi}_{i,k}$, $\triangle\hat{v}_{ti,k}$, $\triangle\hat{r}_{i,k}$ are the latest error estimates available at t_k, k_{ji} are feedback gains to be determined ($j = \overline{1,5}$), $\operatorname{sat}(\sigma, \bar{\sigma}) = \operatorname{sign}(\sigma)\min(|\sigma|, \bar{\sigma})$ is the saturation function.

Similarly (6) the control signal update time t_k is determined by $t_k = t_{k-1} + g_i h$ with g_i to be the minimal positive integer satisfying the following event triggering condition

$$|x_{ei}(t_{k-1} + g_i h) - x_{ei}(t_{k-1})| > \sigma_i |x_{ei}(t_{k-1} + g_i h)|, \quad x_{ei} \triangleq \begin{bmatrix} s_{ei} \ y_{ei} \end{bmatrix}^T, \quad \sigma_i > 0. \quad (6)$$

The structure of the closed-loop system (3), (5) and (6), after some simplification, allows to apply some modifications of algorithms based on vector Lyapunov functions [15,33] for synthesis of feedback gains k_{ij}. Table 1 shows the possible events detected by the leader (L) or follower (F) and the actions to be taken on those events.

Table 1. Events for the formation path-following problem.

AUV role	Event	Action
L, F	Condition (6) is fulfilled	Update control signals using (5)
L	Condition (4) is fulfilled	Broadcast the state of the VT and other data
F	Information from the leader has been received	Update estimates $\triangle\hat{s}_{ei,k}$, $\triangle\hat{y}_{ei,k}$, $\triangle\hat{\psi}_{i,k}$, $\triangle\hat{v}_{ti,k}$, $\triangle\hat{r}_{i,k}$ used in (5)

5   Path-Planning Problem

The path-planning problem is currently being actively studied in literature (see, for example, review papers [20,26]). The main emphasis in these studies is on developing algorithms that provide real-time path planning [12,36]. This section proposes a new event-based approach for reactive path planning in an unknown environment. The designed control system allows the AUV to move alone a reference path represented as a sequence of line and arc segments, avoiding obstacles detected by a forward-looking sonar (FLS). The approach employs a DES to identify situations that require updating the current path. The DES together with path-planning algorithms implements a simple behavioral strategy that can be expressed as follows:

- do not get to a place where it is impossible to get out using standard obstacle avoidance algorithms;

– do not alter once chosen obstacle avoidance direction until the AUV return to the reference path (left or right hand rule);
– rebuild the path only if it is really necessary;

The graphical representation of the designed DES is shown in Fig. 4. The DES has the following states: mPF – following the reference path; mDOL – avoiding the detected obstacle from its left side; mDOR – avoiding the detected obstacle from its right side; mNRF – navigation to the reference path; mSOL – searching for an obstacle on the left; mSOR – searching for an obstacle on the right; mMC – mission completed. State mMC is marked.

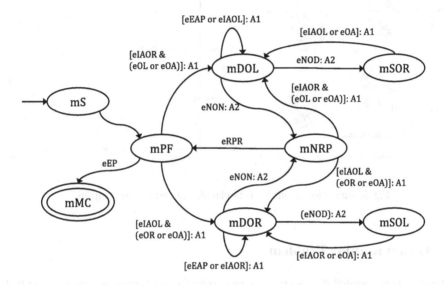

Fig. 4. A graphical representation of DES for path-planning.

We also distinguish the following events: eOP – the AUV has reached the end of the reference path; eOL (eOR) – an obstacle is detected on the left (right) side; eIAOL (eIAOR) – soon the AUV will not be able to avoid approaching the obstacle on the left (right) side; eOA – an obstacle is detected ahead; eNON – there are no obstacles nearby; eNOD – no obstacle detected; eRPR – the AUV has returned on the reference path; eEAP – the AUV has reached the end of the avoidance path.

There are two possible actions when an event is triggered. Action A1 provides construction of a path to avoid a detected obstacle using a modification of algorithms presented in [35]. Action A2 generates a path that leads the AUV to the reference trajectory. In this case the path planner is based on Dubins curves [9,22]. To make the AUV move along the generated path, the path-following controller designed in the previous section is used. All paths generated by the path-planning algorithms is first-order differentiable (smooth) and meet kinematic constraints given by minimum turning radius. The obtained obstacle avoidance paths lie no closer than safe distance from obstacles.

Due to space limitations, here we presented a simplified version of the DES without specifying parameter values (e.g. characterizing the proximity of an obstacle) that affect the logic of the AUV behavior when navigating in obstacle environments. Simulation results for the AUV (with the designed path-planner) moving in an obstacle environment are shown in Fig. 5. The left figure shows the trajectories of the AUV (desired and actual) and the right figure depicts the evolution of the state of the DES in the considered simulation scenario. Points on the graph mark the moments when the current path is rebuilt. In this simulation scenario, the AUV changes its path about 10 times in 600 s.

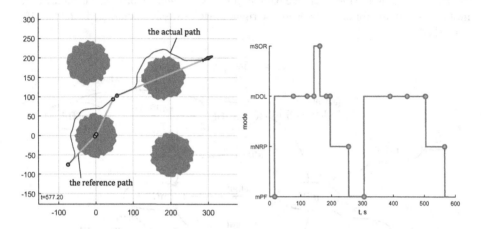

Fig. 5. Simulation results for path-planning in obstacle environment.

6   Group Routing Problem

The use of coordinated AUVs groups seems to be the most promising technology that allows providing operational coverage of big-scaled underwater regions. As the operational field is usually large compared to the vehicles' sensing capabilities, the task allocation and route-planning problems inevitably arise to provide time/resource-effective mission implementation. Therefore, each vehicle in the group must be associated with a series of survey tasks located in different areas of the region. The construction of a feasible and effective group route is a problem of high computational complexity even for the simple classical formulations with one type of constraint.

In this paper, we consider the multi-attribute vehicle routing problem for AUVs environmental monitoring. The group task here is the well-timed inspection of the objects set over a specified period and under a given set of requirements [10]. These requirements may come in various combinations of spatial, temporal or functional constraints [6]. Such routing problems, which combines a whole range of restrictions and requirements of various nature, are primarily classified as rich or multi-attribute routing problems [11]. These extended formulations, aimed at more accurate real-world problems modelling, nowadays are both insufficiently studied and of great scientific research interest [3]. We propose using the decentralized hybrid evolutionary approach for the efficient generation of feasible group routes.

Denote by T the duration of a single required routing period. The value T is usually determined depending on the required frequency of the group communication sessions or as a period towards the nearest expected event to provide the capability to react on any emergent condition changes in a quick and efficient manner. Let the mission include n objects (tasks) located within the connected mission area D. Each task i is assigned a set of parameters describing its inspection requirements:

- object location (x_i, y_i, z_i) within the area D;
- object inspection periodicity p_i as the recommended time-length between its two consecutive inspections by the AUVs of the group;
- the required type of inspections $u_i \in \{0, 1\}$ (1 if early inspection ahead of time p_i is allowed, 0 otherwise);
- minimal number of vehicles l_i to conduct a single object inspection;
- single-inspection duration s_i.

Let the working group consists of m functionally equivalent vehicles, which can still differ by their cruising speed v^i. We define the route of a single vehicle i as a numerical vector $r_i = \langle r_0, r_{i1}, r_{i2}, ..., r_{ih}, r_0 \rangle$ of tasks indexes in the consecutive order of their inspections. On each routing period, the group starts from the specified rendezvous location r_0 and should return there by the time T to provide the cooperative communication session. Such a session involves inner-vehicle data sharing among the group and the current group strategy adjustment (if needed). The group route $R = \{r_1, r_2, ..., r_m\}$ is a set of all single AUV's routes.

During the route implementation, the time t_{ij}, that i-th AUV spends on j-th task inspection, is calculated as $t_{ij} = t_1 + t_2 + t_3$ according to the following rules:

1. The idle time t_1 before the inspection can be started:
   (a) $t_1 = 0$ if $u_j = 1$ or the period since the task's last inspection exceeds p_j;
   (b) In other cases, t_1 is defined as the time until the expiration of p_j.
2. Waiting time $t2$ until the required number of vehicles arrive:
   (a) $t_2 = 0$ if at least l_i vehicles have already gathered to conduct the inspection;
   (b) Otherwise, an AUV has to wait up to the maximum allowed downtime t_W.
3. An actual inspection duration $t3$:
   (a) If $t_2 \leq t_W$, then the inspection is conducted $t_3 = s_i$;
   (b) If downtime has expired $t_2 > t_W$, the i-th AUV leaves the current object.

We use the next function as the objective function to rate the group efficiency:

$$F(R) = \Phi(R) + m \cdot \Psi(R) \rightarrow \min \qquad (7)$$

$\Phi(R)$ here is a penalty function for inspection delays, while $\Psi(R)$ evaluates the "favorableness" of the terminal conditions as being starting conditions for the next routing period. The calculation procedure for both functions is in-detail described in [14].

Currently, no approach can be singled out as a universal method to solve the multi-attribute routing problems for the robotic groups. The most efficient ones combine meta-heuristic structures with intelligent neighbourhood search strategies and exact methods techniques. According to the present-day studies [16,18], hybrid evolutionary algorithms allow finding close-to-optimal solutions with better scalability and in better time (on average) than any other heuristic and meta-heuristic approach.

In this regard, we propose an original modification of the hybrid evolutionary algorithm to solve the route planning problem. The group route R acts as a chromosome here, while (7) is the fitness function. The proposed approach includes several specialized heuristics, advanced local search procedures, and additional solution improvement schemes. The approach demonstrates high efficiency for all data sets without significant quality loss, even in the most complex problem statements. For those test cases, on which we were able to brute-force the global optimum, the proposed evolutionary approach has found solutions that are inferior to the optimal ones by no more than 1.3%. The solution fragment for the case with highly diversified mission condition is illustrated in Fig. 6 (grey circles stand for $u_i = 1$, white for $u_i = 0$).

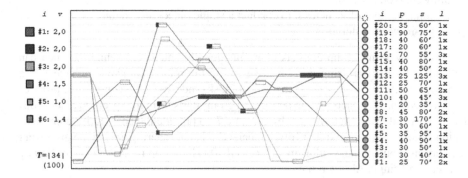

Fig. 6. The fragment of a group route for six vehicles in a mission of twenty objects.

7   Action Planning Problem

Let us present an example of the use of the calculus of PCFs for planning the actions of robots at the upper level. Within the framework developed in this work, the PCF formalization of the problem presented below should be taken as a mission specification for robots. The result of an automatic inference search for that specification will be consecutive action plans that will be transferred to the middle level as task parameters.

Let us consider some working environment for robots in the form of a flat platform, which can be schematically depicted as in Fig. 7: there are three robots, two blocks, and the target area in which it is necessary to move the blocks.

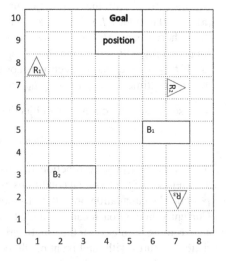

Fig. 7. The initial scene.

We divide the scene into squares and try to find the minimal plans for the movement of robots on the scene from cell to cell at which the blocks will be delivered to the target area.

Plans for the movement of blocks by a group of three robots will be constructed as follows. Any two robots can move blocks, so at first, they are distributed among the blocks, push the blocks to the edge of the scene, and then push them to the target area.

Let us consider the predicates, functional symbols, and their interpretation:

- $R(x,y)$—"x is a robot facing the direction $y \in \{N,E,S,W\}$";
- $B(x)$—"x is the block that needs to be moved";
- $W(x)$—"x is the width of the scene";
- $H(x)$—"x is the height of the scene";
- $Pos(x,y,z)$—"x is a robot or block positioned at coordinates (y,z)";
- $App(x,u,v)$—"x is a robot appointed to the task to move at coordinates (u,v)";
- $Free(x,y)$—"coordinates (x,y) are free for a robot to move there";
- $Ready(x)$—"x is ready to move a block";
- $Move(x)$—"x is moving a block";
- $Bmoved(x,y)$—"x is the block number y that was already moved";
- $\leqslant (x,y)$—"auxiliary computable predicate that is substituted with logic $true$ constant if x is less than y, or equal, and to constant $false$ otherwise";
- $= (x,y), \neq (x,y)$—"the same as above for the treatment of equality and inequality";
- $dist(x,y,u,v)$—is a functional symbol needed to evaluate the distance between cells with coordinates (x,y) and (u,v).

Predicate $Free()$ will also be treated as computable for readability and will be marked with symbol #. The symbol $*$ near the predicate indicates that this predicate will be erased from the base after the successful answer, this is an approach to modeling the obsolescence of facts over time. The base of facts contains atoms that correspond to the initial scene, as depicted in Fig. 7:

$$\exists R(r_1,N),R(r_2,E),R(r_3,S),B(b_1),B(b_2),Pos(r_1,1,8),Pos(r_2,7,7),Pos(r_3,7,2),$$
$$Pos(b_1,6,5),Pos(b_2,2,3),H(10),W(8)$$

The goal is to move blocks B_1,B_2 to the Goal position. It can be formalized with following non-Horn rule:

$$\exists x,yB(x)\&Pos(x,4,10)\&B(y)\&Pos(y,4,9) \lor B(x)\&Pos(x,4,9)\&B(y)\&Pos(y,4,10)$$

Next, we present the rules (questions, in terms of PCFs) on the basis of which a constructive inference of plans will be built to achieve the goal.

The first one, Q_1, corresponds to the distribution of robots by the first part of the plan's goals, which robots will move, which blocks in the north direction.

Q_2 can be read as, if the distance between a robot's next position and the current task position is less than the robot's starting position than it needs to rotate to another direction. Actually, Q_2 is "rotating" the robot eastward, and rotating the rest directions will look similar.

Q_3 corresponds to the movement of robots in the facing direction (this one is for the north). Checking the distance between the current position of the robot and its appointed target is a heuristic to reduce the search space of the inference.

Q_4 is checking the achievement of the currently appointed task. If the robot is positioned near the block and ready to move it, then it needs to rotate north.

Q_5 can be read as, if two different robots are ready to push the block, they simultaneously begin to move.

$Q_{6.1}$ and $Q_{6.1}$ are needed for the verification of reaching the edge of the scene by the robot moving the block.

$Q_{7.1}$ is appointing one of the robots to move to the edge of the block for further pushing to the target. If the current position of the block is on the left side of the scene, i.e. when $x < w/2$, then assign the target position of the robot to the left of the block, otherwise then, and it should be the rule $Q_{7.2}$.

$Q_{8.1}$ and $Q_{8.2}$ are needed to verify the achievement of the second destination and turn robots in the needed direction.

$Q_{9.1}$ and $Q_{9.2}$ are needed to move the robot and the block. These rules will add to the base a fact that completes the inference.

$$Q_1 : \forall b,x,y\, B(b), Pos(b,x,y) - \begin{cases} \exists App(r_1,x,y-1), App(r_2,x,y-1) \\ \exists App(r_2,x,y-1), App(r_3,x,y-1) \\ \exists App(r_1,x,y-1), App(r_3,x,y-1) \end{cases}$$

$$Q_2 : \forall r,x,y,u,v\, R^*(r,N), Pos(r,x,y), Free^\#(x,y+1),$$
$$App(r,u,v), dist(x,y,u,v) < dist(x,y+1,u,v) - \exists R(r,E)$$

$$Q_3 : \forall r,x,y,u,v\, R(r,N), Pos^*(r,x,y), Free^\#(x,y+1),$$
$$App(r,u,v), dist(x,y,u,v) \geqslant dist(x,y+1,u,v) - \exists Pos(r,x,y+1)$$

$$Q_4 : \forall r,x,y,d\, R^*(r,d), Pos(r,x,y-1), App^*(r,x,y) - \exists R(r,N), Ready(r)$$

$$Q_5 : \forall b,x,y,r_1,r_2,x_1,y_1,x_2,y_2\, Ready(r_1), Ready(r_2), Pos^*(r_1,x_1,y_1), Pos^*(r_2,x_2,y_2),$$
$$Pos^*(b,x,y), Free^\#(x_1,y_1-2), Free^\#(x_2,y_2-2), r_1 \neq r_2-$$
$$-\exists Pos(r_1,x_1,y_1+1), Pos(r_2,x_2,y_2+1), Move(r_1,b), Move(r_2,b), Pos(b,x,y+1)$$

$$Q_{6.1} : \forall r,b,x,y,h\, Move^*(r,b), Ready^*(r), Pos(r,x,y), H(h), y=h-1-\exists Bmoved(b,1)$$

$$Q_{6.2} : \forall r,b,x,y,h,n\, Move^*(r,b), Ready^*(r),$$
$$Pos(r,x,y), Bmoved^*(b,n), H(h), y=h-n-\exists Bmoved(b,n+1)$$

$$Q_{7.1} : \forall b,n,x,y,w\, Bmoved(b,n), Pos(b,x,y), W(w), x<w/2 \begin{cases} \exists App(r_1,x-1,y) \\ \exists App(r_2,x-1,y) \\ \exists App(r_3,x-1,y) \end{cases}$$

$$Q_{8.1} : \forall r,b,n,x,y\, Bmoved(b,n), Pos(b,x,y), App^*(r,x-1,y) - \exists Ready(r), R(r,E)$$

$$Q_{8.2} : \forall r,b,n,x,y\, Bmoved(b,n), Pos(b,x,y), App^*(r,x+3,y) - \exists Ready(r), R(r,W)$$

$$Q_{9.1} : \forall b,x,y,r,x_r,y_r\, Ready(r), R(r,E), Pos^*(r,x_r,y_r), Pos^*(b,x,y), Free^\#(x+3,y_r)-$$
$$-\exists Pos(r,x_r+1,y_r), Move(r,b), Pos(b,x+1,y)$$

$$Q_{9.2} : \forall b,x,y,r,x_r,y_r\, Ready(r), R(r,W), Pos^*(r,x_r,y_r), Pos^*(b,x,y), Free^\#(x-1,y_r)-$$
$$-\exists Pos(r,x_r-1,y_r), Move(r,b), Pos(b,x-1,y)$$

It is assumed that with the complete construction of the inference, facts will be accumulated in the refuted bases, reflecting the successive changes in the state of the scene. At the same time, it will be quite easy to extract from there the minimum of the constructed plans. It should be noted that the described setting can be changed and complicated for the movement of different sizes of blocks, the presence of obstacles on the scene. Here we give a minimal formalization to demonstrate the idea of Non-Horn approach to constructive PCF based deductions and inference of plans.

8   Conclusion

The paper presents the general concept of the framework without detailed elaboration of particular issues related, for example, to filtering, processing and dispatching events. We have individually validated the proposed methods to be used on different levels of the suggested architecture. Currently, we are finishing the development of the advanced modelling system combining the software simulator and the real multi-robot testbed platform. Thus, the next step in our study is to conduct integrated simulation and practical experiments with complex scenarios involving different levels of group cooperation.

References

1. Bychkov, I., Davydov, A., Kenzin, M., Maksimkin, N., Nagul, N., Ul'yanov, S.: Hierarchical control system design problems for multiple autonomous underwater vehicles. In: 2019 International Siberian Conference on Control and Communications (SIBCON), pp. 1–6 (2019). https://doi.org/10.1109/SIBCON.2019.8729592
2. Bychkov, I., Davydov, A., Nagul, N., Ul'yanov, S.: Hybrid control approach to multi-AUV system in a surveillance mission. Inf. Technol. Ind. 6(1), 20–26 (2018)
3. Caceres-Cruz, J., Arias, P., Guimarans, D., Riera, D., Juan, A.A.: Rich vehicle routing problem: survey. ACM Comput. Surv. 47(2) (2014). https://doi.org/10.1145/2666003
4. Cassandras, C.G., Lafortune, S.: Introduction to Discrete Event Systems. Springer, Heidelberg (2008). https://doi.org/10.1007/978-0-387-68612-7
5. Chu, X., Peng, Z., Wen, G., Rahmani, A.: Distributed formation tracking of multi-robot systems with nonholonomic constraint via event-triggered approach. Neurocomputing 275, 121–131 (2018). https://doi.org/10.1016/j.neucom.2017.05.007
6. Cissé, M., Yalçındağ, S., Kergosien, Y., Şahin, E., Lenté, C., Matta, A.: Or problems related to home health care: a review of relevant routing and scheduling problems. Oper. Res. Health Care 13–14, 1–22 (2017). https://doi.org/10.1016/j.orhc.2017.06.001
7. Dai, X., Jiang, L., Zhao, Y.: Cooperative exploration based on supervisory control of multi-robot systems. Appl. Intell. 45(1), 18–29 (2016). https://doi.org/10.1007/s10489-015-0741-3
8. Davies, T., Jnifene, A., Davies, T.: Path planning and trajectory control of collaborative mobile robots using hybrid control architecture. J. Syst. Cybern. Inform. 6 (2013)
9. Dubins, E.L.: On curves of minimal length with a constraint on average curvature, and with prescribed initial and terminal positions and tangents. Am. J. Math. (1957)
10. Dunbabin, M., Marques, L.: Robots for environmental monitoring: significant advancements and applications. IEEE Robot. Autom. Mag. 19(1), 24–39 (2012). https://doi.org/10.1109/MRA.2011.2181683
11. Hartl, R., Hasle, G., Janssens, G.: Special issue on rich vehicle routing problems. CEJOR 14, 103–104 (2006). https://doi.org/10.1007/s10100-006-0162-9

12. Hernández, J.D., Vidal, E., Moll, M., Palomeras, N., Carreras, M., Kavraki, L.E.: Online motion planning for unexplored underwater environments using autonomous underwater vehicles. J. Field Robot. **36**(2), 370–396 (2019). https://doi.org/10.1002/rob.21827

13. Ju, C., Son, H.I.: Modeling and control of heterogeneous agricultural field robots based on Ramadge–Wonham theory. IEEE Robot. Autom. Lett. **5**(1), 48–55 (2020)

14. Kenzin, M., Bychkov, I., Maksimkin, N.: Task allocation and path planning for network of autonomous underwater vehicle. Int. J. Comput. Netw. Commun. **10**(2), 33–42 (2018). https://doi.org/10.5121/ijcnc.2018.10204

15. Kozlov, R.I., Kozlova, O.R.: Investigation of stability of nonlinear continuous-discrete models of economic dynamics using vector Iyapunov function. I. J. Comput. Syst. Sci. Int. **48**(2), 262–271 (2009). https://doi.org/10.1134/S1064230709020105

16. Koç, Ç., Bektaş, T., Jabali, O., Laporte, G.: Thirty years of heterogeneous vehicle routing. Eur. J. Oper. Res. **249**(1), 1–21 (2016). https://doi.org/10.1016/j.ejor.2015.07.020

17. Lapierre, L., Soetanto, D.: Nonlinear path-following control of an AUV. Ocean Eng. **34**(11), 1734–1744 (2007). https://doi.org/10.1016/j.oceaneng.2006.10.019

18. Laporte, G., Røpke, S., Vidal, T.: Heuristics for the Vehicle Routing Problem, 2 edn., pp. 87–116. Society for Industrial and Applied Mathematics (2014)

19. Larionov, A., Davydov, A., Cherkashin, E.: The calculus of positively constructed formulas, its features, strategies and implementation. In: International Convention on Information and Communication Technology, Electronics and Microelectronics (MIPRO), Opatija 2013, pp. 1023–1028 (2013)

20. Li, D., Wang, P., Du, L.: Path planning technologies for autonomous underwater vehicles-a review. IEEE Access **7**, 9745–9768 (2019)

21. Liu, X., Ge, S.S., Goh, C., Li, Y.: Event-triggered coordination for formation tracking control in constrained space with limited communication. IEEE Trans. Cybern. **49**(3), 1000–1011 (2019)

22. Manyam, S.G., Casbeer, D.W., Moll, A.V., Fuchs, Z.: Shortest Dubins path to a circle. arXiv, Optimization and Control (2018)

23. Medeiros, A.A.D.: A survey of control architectures for autonomous mobile robots. J. Braz. Comput. Soc. **4** (1998)

24. Metoui, F., Boussaid, B., Abdelkrim, M.N.: Path planning for a multi-robot system with decentralized control architecture. In: Ghommam, J., Derbel, N., Zhu, Q. (eds.) New Trends in Robot Control. SSDC, vol. 270, pp. 229–259. Springer, Singapore (2020). https://doi.org/10.1007/978-981-15-1819-5_12

25. Nair, R.R., Behera, L., Kumar, S.: Event-triggered finite-time integral sliding mode controller for consensus-based formation of multirobot systems with disturbances. IEEE Trans. Control Syst. Technol. **27**(1), 39–47 (2019)

26. Panda, M., Das, B., Subudhi, B., Pati, B.B.: A comprehensive review of path planning algorithms for autonomous underwater vehicles. Int. J. Autom. Comput. **17**(3), 321–352 (2020). https://doi.org/10.1007/s11633-019-1204-9

27. Ramadge, P.J., Wonham, W.M.: Supervisory control of a class of discrete event processes. SIAM J. Control Optim. **25**(1), 206–230 (1987)

28. Ren, W., Sorensen, N.: Distributed coordination architecture for multi-robot formation control. Robot. Autonom. Syst. **56**(4), 324–333 (2008). https://doi.org/10.1016/j.robot.2007.08.005

29. Seatzu, C., Silva, M., van Schuppen, J.H. (eds.): Control of Discrete-Event Systems. Springer, London (2013). https://doi.org/10.1007/978-1-4471-4276-8

30. Shepard, J.T., Kitts, C.A.: A multirobot control architecture for collaborative missions comprised of tightly coupled, interconnected tasks. IEEE Syst. J. **12**(2), 1435–1446 (2018). https://doi.org/10.1109/JSYST.2016.2590430

31. Ulyanov, S., Maksimkin, N.: Formation path-following control of multi-AUV systems with adaptation of reference speed. Math. Eng. Sci. Aerosp. (MESA) **10**(3), 487–500 (2019)
32. Vassilyev, S.N.: Machine synthesis of mathematical theorems. J. Logic Program. **9**(2–3), 235–266 (1990). https://doi.org/10.1016/0743-1066(90)90042-4
33. Vassilyev, S., Ulyanov, S., Maksimkin, N.: A VLF-based technique in applications to digital control of nonlinear hybrid multirate systems. In: AIP Conference Proceedings, pp. 020170(1)–020170(10) (2017). https://doi.org/10.1063/1.4972762
34. Wonham, W.M., Cai, K.: Supervisory Control of Discrete-Event Systems. Springer, Heidelberg (2019). https://doi.org/10.1007/978-3-319-77452-7
35. Yan, Z., Li, J., Zhang, G., Wu, Y.: A real-time reaction obstacle avoidance algorithm for autonomous underwater vehicles in unknown environments. Sensors **18**(2) (2018). https://doi.org/10.3390/s18020438
36. Yao, X., Wang, X., Wang, F., Zhang, L.: Path following based on waypoints and real-time obstacle avoidance control of an autonomous underwater vehicle. Sensors **20**(3) (2020). https://doi.org/10.3390/s20030795

Probability-Based Time-Optimal Motion Planning of Multi-robots Along Specified Paths

Haonan Wang, Biao Hu, and Zhengcai Cao$^{(\boxtimes)}$

Beijing University of Chemical Technology, Beijing 100029, China
wanghnbuct@163.com, hubiao@mail.buct.edu.cn, giftczc@163.com

Abstract. In this paper, we study the time-optimal motion planning problem for multi-robots under their kinodynamic constraints along specified paths. Unlike previous approaches that coordinate their motions only at a road intersection and without considering noise influence, we for the first time use the probabilistic motion model caused by the noise when planning robot motion. By deriving the conflict probability, we design a scheme to sufficiently guarantee conflict probability below our expectation. In particular, we map the potential conflict to the infeasible rectangle region in robot's path-time coordinate plane, which enables us to apply velocity interval propagation to find a time-optimal motion for each robot. Then we integrate our approach into two popular multi-robot motion planners, i.e., conflict-based search and prioritized planning approach. Experimental results demonstrate that, conflict-based search needs more computation time to achieve less motion time than prioritized motion planner.

Keywords: Motion planning · Multi-robot system · Optimization · Probability analysis

1 Introduction

Recent years have witnessed the rapid growth of autonomous multi-robot deployment in many applications such as warehouse automation and manufacturing applications [1]. In multi-robot system, a central problem is how to coordinate robot motions such that they can efficiently reach their goals without collision, and in the meanwhile, their motion time should be minimized because less motion time brings more profit. Towards this aim, a vast majority of motion planning approaches have been presented (see e.g. [1]).

Previous approaches mainly focus on the problem of planning multi-robot motions without specific path constraints and without considering noise influences. Unlike them, we for the first time plan the multi-robot motions along

This work was supported in part by the National Natural Science Foundation of China under Grant 91848103, in part by the Talent Foundation of Beijing University of Chemical University under Grant buctrc201811.

J. Qian et al. (Eds.): ICRRI 2020, CCIS 1336, pp. 138–153, 2020.
https://doi.org/10.1007/978-981-33-4932-2_10

specified paths in a dynamic environment, where the obstacle is not static and robot motions are impacted by stochastic noise. The fixed path constraints may come from scenarios such as traffic intersections or warehouses where robots need to cross multiple lanes or aisles. Along these paths, robots can only accelerate or brake to their goals. To solve the optimal motion planning of multi-robot, we have to first know how to plan an optimal motion for a single robot.

If there is not an acceleration bound, the motion planing of a single robot along a specified path can be formulated as a visibility graph in the path position/time space, as presented in [2]. Based on this formulation, a polynomial-time algorithm is presented in [3] to find an optimal solution with velocity and acceleration bounds in the absence of dynamic obstacle. The influence of dynamic obstacle has been taken into account in [4] that constructs reachable sets in the path-velocity-time space by propagating reachable velocity sets between obstacle tangent points. However, these approaches assume that robot can move strictly as our wish, which is not true in reality due to the uncertainty caused by noise.

The uncertainty of robot motion has been widely studied in the past robotic research. In the book [5], a wealth of techniques and algorithms related to the probabilistic robotics have been introduced, where the measured motion is generally given by the true motion corrupted with noise. The probabilistic motion model also needs to be taken into account when planning robot's motion. In [6], a decentralized probabilistic framework is presented for the path planning of autonomous vehicles by deriving a relation transforming the maximum turn angle into a maximum search angle. In [7], a particle filter framework is proposed to treat the uncertainty in motion planning by updating trajectory candidates, perception measurement, trajectory selection and resampling motion goal. Via assigning probabilities to the generated trajectories according to their likelihood of obeying the driving requirements, particle filtering approach is also used in [8] that formulates the motion planning as a nonlinear non-Gaussian estimation problem. In the same line with them, this paper also builds the motion corrupted with noise as a probabilistic motion model.

For multi-robot systems, the optimal coordination is a well-known NP-hard problem [9], even in static environment. There are already some optimal approaches that solve this problem in a graph-based map. Many of them, such as [10–12], first plan optimal path individually for each robot and then coordinate them when necessary. Analogously, the robot motion coordination with specified paths also relies on the individually optimal motion. For example, the approach in [13] first finds out the minimum and maximum traversal times for each path segment of each robot, then formulates their coordination as a mixed integer nonlinear programming problem that combines collision avoidance constraints for pairs of robots. To handle the non-convex challenge in this formulation, another more advanced approach in [14] successfully transforms it into two linear subproblems by introducing some additional constraints. As a result, the general-purpose solver such as mixed integer linear programming can be applied to find out an approximated optimal solution in a reasonable time. However, all these approaches assume that robot's motion is deterministic. It is true in video game or computer simulation, but not in practical implementation.

In this paper, we study how to optimally coordinate multi-robot motion with uncertain noise influence and under kinodynamic constraints along specified paths. The motion uncertainty brings the difficulty in conflict detection. For this problem, we develop a conflict detection scheme based on the probability analysis that can ensure the conflict probability below our expectation at any time. In order to plan motion in the path-time coordinate plane, we simplify dynamic obstacle or robot obstacle to rectangle and then apply the velocity interval propagation to plan the time-optimal motion of individual robot. Then, inspired by the conflict-based scheme proposed in [15], we apply a two-level scheme to resolve conflicts. In particular, a conflict tree is constructed to store all coordination options at the high-level algorithm, and optimal solution is individually found out for each coordination option. After enumerating all possible options, the final optimal motion coordination can be established. As a comparison, we also integrate our probabilistic conflict analysis into a lightweight heuristic planner, i.e., prioritized motion planning, to reduce its computation time.

The structure of this paper is as follows: in next section we present the problem formulation. In Sect. 3, we present how to plan an optimal motion for a single robot based on the probabilistic motion model. Section 4 presents the optimal multi-robot coordination scheme. In Sect. 5, experimental results are demonstrated that compare the effectiveness of our proposed probabilistic conflict-based search with prioritized path planning. Section 6 provides the conclusion of this paper.

2   Problem Formulation

We start by formally formulating the multi-robot path planning problem. Suppose there are n robots denoted as r_1, r_2, .., r_n that need to operate in a region $\mathcal{W}$ of 2-d Euclidean space. Each robot r_i is assigned a path to move from its start point s_i to its goal point g_i where $s_i \in \mathcal{W}$, $g_i \in \mathcal{W}$, where its path can be represented as an arc-length function $q_i(p) : [0, p_i^{\max}] \mapsto \mathcal{W}$. Our aim is to determine $\pi_i = (p_i, v_i, t_i)$, $\forall i \leq n$ in the path-velocity-time (PVT) state space. We assume that robot r_i starts its movement at time 0 and ends at time $t_i^{\max}$. Then we have $\pi_i(0) = s_i$, $\pi_i(t_i^{\max}) = g_i$, and $\pi_i(t) \in \mathcal{W}\,(0 \leq t \leq t_i^{\max})$. The robot dynamics must be under its velocity and acceleration bounds [1]:

$$\forall 0 \leq i \leq n, \quad \dot{p}_i = v_i + \delta_i$$
$$v_i \in [\underline{v}_i, \overline{v}_i] \tag{1}$$
$$\dot{v}_i \in [\underline{a}_i, \overline{a}_i]$$

where δ_i is a noise that follows a normal distribution $N(0, \sigma_i^2)$, and $\underline{v}_i \geq 0$ and $\underline{a}_i < 0 < \overline{a}_i$. The robot trajectory with respect to space-time is denoted as π_i, i.e., $\pi_i(t_i) = (p_i(t_i), v_i(t_i), t_i)$. Note that a robot path $q_i(p)$ has been determined

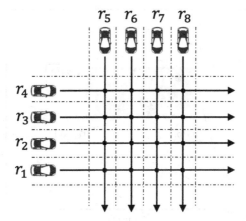

Fig. 1. An eight-robot system, where the path of each robot has been marked out with arrow.

and the remained problem is to decide the one-to-one correspondence between its path position $p_i(t)$ and time t. Besides, we denote that $\Pi(\pi_i, \pi_j)$ stands for the trajectory collision part between π_i and π_j. Then, a feasible solution should satisfy that $\Pi(\pi_i, \pi_j) = \emptyset, \forall i \neq j$, which means no collision between any two robots. We further denote $T^{\max} = \max\{t_i^{\max}, i = 1..n\}$. Then we can formally define our problem as

$$\min \ T^{\max}$$

subject to

$$(1), \tag{2}$$

$$\text{and } \pi_i(0) = s_i, \ \pi_i(t_i^{\max}) = g_i, \ i = 1..n,$$

$$\text{and } \forall i \neq j, \ \Pi(\pi_i, \pi_j) = \emptyset.$$

An example of this problem is shown in Fig. 1, where 8 robots need to move in the arrow direction. We assume that the path for each robot has been decided, but their positions with respect to time are unknown. Our goal is to decide the relationship between their positions and time such that their motions are strictly under the velocity and acceleration bounds, and they do not collide with each other, and the makespan of completing all motions is minimized.

3   Time-Optimal Motion for a Single Robot

We start by building the time-optimal motion for a single robot in a dynamic environment. This problem has been partly addressed in paper [4] that relies on the deterministic motion model to find the optimal bounded-acceleration trajectory for a single robot moving along a fixed, given path with dynamic obstacles. In this paper, we extend this approach to handle probabilistic motion model.

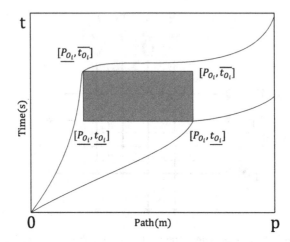

Fig. 2. An exemplified diagram of velocity interval propagation.

3.1 Velocity Interval Propagation

Velocity interval propagation(VIP) is the backbone of the single robot's planner. In the path-time coordinate plane, dynamic obstacles can be formulated as a list of rectangles $O_i = [\underline{p}_{o_j}, \overline{p}_{o_j}] \times [\underline{t}_{o_j}, \overline{t}_{o_j}], j = 1, ..., m$, as shown in Fig. 2. The robot's initial state is $(p_i(0), v_i(0), 0)$, where $v_i(0) \in [\underline{v}_i, \overline{v}_i]$ and the terminal condition is $(p_i(t_i^{\max}), v_i(t_i^{\max}), t_i^{\max})$, where $v_i(t_i^{\max}) \in [\underline{v}_i, \overline{v}_i]$. Besides, the acceleration $a_i \in [\underline{a}_i, \overline{a}_i]$. The planner's task is to find a set of collision-free trajectory π_i over above conditions in the path-velocity-time state space.

Ignoring obstacle's influence and assuming the initial velocity $v_i(0)$ is $v_i^1(0)$, the robot's reachable set $R(t : (p_i(0), v_i(0), 0))$ of velocities at time t is proved to be a convex region in the path-velocity plane, where $(p_i(0), v_i^1(0), 0)$ is the initial state of the robot. The set of velocities $V(t_i^{\max})$ attainable a target point $(p(t_i^{\max}), t(t_i^{\max}))$ is the intersection $R(t_i^{\max} : (p_i(0), v_i(0), 0)) \cap \{(p_i, v_i)|p_i = p_i(t_i^{\max}), v_i \in [\underline{v}_i, \overline{v}_i]\}$.

On the basis of the above, VIP computes the minimum and maximum starting velocity $\underline{v}_i$ and $\overline{v}_i$'s terminal velocity intervals V_1 and V_2, respectively. Then VIP minimizes and maximizes the terminal velocity without input terminal velocity constraint, and generates terminal intervals V_1 and V_2 by constructing a parabolic trajectory with acceleration $\underline{a}_i$ and $\overline{a}_i$ that interpolates $(p_i(0), 0)$ and $(p_i(t_i^{\max}), t_i^{\max})$. The final terminal velocity intervals $[\underline{v}_i, \overline{v}_i]$ is the smallest interval containing V_1, V_2, V_3, V_4.

Meanwhile, there are only two cases of collision free trajectory. The first one connects directly to the goal and the next pass tangentially along either the upper-left or lower-right vertex of one or more path-time obstacles. So the planner is designed to find the collision-free trajectories that pass through combinations of upper-left and lower-right obstacle vertices. When there is only one obstacle O, according to the obstacle's location in path-time coordinate plane,

---

Algorithm 1. Time-Optimal Motion Planning of Single Robot

1: **function** FORWARD(V_i, S_i)
2:     **for** j **do** = 1,...,2n
3:         **for** i **do** = 0,...,j-1
4:             For each disjoint interval [a,b] in V_i
5:             Call Propagate($[a, b], i, j$)
6:         **end for**
7:         $V_j \leftarrow Merge(S_j)$ For each disjoint interval $[a, b]$ in V_j:
8:         Call PropagateGoal($[a, b], j$)
9:     **end for**
10: **end function**
11: **function** PROPAGATE($[a, b], i, j$)
12:     $[a', b'] \leftarrow VIP((p_i, t_i), [a, b], (p_j, t_j))$
13:     If $[a', b']$ is empty, return
14:     Let $L(t)$ and $U(t)$ be the lower and upper examples.
15:     $B_L \leftarrow Channel(L(t))$, $B_U \leftarrow Channel(U(t))$
16:     If $B_L = B_U \neq nil$, Insert($(([a', b'], B_L), S_j$)
17:     Else
18:     If $B_L \neq nil$, Insert($(([a'], B_L), S_j$)
19:     If $B_L \neq nil$, Insert($(([b'], B_U), S_j$)
20: **end function**
21: **function** MERGE(S)
22:     Repeat until S is unchanged
23:     If $\exists([a, b], B), ([a', b'], B')$ such that $Suffix?(B, B')$
24:     Remove $([a, b], B), ([a', b'], B')$
25:     Add $([\min(a, a'), \max(b, b'), B'])$ to S
26:     Output $V = \bigcup_{([a,b],B) \in S, b-a \geq \epsilon}[a, b]$
27: **end function**

---

we just have to combine O and the interval $[\underline{v}_i, \overline{v}_i]$ to calculate the collision-free trajectory π_i, as shown in an example of Fig. 2. By connecting multiple obstacles' upper-left and lower-right obstacle vertices, the algorithm can be easily extended to multiple obstacle environments.

Then the time-optimal motion planning approach for a single robot with dynamic obstacles can be developed as Algorithm 1. In the first step, we sort the upper left and lower right obstacle vertices in the ascending order of their time into the list $(p_1, t_1), ..., (p_{2n}, t_{2n})$. Besides, we use a set S_i to store velocity interval and singleton velocity for each element in this list. Note that we only highlight its main part and ignore details, because you can read [4] for its complete information.

3.2 Probability-Based Motion Planning

The approach of velocity interval propagation assumes that the robot strictly moves as planned. However, due to the noise influence, there may be some deviations with the planned motion, which may lead to the robot collision in reality. We thus have to take these deviations into account in order to plan a safe motion.

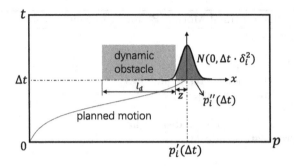

Fig. 3. An illustration of robot position influenced by noise.

Suppose that robot r_i starts at $t = 0$ and moves as planned for Δt. Then the robot position is:

$$p_i(\Delta t) = \int_0^{\Delta t} v_i(t)\, dt + \int_0^{\Delta t} \delta_i\, dt = p_i'(\Delta t) + p_i''(\Delta t), \tag{3}$$

where $p_i'(\Delta t)$ is the part that moves exactly as our plan, and $p_i''(\Delta t)$ is the deviation part caused by the noise. Because the noise δ_i is an independent input that follows normal distribution $N(0, \sigma_i^2)$, we can conclude that $p_i''(\Delta t)$ follows $N(0, \Delta t \cdot \sigma_i^2)$. As shown in Fig. 3, the red curve represents the planned motion. At time Δt, the robot is planned to move to $p_i'(\Delta t)$. The deviation at this moment is represented as $p_i''(\Delta t)$, from which we find that the small overlapping white area between dynamic obstacle and normal distribution represents the probability that robot collide with dynamic obstacle. This probability can be calculated by

$$\mathcal{P}_i(\Delta t) = \int_{-z-l_d}^{-z} \frac{1}{\sqrt{2\pi \Delta t \sigma_i}} e^{-\frac{x^2}{2\Delta t \sigma_i^2}}\, dx. \tag{4}$$

If this probability is small enough, it can be inferred that the planned path is safe. We use $\mathcal{P}_\epsilon$ as the probability threshold, i.e., a safe path should meet $\mathcal{P}_i(\Delta t) \leq \mathcal{P}_\epsilon$. By this means, we can get the minimum $z^*(\Delta t)$ that meets this condition:

$$z^*(\Delta t) = \min\left\{z \,|\, \mathcal{P}_i(\Delta t) \leq \mathcal{P}_\epsilon\right\}. \tag{5}$$

Since $\mathcal{P}_i(\Delta t)$ is a monotonously decreasing function with z, $z^*(\Delta t)$ can be easily found out by solving

$$\mathcal{P}_i(\Delta t) = \int_{-z-l_d}^{-z} \frac{1}{\sqrt{2\pi \Delta t \sigma_i}} e^{-\frac{x^2}{2\Delta t \sigma_i^2}}\, dx = \mathcal{P}_\epsilon. \tag{6}$$

Suppose that this dynamic obstacle obstructs the path from t_1 to t_2. Then for $\Delta t \in [t_1, t_2]$, its corresponding $z^*(\Delta t)$ can be determined. As shown in Fig. 4, $z^*(\Delta t)$ forms two curves $z_l^*(\Delta t)$ and $z_r^*(\Delta t)$, which are at the left and right side of dynamic obstacle, respectively. It is easy to prove that the extended dynamic obstacle is centerline symmetry because

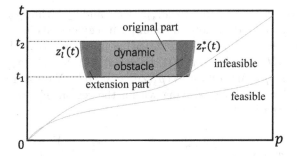

Fig. 4. The extension of dynamic obstacle for collision avoidance.

$$\mathcal{P}_i(\Delta t) = \int_{-z-l_d}^{-z} \frac{1}{\sqrt{2\pi\Delta t\sigma_i}} e^{-\frac{x^2}{2\Delta t\sigma_i^2}} \, dx \tag{7}$$

$$= \int_{z}^{z+l_d} \frac{1}{\sqrt{2\pi\Delta t\sigma_i}} e^{-\frac{x^2}{2\Delta t\sigma_i^2}} \, dx = \mathcal{P}_\epsilon, \tag{8}$$

where (7) denotes the right-side curve and (8) denotes the left-side curve. With the extended dynamic obstacle, a feasible path is the one that does not cross over it. In this way, we can apply the approach of velocity interval propagation to find the time-optimal path, only if it takes the extended dynamic obstacle into account.

4   Time-Optimal Motion for Multiple Robots

It is difficult to plan time-optimal motions for multi-robots because each robot has numerous motion options to reach their goals. The time-optimal motion planner should take all potential options into account and find out the best one from them. We have already proposed an approach to plan the time-optimal motion for a single robot in dynamic environment. To coordinate the multi-robot motions, we can either apply a conflict-based search scheme to spot all conflicts and resolve them by constructing a conflict tree, or prioritize the motion order of all robots and plan their motions successively according to this order.

4.1   Conflict Probability Between Robots

We have derived how to get the conflict probability between a robot and a dynamic obstacle. Analogously, we can derive the conflict probability between robots, by replacing deterministic moving obstacle with non-deterministic moving robot. Suppose that there are two robots r_1 and r_2 whose paths and motion have been given. Their paths intersect at point v, as shown in Fig. 5(a). The conflict is defined to happen when both robots are very near v simultaneously.

To quantitatively provide the conflict probability, we define that the intersection point v is located at p_1^v of r_1's path and p_2^v of r_2's path. The collision happens when r_1 stays within the region $[p_1^v - d_\epsilon, p_1^v + d_\epsilon]$ and r_2 within $[p_2^v - d_\epsilon, p_2^v + d_\epsilon]$ at the same time, where d_ϵ denotes a small path segment. We have derived that $p_1(\Delta t) = p_1'(\Delta t) + p_1''(\Delta t)$ in (3) where $p_1'(\Delta t)$ is given and $p_1''(\Delta t)$ follows normal distribution $N(0, \Delta t \cdot \sigma_1^2)$. Thus, $p_1(\Delta t)$ follows $N(p_1'(\Delta t), \ \Delta t \cdot \sigma_1^2)$. In a similar way, we know that $p_2(\Delta t)$ follows $N(p_2'(\Delta t), \ \Delta t \cdot \sigma_2^2)$. We can further get the probability of robot r_1 and r_2 staying within the region $[p_1^v - d_\epsilon, p_1^v + d_\epsilon]$ and $[p_2^v - d_\epsilon, p_2^v + d_\epsilon]$ at time $t = \Delta t$,

$$\mathcal{P}_1^v(\Delta t) = \int_{p_1^v - d_\epsilon}^{p_1^v + d_\epsilon} \frac{1}{\sqrt{2\pi \Delta t}\sigma_1} e^{-\frac{\left(x - p_1'(\Delta t)\right)^2}{2\Delta t \sigma_1^2}} \, dx, \tag{9}$$

$$\mathcal{P}_2^v(\Delta t) = \int_{p_2^v - d_\epsilon}^{p_2^v + d_\epsilon} \frac{1}{\sqrt{2\pi \Delta t}\sigma_2} e^{-\frac{\left(x - p_2'(\Delta t)\right)^2}{2\Delta t \sigma_2^2}} \, dx. \tag{10}$$

So the conflict probability at Δt between r_1 and r_2 is $\mathcal{P}_{1,2}^v(\Delta) = \mathcal{P}_1^v(\Delta t) \cdot \mathcal{P}_2^v(\Delta t)$. Our aim is to let

$$\forall \Delta t > 0, \ \mathcal{P}_{1,2}^v(\Delta t) \leq \mathcal{P}_\epsilon. \tag{11}$$

4.2   Conflict Resolution

To coordinate the motion of r_1 and r_2 at v, we can let one of them move freely, and the other one considers it as a dynamic obstacle. Suppose that r_1 has the time-optimal motion without considering r_2's motion. Its motion will be considered as a dynamic obstacle by r_2, as presented in Fig. 5(b). To apply VIP to find r_2's time-optimal motion, we should get the feasible area in its path-time coordinate plane. Because r_1's motion is given, $\mathcal{P}_1^v(\Delta t)$ is known, and we can find probability requirement on r_2 passing $[p_2^v - d_\epsilon, p_2^v + d_\epsilon]$ by solving

$$\mathcal{P}_2^v(\Delta t) = \int_{p_2^v - d_\epsilon}^{p_2^v + d_\epsilon} \frac{1}{\sqrt{2\pi \Delta t}\sigma_2} e^{-\frac{\left(x - p_2'(\Delta t)\right)^2}{2\Delta t \sigma_2^2}} \, dx < \frac{\mathcal{P}_\epsilon}{\mathcal{P}_1^v(\Delta t)}, \tag{12}$$

as shown in Fig. 5(c). Then we can get r_2's conflict region in its path-time coordinate plane. As shown in Fig. 5(d), we use C_{r_1, r_2}^v to denote it. On the contrary, when r_2 moves freely, C_{r_2, r_1}^v denotes the conflict region that r_1 needs to avoid.

4.3   Obstacle Simplification

It should be noted that VIP can only handle the dynamic obstacle whose shape is rectangle in the path-time coordinate plane, while the shape of our obstacle based on probability is not. In order to apply VIP to solve our problem, we have to simplify the shape of our obstacle to rectangle. We discuss the simplification scheme of dynamic obstacle and robot obstacle, respectively.

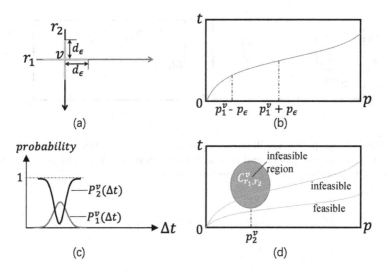

Fig. 5. A two-robot motion coordination, (a) r_1 and r_2 intersects at v, and red lines represent conflict segment, (b) r_1's optimal motion is given, (c) r_1's probability at conflict segment and probability requirement on r_2, both with respect to time, (d) the infeasible region on r_2's path-time coordinate plane. (Colour figure online)

Dynamic Obstacle Simplification. We suppose that the exact region of its original dynamic obstacle is known, as shown in gray part of Fig. 6(a). The extension part can be obtained by applying (7) and (8) for $\Delta t \in (t_1, t_2)$, as shown in the dark area of Fig. 6(a), where $z^{\max}$ denotes its maximum path length. Then we can use a rectangle with length $l^d + 2z^{\max}$ and width $t_2 - t_1$ to cover it, and this rectangle can be applied for calling VIP to obtain the optimal motion.

Robot Obstacle Simplification. For the robot obstacle, we suppose that it is covered by a rectangle with a top left vertex (t_u, p_u) and bottom right vertex (t_d, p_d). Our job is to decide the two vertexes. From (12), we know that if $\mathcal{P}_1^v(\Delta t)$ is smaller than $\mathcal{P}_\epsilon$, the required probability of $\mathcal{P}_2^v(\Delta t)$ is less than 1. In this case there is no limitation on robot path. However, when $\mathcal{P}_1^v(\Delta t)$ is equal to $\mathcal{P}_\epsilon$, the limitation comes. Thus, we can conclude that the robot obstacle comes into being when $\mathcal{P}_1^v(\Delta t) = \mathcal{P}_\epsilon$. Then we can get two Δt to meet this condition, where the large one is t_u and the small one is t_d, i.e.,

$$\mathcal{P}_1^v(t_u) = \mathcal{P}_1^v(t_d) = \mathcal{P}_\epsilon \tag{13}$$

Next, we need to decide the width of this rectangle. It can be found that the width of robot obstacle is negatively proportional to the required probability. The smaller required probability is, the larger obstacle width should be. According to (12), we can infer that the smallest required probability corresponds to the largest $\mathcal{P}_1^v(\Delta t)$, i.e.,

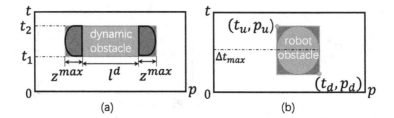

Fig. 6. Obstacle simplification illustration, (a) gray part is the original dynamic obstacle, dark part is the extension, and rectangle is its simplified form, (b) gray part is the robot obstacle, and rectangle is its simplified form.

$$\mathcal{P}_{1,\max}^v = \mathcal{P}_1^v(\Delta t_{\max}) = \max\left\{\mathcal{P}_1^v(\Delta t), \forall \Delta t > 0\right\},$$
$$\mathcal{P}_{2,\min}^v = \frac{\mathcal{P}_\epsilon}{\mathcal{P}_{1,\max}^v}, \tag{14}$$

where $\Delta t_{\max}$ denotes the moment of maximizing $\mathcal{P}_1^v$. Then, we can also get two $p_2'(\Delta t)$ to meet (14), where the small one is p_u and the large one is p_d, i.e.,

$$\mathcal{P}_{2,\min}^v = \int_{p_2^v - d_\epsilon}^{p_2^v + d_\epsilon} \frac{1}{\sqrt{2\pi}\Delta t_{\max}\sigma_2} e^{-\frac{\left(x - p_u\right)^2}{2\Delta t_{\max}\sigma_2^2}} dx,$$
$$= \int_{p_2^v - d_\epsilon}^{p_2^v + d_\epsilon} \frac{1}{\sqrt{2\pi}\Delta t_{\max}\sigma_2} e^{-\frac{\left(x - p_v\right)^2}{2\Delta t_{\max}\sigma_2^2}} dx. \tag{15}$$

Hence, to find the rectangle that can cover robot obstacle, it is not necessary to compute all parts of robot obstacle. By simply calling (13), (14) and (15), the simplified rectangle can be built.

4.4   Probabilistic Conflict-Based Search Scheme

The conflict-based search (CBS) is originally presented to minimize the sum of all robots' cost in an undirected graph whose edges have been assigned specific cost. In this section we extend it to find the time-optimal motion for multi-robots with specific paths. CBS works at two levels, where at the low level a time-optimal motion is planned for each individual robot under dynamic obstacles and high-level constraints, and at the high level conflicts are spotted and resolved at their earliest start time. Conflict is resolved by adding two successor nodes in the conflict tree. Each robot involved by this conflict is assigned an additional constraint at the low level.

We have presented how to plan time-optimal motion for a single robot with dynamic obstacle, as presented as Algorithm 1. This acts as the low-level algorithm in the probabilistic conflict-based search (P-CBS) scheme. We extend the high-level search to incorporate the conflict based on probability, as presented in

Algorithm 2. Probabilistic Conflict-based Search Scheme
___
1: **function** HIGHLEVELSEARCH
2:     $\mathcal{R}.constraints = \emptyset$
3:     $\mathcal{R}.solution =$ find individual motion by Algo. 1
4:     $\mathcal{R}.cost =$ the maximum cost of all individual paths in $\mathcal{R}.solution$
5:     Insert $\mathcal{R}$ to $Open$
6:     **while** $Open \neq \emptyset$ **do**
7:         $\mathcal{P} \leftarrow$ best node from $Open$ // lowest solution cost
8:         Validate the paths in $\mathcal{P}$ until a conflict occurs
9:         **if** $\mathcal{P}$ has no conflict **then**
10:             **Return** $\mathcal{P}.solution$
11:         **end if**
12:         $\mathcal{C} \leftarrow$ conflict between r_i and r_j in $\mathcal{P}$
13:         **for** robot r_i in $\mathcal{C}$ **do**
14:             $\mathcal{A} \leftarrow$ new node
15:             $\mathcal{A}.constraints \leftarrow \mathcal{P}.constraints + C_{r_j,r_i}^v$
16:             $\mathcal{A}.solution \leftarrow \mathcal{P}.solution$
17:             Update $\mathcal{A}.solution$ by by Algo. 1
18:         **end for**
19:     **end while**
20: **end function**
___

Algorithm 2. At the beginning, all robots are given their time-optimal motions by solely considering dynamic obstacles. Then, the first conflict between two robots is spotted, as shown in line 12, where the two robots are denoted as r_i and r_j. For each of them, a new node is added to the conflict tree, and a new solution is found. This procedure continues until all conflicts have been resolved.

Similar to CBS, P-CBS is also a complete and optimal algorithm with respect to the motion time, because the search of P-CBS has enumerated all possible situations for the time-optimal solution.

4.5   Prioritized Motion Planning

Intuitively, priority-based approach assigns each robot an unique priority, and robots' motion is planned in the descending order of their priorities [16,17]. The collision can thus be avoided with this scheme because low-priority robot consider high-priority robot as dynamic obstacle. In this approach, priority assignment plays a significant role in its performance. We assign the priority to robots according to their Euclid distance of start points and goal points, with a long Euclid distance leading to a high priority, because robots with long distance should avoid interference as much as possible to shorten the maximum of motion time. Its basic procedures are presented in Algorithm 3.

Algorithm 3. Basic procedures of prioritized planning

1: **function** PRIORITIZEDPLANNING
2:     Assign the priority in the descending order of l_i
3:     $O = \emptyset$
4:     **for** $i \leftarrow 1..n$ **do**
5:         $\mathcal{R}.solution$ = find individual motion by Algo. 1
6:         **if** not found **then**
7:             Report failure and terminate
8:         **end if**
9:         $O = O \cup \mathcal{R}.solution$
10:     **end for**
11: **end function**

5   Experiments and Results

5.1   Environments and Metrics

The experimental environments are set up like Fig. 1, where there are $n \times n$ robots moving in horizontal and vertical directions with path length = 70 m. Thus, there are $2n$ robots in total. Here we evaluate the performance of motion planning approaches in three configurations, which are $n = 2, 4, 6$. A robot is configured to be with width = 1 m and length = 2 m. Its velocity and acceleration bound is set to $\underline{v}_i = 0$, $\overline{v}_i = 3$ and $\underline{a}_i = -3$, $\overline{a}_i = 2$, respectively.

In general, we investigate the influence of noise intensity and safety requirement on their performance. The noise intensity is represented by its standard variance, i.e., σ_i, and safety requirement means the conflict probability threshold, i.e., $\mathcal{P}_\epsilon$. Here we vary σ_i from 0.1 to 0.9 and vary $\mathcal{P}_\epsilon$ from 10^{-4} to 10^{-1}, respectively. We set the maximum of their motion time as our optimized goals. In addition, we compare the computation time of compared approaches to evaluate their computation complexity.

5.2   Results

First, we investigate the influence of noise intensity on the performance of compared approaches by fixing conflict probability threshold to 0.01. The maximum of motion time of all robots and computation time in finding solution are presented in Fig. 7, where PRIOR denotes the prioritized motion planning and P-CBS denotes the probabilistic conflict-based search. From Fig. 7(a), we find that, with the increase of noise intensity, the maximum of motion time also increases for both approaches, because an intensive noise brings more uncertainty and forces robots to take more conservative actions to keep safety. It can also be observed that P-CBS achieved a lower motion time compared to PRIOR, especially for a large number of robots. From Fig. 7(b), we observe that the noise intensity has a minor impact on the computation time when robot count is small, but clearly increases it when robot count is large, such as $n = 6$. Besides, P-CBS needs more computation time than PRIOR in general because P-CBS needs to

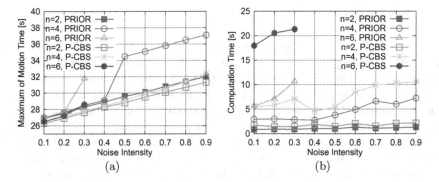

Fig. 7. Performance of compared approaches w.r.t. noise intensity, where (a) is the maximum of motion time, (b) is the computation time, and the conflict probability threshold $\mathcal{P}_\epsilon = 0.01$

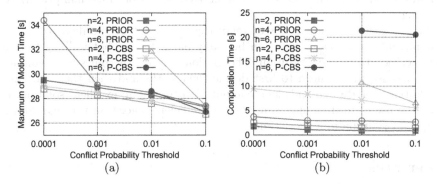

Fig. 8. Performance of compared approaches w.r.t. conflict probability threshold, where (a) is the maximum of motion time, (b) is the computation time, and the noise intensity $\sigma_i = 0.3$

check more motion options, while there is only one option for PRIOR. Last, Fig. 7 does not show the results of 6 robots when noise intensity is large, because there is not a feasible solution in these cases. This tells us that a large number of robot or an intensive noise may result in the failure of P-CBS and PRIOR.

Second, we investigate the influence of conflict probability threshold on the performance of compared approaches by fixing noise intensity to 0.3. The results of motion time and computation time are presented in Fig. 8. We find that the motion time decreases with the increase of conflict probability threshold, because a large conflict probability threshold relieves safety requirement and provides a large freedom in motion planning. Regarding to its influence on the computation time, the planner needs more computation time when conflict probability threshold is small. Last, Fig. 8 also demonstrates that P-CBS achieves smaller motion time and needs more computation time than the prioritized motion planning approach.

6   Conclusions

In this paper, we study the problem how to plan time-optimal motion for multi-robots moving along specified paths. Unlike previous approaches that plan robots' motion without considering their stochastic noise, we build our approach on the probabilistic motion robot. Towards the aim of minimizing the motion time, we present the conflict probability analysis to ensure the robot's safety. Because the shape of dynamic obstacle and robot obstacle is irregular in the path-time coordinate plane, we present a simplification scheme to transform them to rectangle, such that the velocity interval propagation can be used to find the time-optimal motion of each individual robot. In the end, we integrate our probability analysis with two popular multi-robot planners, i.e., conflict-based search and prioritized planning, and compare their performance in solution quality and computation time. Experimental results show that conflict-based search can find better solution, while incurring more computation time than prioritized planning.

In the future, we are going to relax the constraint of specified path. For example, in a free space, there can be many paths for a robot walking from a start position to the goal position. The motion planning becomes more complicated because robots need to first decide their paths and then their motions. In addition, P-CBS demands long computation to find an excellent solution. It would be useful to develop a simplified but efficient algorithm that can find a competent solution in a short time.

References

1. Mohanan, M.G., Salgoankar, A.: A survey of robotic motion planning in dynamic environments. Robot. Auton. Syst. **100**, 171–185 (2018)
2. Kant, K., Zucker, S.W.: Toward efficient trajectory planning: the path-velocity decomposition. Int. J. Robot. Res. **5**(3), 72–89 (1986)
3. O'Dunlaing, C.: Motion planning with inertial constraints. Algorithmica **2**(1–4), 431–475 (1987)
4. Johnson, J., Hauser, K.: Optimal acceleration-bounded trajectory planning in dynamic environments along a specified path. In: IEEE International Conference on Robotics & Automation, pp. 2035–2041 (2012)
5. Sebastian, T., Wolfram, B., Dieter, F., Ronald, C.A.: Probabilistic Robotics. MIT Press (2005)
6. Kamal, W., Gu, D.-W., Postlethwaite, I.: A decentralized probabilistic framework for the path planning of autonomous vehicles. IFAC Proc. Vol. **38**(1), 37–42 (2005)
7. Kim, J., Jo, K., Lim, W., Sunwoo, M.: A probabilistic optimization approach for motion planning of autonomous vehicles. Proc. Inst. Mech. Eng. Part D J. Automob. Eng. **232**, 632–650 (2018)
8. Karl, B., Tru, H., Stefano, D.-C.: Motion planning of autonomous road vehicles by particle filtering. IEEE Trans. Intell. Veh. **4**(2), 197–210 (2019)
9. Dasler, P., Mount, D.M.: On the complexity of an unregulated traffic crossing. In: Dehne, F., Sack, J.-R., Stege, U. (eds.) WADS 2015. LNCS, vol. 9214, pp. 224–235. Springer, Cham (2015). https://doi.org/10.1007/978-3-319-21840-3_19

10. Standley, T.: Finding optimal solutions to cooperative pathfinding problems. In: Twenty-Fourth AAAI Conference on Artificial Intelligence (2010)
11. Wagner, G., Choset, H.: M*: a complete multirobot path planning algorithm with performance. In: IEEE/RSJ International Conference on Intelligent Robots & Systems, pp. 3260–3267 (2011)
12. Wagner, G., Choset, H.: Subdimensional expansion for multirobot path planning. Artif. Intell. **219**(3), 1–24 (2015)
13. Peng, J., Akella, S.: Coordinating multiple robots with kinodynamic constraints along specified paths. Int. J. Robot. Res. **24**(4), 295–310 (2005)
14. Altché, F., Qian, X., Fortelle, A.D.L.: Time-optimal coordination of mobile robots along specified paths. In: IEEE/RSJ International Conference on Intelligent Robots & Systems, pp. 5020–5026 (2016)
15. Sharon, G., Stern, R., Felner, A., Sturtevant, N.R.: Conflict-based search for optimal multi-agent pathfinding. Artif. Intell. **219**, 40–66 (2015)
16. Van Den Berg, J.P., Overmars, M.H.: Prioritized motion planning for multiple robots. In: IEEE/RSJ International Conference on Intelligent Robots & Systems, pp. 430–435 (2005)
17. Hu, B., Cao, Z.: Minimizing task completion time of prioritized motion planning in multi-robot systems. In: IEEE International Conference on Systems, Man and Cybernetics, pp. 1018–1023 (2019)

Robot Design and Control

Dynamic Precision Analysis
of a Redundant Sliding Manipulator

Yuchuang Tong[1,2,3], Jinguo Liu[1,2(✉)], Zhaojie Ju[1,2,4], Yuwang Liu[1,2],
and Lingshen Fang[2,5]

[1] Stata Key Laboratory of Robotics, Shenyang Institute of Automation,
Chinese Academy of Sciences, Shenyang 110016, China
liujinguo@sia.cn
[2] Institutes for Robotics and Intelligent Manufacturing,
Chinese Academy of Sciences, Shenyang 110169, China
[3] University of the Chinese Academy of Sciences, Beijing 100049, China
[4] School of Computing, University of Portsmouth, Portsmouth PO1 3HE, UK
[5] Kunshan Intelligent Equipment Research Institute, Shenyang Institute
of Automation, Chinese Academy of Sciences, Kunshan 215347, China

Abstract. This paper deals with the dynamic precision analysis of a
linear-type rail redundant sliding manipulator. Based on the analysis of
the mechanism structure, kinematics, stiffness performance and dynamic
characteristics of the redundant sliding manipulator, the dynamic error
caused by elastic deformation of links is analyzed. A mathematical model
of the dynamic precision of the redundant sliding manipulator system
is established in this paper by the geometric method. The influence of
parameters such as equivalent stiffness and moment of inertia on the
dynamic error of the manipulator is analyzed and parameter optimization schemes are proposed to reduce the dynamic error. Simulation analysis and experiment verify the dynamic precision of the redundant sliding
manipulator. This paper provides a theoretical guidance for the high-
precision operations of the redundant sliding manipulator.

Keywords: Dynamic precision analysis · Dynamic error · Redundant
manipulator · Sliding manipulator

1 Introduction

For decades, as redundant manipulators have huge application prospects in
aerospace, military industry, manufacturing, medical, civil and other fields, and
will play an increasingly important role in the future [1–3]. Thus the study of
their dynamic errors can improve the theoretical study of dynamic accuracy.
With the progress of exploration and the change of experiment demand, more
and more scientific experiments require the redundant manipulators to simultaneously possess the ability to carry out autonomous and precise operations to
improve the efficiency and operational safety [4,5]. Therefore, research on the

© Springer Nature Singapore Pte Ltd. 2020
J. Qian et al. (Eds.): ICRRI 2020, CCIS 1336, pp. 157–171, 2020.
https://doi.org/10.1007/978-981-33-4932-2_11

dynamic accuracy of redundant manipulators has become an important part of robotic technology [6, 7].

Redundant manipulators are nonlinear and strongly coupled dynamic systems [8, 9]. Due to the limitation of mechanical manufacturing technology level and manufacturing cost, actual structural parameters and motion parameters inevitably have deviations [10]. In addition, during the high precision technical operation process of the redundant manipulator, some of their own error sources, elastic deformation and vibration caused by inertial forces during high-speed movement, and errors caused by external interference are uncertain and random [11]. Fang et al. [12] proposed a new set of kinematic indicators that can evaluate redundant parallel manipulators. Guo et al. [13] proposed a dynamic analysis method for spherical motors. Tong et al. [14] proposed a Gough-Stewart parallel manipulator with linear orthogonality to achieve dynamic decoupling. Tsai et al. [15] integrated the existing uncertainty models for the practical application of serial and parallel manipulators. Yang et al. [16] used a hypothetical mode method and Lagrange method to describe the dynamic model of a flexible manipulator with unknown interference. At present, most of the researchers' research on the dynamic precision of manipulators is for parallel manipulators or the flexible manipulators. However, there are few studies on the dynamic accuracy of the redundant rail-type sliding manipulator which is used for high-precision operations. Therefore, research on the dynamic accuracy of the redundant sliding manipulator is the top priority for the development of robotic technology.

In order to resolve the appeal issue, this paper deals with the dynamic precision analysis of a linear-type rail redundant sliding manipulator. The dynamic error of the redundant sliding manipulator caused by elastic deformation of links is analyzed, and parameter optimization schemes are proposed to reduce the dynamic error. This paper provides a theoretical guidance for the high-precision operations of the redundant sliding manipulator.

The remainder of this paper is organized as follows. Section 2 outlines the redundant sliding manipulator studied in this paper. Section 3 establishes the mathematical model of dynamic precision of the redundant sliding manipulator by the geometric method. Section 4 analyzes the influence of parameters on dynamic errors. Section 5 verifies the dynamic accuracy through simulation and experiment, and proposes parameter optimization schemes. Section 6 summarizes the innovation in this paper.

2   Overview of the Redundant Sliding Manipulator

In this paper, we have carried out the development of a linear-type rail redundant sliding manipulator. The redundant sliding manipulator is mainly composed of macro manipulator, a micro manipulator platform, micro gripper, micro vision system and control system. Figure 1 shows the prototype of the macro-micro manipulator. The dexterous manipulator provides a wide range of accessible space and the dexterous posture of the end-effector, the micro manipulator performs the micro-operation, and the macro manipulator performs the macro-operation. The whole precision manipulator has good spatial accessibility and

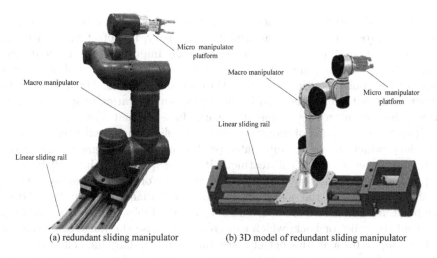

(a) redundant sliding manipulator   (b) 3D model of redundant sliding manipulator

Fig. 1. The linear-type rail redundant sliding manipulator

dexterity, as well as the characteristics of precise positioning, rapid response and easy operation. Combined with the measuring instruments such as the microscope, the micro-operation task with high precision can be realized.

The redundant sliding manipulator studied in this paper is composed of different types of sliding rails as a mobile base and a six degrees-of-freedom (DOF) redundant manipulator. Due to its unique structural features, it has the advantages and characteristics of being both a redundant manipulator and a mobile manipulator. On the one hand, as a redundant manipulator, the DOFs of the redundant manipulator is greater than the DOFs required to achieve the primary task defined by its end effector, so it has received great attention. The existence of redundancy allows the manipulator to complete secondary tasks and primary tasks simultaneously. Specifically, due to redundancy, the redundant manipulator has multiple control solutions for the tasks that can be completed, from which we can choose the best solution to complete the secondary tasks. On the other hand, with the development of complex technological society and the introduction of new concepts and innovative theoretical tools in the field of intelligent systems, the mobile manipulator has attracted great interest in the industrial, military and public service fields. Compared with fixed robots, they have large-scale maneuverability and control capability.

Therefore, the redundant sliding manipulator studied in this paper has the advantages of large working space, light weight, high fault tolerance, strong robustness, high operating accuracy, easy replacement of different types of rails to complete various operating tasks and so on. Compared with the traditional redundant manipulator, its configuration has larger workspace, more flexible task operation and higher fault tolerance. Moreover, compared with the traditional mobile manipulator, its controllability is stronger and the precision is higher.

Based on the above advantages, redundant sliding manipulators have huge application prospects in aerospace, military industry, manufacturing, medical, civil and other fields, and will play an increasingly important role in the future. In the military field, it can meet the operational requirements of military operations such as mobile shooting, pursuit of the enemy, and material transmission. In the aerospace field, it can be used for high-precision operation of space on-orbit service, thereby improving astronauts' work efficiency and operating accuracy simultaneously. In terms of industry, redundant sliding manipulators can complete the production work of automated production lines, thus greatly liberating manpower. In terms of medical treatment, it can be applied in minimally invasive surgery and surgical surgery to improve the accuracy of surgery, thereby improving the success rate of the operation. In terms of civilian use, redundant sliding manipulators can be used as supernumerary robotic limbs. The wearer can wrap it around the waist or back, which can greatly increase the range of movement of the manipulator and thus better assist the people in daily life. Therefore, the redundant sliding manipulator in this paper has great application prospects, and plays an important role in practical applications. Therefore, the research on the precision of the redundant sliding manipulator is the most important part of the robot technology.

3 Dynamic Precision Modeling of the Redundant Sliding Manipulator

3.1 Establish Vibration Differential Equations

This paper assumes that the link of the manipulator is a cantilever beam fixed at one end, and the flexibility of the joint and the link itself merges into equal flexibility.

In general, linear vibration of mechanical system can be described by linear differential equation as follows:

$$m\ddot{x} + c\dot{x} + kx = f(t) \tag{1}$$

where m is the vibration mass, c is the damping coefficient, k is the stiffness, x is the displacement, $\dot{x}$ is the speed, $\ddot{x}$ is the acceleration, $f(t)$ is the disturbance force or excitation force, and t is time.

The Lagrangian equation method is to solve the vibration differential equation of the vibration system by establishing the kinetic energy T, potential energy U, and energy loss function D of the vibration system, i.e.,

$$\frac{d}{d}\left(\frac{\partial T}{\partial \dot{q}_i}\right) - \frac{\partial T}{\partial q_i} + \frac{\partial U}{\partial q_i} + \frac{\partial D}{\partial \dot{q}_i} = F_i(t) \quad (i = 1, 2, 3, \cdots, 7) \tag{2}$$

where q_i is generalized coordinate, $\dot{q}_i$ is a generalized velocity, and $F_i(t)$ is a generalized excitation force.

The vibration model discussed in this paper is the free vibration under the ideal state, so the energy dissipation function and the generalized excitation force are ignored, then the Laplace equation (2) can be simplified as:

$$\frac{d}{dt}\left(\frac{\partial T}{\partial \dot{q}_i}\right) - \frac{\partial T}{\partial q_i} + \frac{\partial U}{\partial q_i} = 0 \quad (i = 1, 2, 3, \cdots, 7) \tag{3}$$

which gives

$$T = \sum_{i=1}^{7} \frac{1}{2} J_i \Delta \dot{\theta}_i^2$$
$$V = \sum_{i=1}^{7} \frac{1}{2} k_i \left(l_i \Delta \theta_i\right)^2 \tag{4}$$

where l_i, m_i, J_i and k_i are the length, the mass, moment of inertia and equivalent stiffness of the link i, respectively.

For the link of the manipulator, only the bending stiffness and tensile stiffness are considered, and the stiffness of the joint and the link is approximated to a stiffness, which is called the equivalent bending stiffness. The equivalent bending stiffness and tensile stiffness are denoted by $k = 3EI/l^3$ and $k = EA/l$, where E is the elastic modulus, I is the moment of inertia, and l is the link length, A is the cross-sectional area.

The moment of inertia is a measure of inertia representing the rotation of an object and a property of the object itself. The moment of inertia of an object on any axis c is defined as $J_c = \sum_{i=1}^{n} m_i r_i^2$. It can be seen that the moment of inertia is determined by the mass and its distribution, and its unit is $kg \cdot m^2$.

By substituting the kinetic energy T and the potential energy V into (3), it can be obtained as follows:

$$0 = J_i \Delta \ddot{\theta}_i + k_i l_i^2 \Delta \theta_i \tag{5}$$

In the form of matrix, the following can be obtained:

$$\begin{bmatrix} 0 \\ \vdots \\ 0 \end{bmatrix} = \begin{bmatrix} J_1 & \ldots & 0 \\ \vdots & \ddots & \vdots \\ 0 & \ldots & J_7 \end{bmatrix} \begin{bmatrix} \Delta \ddot{\theta}_1 \\ \vdots \\ \Delta \ddot{\theta}_7 \end{bmatrix} + \begin{bmatrix} k_1 l_1^2 & \ldots & 0 \\ \vdots & \ddots & \vdots \\ 0 & \ldots & k_7 l_7^2 \end{bmatrix} \begin{bmatrix} \Delta \theta_1 \\ \vdots \\ \Delta \theta_7 \end{bmatrix} \tag{6}$$

where $\begin{bmatrix} \Delta \ddot{\theta}_1 \ldots \Delta \ddot{\theta}_7 \end{bmatrix}^T$ is the angular acceleration array and $\begin{bmatrix} \Delta \theta_1 \ldots \Delta \theta_7 \end{bmatrix}^T$ is the angular displacement array. Further simplified (6) as the following equation:

$$[\mathbf{M}] [\Delta \ddot{\theta}] + [\mathbf{K}][\Delta \theta] = [0] \tag{7}$$

where $[\mathbf{M}]$ is the mass matrix and $[\mathbf{K}]$ is the stiffness matrix. Equation (7) is the free vibration differential equation of the manipulator.

3.2   Natural Frequency and Main Vibration of the System

The angular displacements $\Delta\theta_i (i \in [1, 2, 3, 4, 5, 6, 7])$ are used to describe the deviation between the deformed position and the ideal position of the links. Moreover, the angular displacements are used to qualitatively describe the dynamic precision caused by the elastic deformation of the manipulator. $\Delta\theta_i$ are assumed as harmonic vibration at the same frequency and phase but different amplitude, i.e.,

$$\Delta\theta_i = A_i \sin(\omega t + \varphi) \tag{8}$$

where A_i are different amplitude.

Substitute (8) into (6), we can obtain:

$$\begin{bmatrix} k_1 l_1^2 - J_1 \omega^2 \cdots & & 0 \\ \vdots & \ddots & \vdots \\ 0 & & \cdots k_7 l_7^2 - J_7 \omega^2 \end{bmatrix} \begin{bmatrix} A_1 \\ \vdots \\ A_7 \end{bmatrix} = \begin{bmatrix} 0 \\ \vdots \\ 0 \end{bmatrix} \tag{9}$$

The characteristic equation of (9) is as follows:

$$|B| = |[K] - \omega^2[M]| = \begin{vmatrix} k_1 l_1^2 - J_1 \omega^2 \cdots & & 0 \\ \vdots & \ddots & \vdots \\ 0 & & \cdots k_7 l_7^2 - J_7 \omega^2 \end{vmatrix} = 0 \tag{10}$$

Further extending the determinant in (10) can obtain that:

$$(k_1 l_1^2 - J_1 \omega^2) \ldots (k_i l_i^2 - J_i \omega^2) = 0 \tag{11}$$

Then, (11) can be regarded as a quadratic equation with the frequency ω_i^2 as the unknown, and the solution can be expressed as:

$$\omega_i^2 = \frac{k_i l_i^2}{J_i} \tag{12}$$

It can be seen that the natural frequency problem can be solved by solving the eigenvalues. Theoretically, the vibration system has n DOFs, and n natural frequencies can be obtained. Since these frequencies are related only to the natural parameters of the vibration system (such as mass and stiffness) and not to the initial energy of the vibration, they are called the natural frequencies of the vibration system.

Differential equations cannot directly reveal the relationship between the amplitude of the link and the amplitude of the end-effector in the vibration process. Assuming that the initial condition of the vibration is: $t = 0, \Delta\theta_i = \Delta\theta_{i0}, \Delta\dot{\theta}_i = \Delta\dot{\theta}_{i0}$, then the expression for amplitude A_i is $A_i = \sqrt{\Delta\theta_{i0}^2 + \left(\frac{\Delta\dot{\theta}_{i0}}{\omega}\right)^2}$. Corresponding to different natural frequencies, the expression of the amplitude A_i is as follows:

$$A_i^{(n)} = \sqrt{\Delta\theta_{i0}^2 + \left(\frac{\Delta\dot{\theta}_{i0}}{\omega_n}\right)^2} \quad n \in (1, 2, \cdots, 7) \tag{13}$$

where $A_i^{(n)}$ is the amplitude when the natural frequency is ω_n.

Thus, the main vibration of each order n of the system is:

$$\begin{cases} \Delta\theta_1^{(n)} = A_1^{(n)} \sin(\omega_n t + \varphi_n) \\ \vdots \\ \Delta\theta_7^{(n)} = A_7^{(n)} \sin(\omega_n t + \varphi_n) \end{cases} \tag{14}$$

In general, the free vibration of 7-DOF is synthesized by the main vibration of 7 different frequencies, and the result of the synthesis is not necessarily simple harmonic vibration. Therefore, the general solution of the differential equation (7) is the superposition of the above seven main vibrations, that is:

$$\Delta\theta_i = \Delta\theta_i^{(1)} + \cdots + \Delta\theta_i^{(7)} = A_i^{(1)} \sin(\omega_1 t + \varphi_1) + \cdots + A_i^{(7)} \sin(\omega_7 t + \varphi_7) \tag{15}$$

Equation (15) gives the general vibration solution of the 7-DOF free vibration model, which provides a theoretical basis for establishing the mathematical model of dynamic accuracy.

3.3   Establish Mathematical Model of Dynamic Precision

When discussing the error of the end-effector, the maximum value of the angular displacement is selected as the research object, so the maximum vibration error of 7-DOF is calculated as follows:

$$\Delta\theta_i = A_i^{(1)} + \cdots + A_i^{(7)} \tag{16}$$

The most commonly used maximum error in the error analysis selected in (16) can simplify the analysis process and make the results relatively accurate.

The linear sliding rail of the redundant sliding manipulator is considered as a revolute joint in kinematic modeling, so the manipulator is considered as a 7-DOF redundant sliding manipulator. The kinematic model of the redundant sliding manipulator is shown in Fig. 2. According to the kinematics of the manipulator, the position of the end-effector can be obtained on the Cartesian coordinates, as shown below:

$$\begin{aligned} p_x &= l_3 * \cos(\theta_1 + \theta_2 + \theta_3) * \sin(\theta_4 + \theta_5) - l_4 * \cos(\theta_7) * (\sin(\theta_1 + \theta_2 + \theta_3)* \\ &\quad \sin(\theta_6) - \cos(\theta_1 + \theta_2 + \theta_3) * \cos(\theta_4 + \theta_5) * \cos(\theta_6)) - l_0 * \cos(\theta_1 + \theta_2) + l_2* \\ &\quad \cos(\theta_1 + \theta_2 + \theta_3) * \cos(\theta_4) - l_4 * \cos(\theta_1 + \theta_2 + \theta_3) * \sin(\theta_4 + \theta_5) * \sin(\theta_7) \\ p_y &= l_4 * \cos(\theta_7) * (\cos(\theta_1 + \theta_2 + \theta_3) * \sin(\theta_6) + \sin(\theta_1 + \theta_2 + \theta_3) * \cos(\theta_4+ \\ &\quad \theta_5) * \cos(\theta_6)) - l_0 * \sin(\theta_1 + \theta_2) + l_3 * \sin(\theta_1 + \theta_2 + \theta_3) * \sin(\theta_4 + \theta_5) + l_2* \\ &\quad \sin(\theta_1 + \theta_2 + \theta_3) * \cos(\theta_4) - l_4 * \sin(\theta_1 + \theta_2 + \theta_3) * \sin(\theta_4 + \theta_5) * \sin(\theta_7) \\ p_z &= l_1 + l_3 * \cos(\theta_4 + \theta_5) - l_2 * \sin(\theta_4) - (l_4 * \cos(\theta_6 + \theta_7) * \sin(\theta_4+ \\ &\quad \theta_5))/2 - l_4 * \cos(\theta_4 + \theta_5) * \sin(\theta_7) - (l_4 * \cos(\theta_6 - \theta_7) * \sin(\theta_4 + \theta_5))/2 \end{aligned} \tag{17}$$

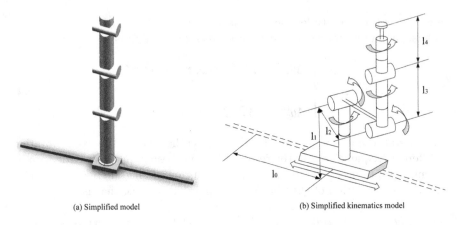

(a) Simplified model   (b) Simplified kinematics model

Fig. 2. Simplified model and simplified kinematics model of the redundant sliding manipulator

Furthermore, the position error of the end-effector of the redundant sliding manipulator can be obtained as follows:

$$
\begin{aligned}
e_x &= l_3 * \cos(\Delta\theta_1 + \Delta\theta_2 + \Delta\theta_3) * \sin(\Delta\theta_4 + \Delta\theta_5) - l_4* \\
&\quad \cos(\Delta\theta_7) * (\sin(\Delta\theta_1 + \Delta\theta_2 + \Delta\theta_3) * \sin(\Delta\theta_6) - \cos(\Delta\theta_1 + \Delta\theta_2 + \Delta\theta_3)* \\
&\quad \cos(\Delta\theta_4 + \Delta\theta_5) * \cos(\Delta\theta_6)) - l_0 * \cos(\Delta\theta_1 + \Delta\theta_2) + l_2 * \cos(\Delta\theta_1 + \Delta\theta_2 + \\
&\quad \Delta\theta_3) * \cos(\Delta\theta_4) - l_4 * \cos(\Delta\theta_1 + \Delta\theta_2 + \Delta\theta_3) * \sin(\Delta\theta_4 + \Delta\theta_5) * \sin(\Delta\theta_7) \\
e_y &= l_4 * \cos(\Delta\theta_7) * (\cos(\Delta\theta_1 + \Delta\theta_2 + \Delta\theta_3) * \sin(\Delta\theta_6) + \sin(\Delta\theta_1 + \\
&\quad \Delta\theta_2 + \Delta\theta_3) * \cos(\Delta\theta_4 + \Delta\theta_5) * \cos(\Delta\theta_6)) - l_0 * \sin(\Delta\theta_1 + \Delta\theta_2) + \\
&\quad l_3 * \sin(\Delta\theta_1 + \Delta\theta_2 + \Delta\theta_3) * \sin(\Delta\theta_4 + \Delta\theta_5) + l_2 * \sin(\Delta\theta_1 + \Delta\theta_2 + \\
&\quad \Delta\theta_3) * \cos(\Delta\theta_4) - l_4 * \sin(\Delta\theta_1 + \Delta\theta_2 + \Delta\theta_3) * \sin(\Delta\theta_4 + \Delta\theta_5) * \sin(\Delta\theta_7) \\
e_z &= l_3 * \cos(\Delta\theta_4 + \Delta\theta_5) - l_2 * \sin(\Delta\theta_4) - (l_4* \\
&\quad \cos(\Delta\theta_6 + \Delta\theta_7) * \sin(\Delta\theta_4 + \Delta\theta_5))/2 - l_4 * \cos(\Delta\theta_4 + \\
&\quad \Delta\theta_5) * \sin(\Delta\theta_7) - (l_4 * \cos(\Delta\theta_6 - \Delta\theta_7) * \sin(\Delta\theta_4 + \Delta\theta_5))/2
\end{aligned}
$$

$$(18)$$

It can be seen from (18) that the errors e_x, e_y and e_z of the end-effector are functions of many parameters (such as equivalent stiffness k_i and moment of inertia J_i). Furthermore, the dynamic error can be minimized by analyzing the influence of these specific parameters on the dynamic errors, which is of guiding significance for the high-precision operation of redundant sliding manipulator.

4   Dynamic Precision Analysis of the Redundant Sliding Manipulator

4.1   Preliminaries of Dynamic Precision Analysis

Parametric Analysis Theory. It is known that the dynamic errors $e_{x/y/z}$ are multi-independent and complex function variables. In the analysis process, dynamic errors $e_{x/y/z}$ are taken as the research objects. Taking only the research object as an independent variable and other variables as constants, we can obtain the following functions:

$$e_{x/y/z} = C_{x/y/z} \cdot f(\beta) \tag{19}$$

where β is the variable under study and $C_{x/y/z}$ is the constant quantity corresponding to $e_{x/y/z}$.

Influence Coefficient Analysis Theory. The error influence coefficient is the ratio between the error increment and the increment of the independent variable causing the increment. Partial deviations of dynamic errors $e_{x/y/z}$ relative to each parameter are defined as the error influence coefficient of each parameter, that is:

$$Y_{x/y/z} = \frac{\partial e_{x/y/z}}{\partial \beta} \tag{20}$$

where β is the study variable, and $Y_{x/y/z}$ is the influence coefficient for $e_{x/y/z}$ corresponding to β.

The error influence coefficient is an important basic theory in the study of error and its compensation. The two-dimensional parametric analysis method (19) and the error influence coefficient method (20) are used to analyze the influence of each parameter on the dynamic error, which can not only make the two independent analysis results complement each other, but also make the overall analysis results more complete and accurate.

4.2   Stiffness and Rotational Inertia Analysis of the Redundant Sliding Manipulator

Influence of Stiffness and Rotational Inertia Analysis. According to (19), the influence of stiffness k_i change and the moment of inertia J_i change on the dynamic errors $e_{x/y/z}$ is analyzed, and the functional relationships between $e_{x/y/z}$ and the independent variables k_i and J_i are as follows:

$$\begin{aligned} e_{x/y/z} &= C_{x/y/z} \cdot f(k_i) \\ e_{x/y/z} &= C_{x/y/z} \cdot f(J_i) \end{aligned} \tag{21}$$

Influence Coefficients of Stiffness and Rotational Inertia Analysis. The dynamic error influence coefficients of the equivalent stiffness k_i and the moment of inertia J_i are defined as the ratio of the increment of the dynamic errors $e_{x/y/z}$ to the increment of the equivalent stiffness k_i and the moment of inertia J_i that caused the increment. It can be obtained as follows:

$$Y_{k_i} = \frac{\partial e_{x/y/z}}{\partial k_i}$$
$$Y_{J_i} = \frac{\partial e_{x/y/z}}{\partial J_i}$$
(22)

5   Simulation Analysis and Experiment

5.1   Simulation Analysis

The links of the 7-DOF linear-type rail redundant sliding manipulator are made of alloy steel, whose cross section is a circle with radius $r = 20\,\mathrm{mm}$. The preliminary dimensions of each link are $l_0 = 236\,\mathrm{mm}$, $l_1 = 208\,\mathrm{mm}$, $l_2 = 166\,\mathrm{mm}$, $l_3 = 166\,\mathrm{mm}$, and $l_4 = 47\,\mathrm{mm}$. According to the equivalent stiffness theory and the configuration of the redundant sliding manipulator, it can be obtained that the link 1, link 3, link 4 and link 6 are equivalent bending stiffness, and the link 2, link 5 and link 7 are equivalent tensile stiffness.

Based on the known material and the design dimensions of each link, the required parameters are calculated as follows: $m_1 = 2.327\,\mathrm{kg}$, $m_2 = 2.051\,\mathrm{kg}$,

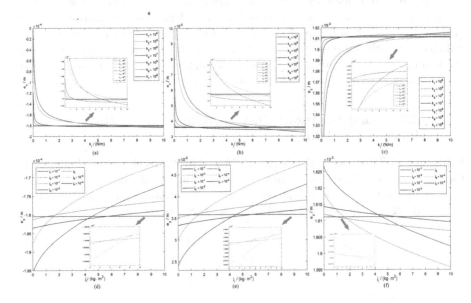

Fig. 3. Dynamic errors e_x, e_y and e_z corresponding to different equivalent stiffness k_i and equivalent rotational inertia J_i.

$m_3 = 1.637\,\text{kg}$, $m_4 = m_5 = 1.637\,\text{kg}$, $m_6 = m_7 = 0.463\,\text{kg}$, $J_1 = 0.452\,\text{kg} \cdot \text{m}^2$, $J_2 = 2.315 \times 10^{-3}\,\text{kg} \cdot \text{m}^2$, $J_3 = 0.159\,\text{kg} \cdot \text{m}^2$, $J_4 = 5.700 \times 10^{-2}\,\text{kg} \cdot \text{m}^2$, $J_5 = 1.025 \times 10^{-3}\,\text{kg} \cdot \text{m}^2$, $J_6 = 6.261 \times 10^{-4}\,\text{kg} \cdot \text{m}^2$, $J_7 = 1.582 \times 10^{-4}\,\text{kg} \cdot \text{m}^2$, $k_1 = 5.905 \times 10^6\,\text{N/m}$, $k_2 = 1.244 \times 10^9\,\text{N/m}$, $k_3 = 8.626 \times 10^6\,\text{N/m}$, $k_4 = 1.697 \times 10^7\,\text{N/m}$, $k_5 = 1.559 \times 10^9\,\text{N/m}$, $k_6 = 7.476 \times 10^8\,\text{N/m}$, $k_7 = 5.505 \times 10^9\,\text{N/m}$. The initial conditions of vibration are $A_1 = \cdots = A_7 = 0.1\,\text{mm}$, $\Delta\theta_{10} = \cdots = \Delta\theta_{70} = 0.1\,\text{mm}$, $\Delta\dot{\theta}_{10} = \cdots = \Delta\dot{\theta}_{70} = 5\,\text{mm/s}$.

It can be seen from Fig. 3 that the change of equivalent stiffness k_i and the moment of inertia J_i will cause changes in dynamic errors and exhibit different curve relationships. The changes of equivalent stiffness k_7 and the moment of inertia J_3 have the greatest influence on the dynamic errors, while the changes of equivalent stiffness k_1 and the moment of inertia J_7 have the smallest influence on the dynamic errors. That means that the design of the redundant sliding manipulator has a great advantage for improving the dynamic accuracy due to the unique configuration of its slide rail. In other words, if the dynamic error is guaranteed to meet certain conditions, the equivalent stiffness of link 7 and the moment of inertia of link 3 are required the most in the design process, that is, the equivalent stiffness of link 7 and the moment of inertia of link 3 are first considered.

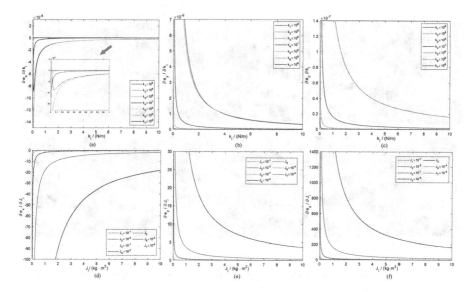

Fig. 4. The stiffness influence coefficient and the rotation inertia influence coefficient of dynamic errors e_x, e_y and e_z.

Figure 4 shows the stiffness and rotational inertia influence coefficient of the dynamic errors e_x, e_y and e_z. Comprehensive analysis of the dynamic error influence coefficient curves of each equivalent stiffness k_i and moment of inertia J_i shows that the equivalent stiffness of link 6 and link 7 and the moment of inertia of link 3 and link 4 have the greatest influence on the dynamic errors of the manipulator in the whole working space. Generally speaking, when the equivalent stiffness and the moment of inertia are small, the dynamic errors are not very stable and may change rapidly. When designing, we must pay attention to the instability of dynamic errors within this range.

Remark: From the above analysis, the parameter optimization scheme for reducing the dynamic errors can be obtained as follows:

(1) Equivalent stiffness: The bending deformation of the link has a much greater influence on the dynamic error than the tensile deformation. Therefore, the adjustment of the equivalent bending stiffness is more important for reducing the dynamic errors of the system.
(2) Moment of inertia: The size and distribution of the mass directly affect the magnitude of the moment of inertia, and thus directly affects the magnitude of the dynamic error of the manipulator.
(3) When adjusting the moment of inertia, the cross-sectional area of each link may be changed, so the change of the equivalent stiffness must also be considered. Therefore, the moment of inertia and the equivalent stiffness can be adjusted at the same time, so as to reduce the dynamic errors of the system.

Fig. 5. Configuration changes of the manipulator during operation.

5.2   Implementation and Verification Experiment

The experimental platform in this paper is the linear-type rail redundant sliding manipulator introduced in Sect. 2, whose dimensions and D-H parameters are the same as the simulation. In this experiment, a set of joint angles is input to the manipulator, and the configuration changes of the manipulator during operation is as shown in Fig. 5.

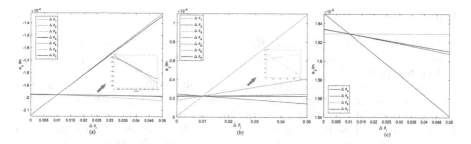

Fig. 6. Dynamic errors e_x, e_y and e_z corresponding to angular displacements $\Delta\theta_i$.

By changing the values of the angular displacements $\Delta\theta_i$ in real time during the movement of the redundant sliding manipulator, the corresponding errors are shown in Fig. 6. We can find that the change of angular displacements $\Delta\theta_5$, $\Delta\theta_3$ and $\Delta\theta_4$ has the greatest impact on dynamic errors e_x, e_y and e_z, respectively. The rule of error variation in the experiment is consistent with the simulation, which further proves the accuracy and practicability of the dynamic error analysis method in this paper.

6   Conclusion

This paper deals with the dynamic precision analysis of a linear-type rail redundant sliding manipulator. Based on the analysis of the mechanism structure, kinematics, stiffness performance and dynamic characteristics of the redundant sliding manipulator, the dynamic error caused by elastic deformation of links is analyzed. A mathematical model of the dynamic precision of the redundant sliding manipulator system is established by the geometric method. The influence of parameters such as equivalent stiffness and moment of inertia on the dynamic error of the redundant sliding manipulator is analyzed and parameter optimization schemes are proposed to reduce the dynamic error. Simulation analysis and experiment verify the dynamic precision of the redundant sliding manipulator. This paper provides a theoretical guidance for the high-precision operations of the redundant sliding manipulator.

Acknowledgments. This work was supported in part by the National Key R&D Program of China (Grant No. 2018YFB1304600), the Natural Science Foundation of China (Grant No. 51775541), CAS Interdisciplinary Innovation Team (Grant No. JCTD-2018-11).

References

1. Flores-Abad, A., Ma, O., Pham, K., Ulrich, S.: A review of space robotics technologies for on-orbit servicing. Prog. Aerosp. Sci. **68**, 1–26 (2014). https://doi.org/10.1016/j.paerosci.2014.03.002
2. Li, W.J., Cheng, D.Y., Liu, X.G., Wang, Y.B., Shi, W.H., Tang, Z.X., et al.: On-orbit service (OOS) of spacecraft: a review of engineering developments. Prog. Aerosp. Sci. **108**, 32–120 (2019). https://doi.org/10.1016/j.paerosci.2019.01.004
3. Opromolla, R., Fasano, G., Rufino, G., Grassi, M.: A review of cooperative and uncooperative spacecraft pose determination techniques for close-proximity operations. Prog. Aerosp. Sci. **93**, 53–72 (2017). https://doi.org/10.1016/j.paerosci.2017.07.001
4. Chang, C.G., Liu, J.G., Ni, Z.Y., Qi, R.L.: An improved kinematic calibration method for serial manipulators based on POE formula. Robotica **36**(8), 1244–1262 (2018). https://doi.org/10.1017/S0263574718000280
5. Zhang, X., Liu, J.G., Feng, J.K., Liu, Y.W., Ju, Z.J.: Effective capture of non-graspable objects for space robots using geometric cage pairs. IEEE/ASME Trans. Mechatron. **25**(1), 95–107 (2020). https://doi.org/10.1109/TMECH.2019.2952552
6. Liu, J.G., Tong, Y.C., Ju, Z.J., Liu, Y.W.: Novel method of obstacle avoidance planning for redundant sliding manipulators. IEEE Access **8**, 78608–78621 (2020). https://doi.org/10.1109/ACCESS.2020.2990555
7. Jia, S.Y., Shan, J.J.: Finite-time trajectory tracking control of space manipulator under actuator saturation. IEEE Trans. Ind. Electron. **67**(3), 2086–2096 (2020). https://doi.org/10.1109/tie.2019.2902789
8. Xu, W.F., Meng, D.S., Liu, H.D., Wang, X.Q., Liang, B.: Singularity-free trajectory planning of free-floating multiarm space robots for keeping the base inertially stabilized. IEEE Trans. Syst. Man Cybern.-Syst. **49**(12), 2464–2477 (2019). https://doi.org/10.1109/tsmc.2017.2693232
9. Liu, J.G., Tong, Y.C., Liu, Y.J., Liu, Y.W.: Development of a novel end-effector for an on-orbit robotic refueling mission. IEEE Access **8**, 17762–17778 (2020). https://doi.org/10.1109/access.2020.2964641
10. Dong, J., He, B., Ming, M., Zhang, C.H., Li, G.: Design of open-closed-loop iterative learning control with variable stiffness for multiple Flexible Manipulator Robot Systems. IEEE Access **7**, 23163–23168 (2019). https://doi.org/10.1109/access.2019.2898266
11. Liang, D., Song, Y., Sun, T.: Nonlinear dynamic modeling and performance analysis of a redundantly actuated parallel manipulator with multiple actuation modes based on FMD theory. Nonlinear Dyn. **89**(1), 391–428 (2017). https://doi.org/10.1007/s11071-017-3461-x
12. Fang, H.L., Tang, T.F., Zhang, J.: Kinematic analysis and comparison of a 2R1T redundantly actuated parallel manipulator and its non-redundantly actuated forms. Mech. Mach. Theory **142**, (2019). https://doi.org/10.1016/j.mechmachtheory.2019.103587
13. Guo, X.W., Li, S., Wang, Q.J., Wen, Y., Gong, N.W.: Dynamic analysis and current calculation of a permanent magnet spherical motor for point-to-point motion. IET Electr. Power Appl. **13**(4), 426–434 (2019). https://doi.org/10.1049/iet-epa.2018.5149
14. Tong, Z.Z., Gosselin, C., Jiang, H.Z.: Dynamic decoupling analysis and experiment based on a class of modified Gough-Stewart parallel manipulators with line orthogonality. Mech. Mach. Theory **143** (2020). https://doi.org/10.1016/j.mechmachtheory.2019.103636

15. Tsai, Y.K., Chan, K.Y.: Investigation on the impact of nongeometric uncertainty in dynamic performance of serial and parallel robot manipulators. Proc. Inst. Mech. Eng. Part C-J. Mech. Eng. Sci. **233**(10), 3487–3511 (2019). https://doi.org/10.1177/0954406218815518

16. Yang, X.X., Ge, S.S., He, W.: Dynamic modelling and adaptive robust tracking control of a space robot with two-link flexible manipulators under unknown disturbances. Int. J. Control **91**(4), 969–988 (2018). https://doi.org/10.1080/00207179.2017.1300837

Kinematic Optimal Design of Ball-Screw Transmission Mechanisms Based on NoCuSa

Zhenpeng Ge[2] , Shuzhi Gao[1(✉)] , and Tianchi Li[1]

[1] Equipment Reliability Institute, Shenyang University of Chemical Technology,
Shenyang 110142, China
szg6868@126.com
[2] School of Information Engineering, Shenyang University of Chemical
Technology, Shenyang 110142, China

Abstract. This paper proposes a variable speed optimal designing method of ball-screw transmission mechanisms based on NoCuSa (Nonhomogeneous Cuckoo Search Algorithm). The optimal function of the input speed was designed, without changing the mechanical structure of the original ball-screw driving mechanism. It can reduce the peak output of the ball-screw and improve output performance. In this paper, kinematic dimensionless equations of rotation angle, velocity, acceleration, and jerk were established for the optimization problem-solving. the mathematical model of the optimization based on seventh-order polynomial was given. The NoCuSa was used to optimize the design function objective function and get the optimal speed function. Finally, the optimized speed function was brought into the dimensionless equation to obtain the optimal output and compared with the GA (Genetic Algorithm) optimization, Liu's optimal design method, and the constant speed input method. And a set of actual parameters of the ball-screw is selected to simulate the actual output performance.

Keywords: Ball-screw transmission · Cuckoo search algorithm · Variable speed design · Kinematic optimization

1 Introduction

The variable input speed design method of the transmission mechanism originated from the design of the cam mechanism. In order to improve the output performance of the system, the traditional method can only redesign the mechanical structure of transmission system. For the traditional constant input design method, Mills et al. [1] studied the effects of various structural dynamic parameters on the peak acceleration and the peak velocity of the cam mechanism, proposed an optimized design method (1993). Yu et al. [2] proposed a cam curvature optimization method based on parametric polynomial and studied the relationship between cam profile and peak acceleration (1996). Qiu et al. [3] proposed a generalized cam curve design optimization method that can simultaneously multiple targets for kinematics or dynamics optimization (2005). In recent years, Kailash & Chaudhary [4] have used the method of seeking the optimal mass distribution of the crank-sliding mechanism to reduce vibration and noise, and compared the optimization results of TLBO and GA (2017).

© Springer Nature Singapore Pte Ltd. 2020
J. Qian et al. (Eds.): ICRRI 2020, CCIS 1336, pp. 172–187, 2020.
https://doi.org/10.1007/978-981-33-4932-2_12

After 1990, some people began to pay attention to the research of variable speed mechanism. For the variable speed input method, it is not necessary to redesign the system mechanical structure. Simply optimize the design of the system input speed can get better output characteristics and save economic costs. However, it is difficult for conventional motors to realize complex variable speed input methods, so the rapid development and popularization of servo motors provide a guarantee variable speed input control. For example, in the 1990s, it emerged an electronic cam [5] composed of servo motor system, that is, the cam curve constructed was used to simulate the mechanical cam. Yan et al. [6] proposed a method to reduce the peak acceleration of follower through the design of cam speed. At the same time, the polynomial speed function was used to optimize the design and verify the feasibility of the variable speed input method (1996). Yao et al. [7] proposed the concept of 'active control of cam mechanism', proved that the output performance of servo system can be greatly improved by applying the optimal control theory (2000). Yan et al. [8] applied B-spline curve to input speed function design for improving the output performance of watt press (2000). Yan et al. [9] applied the variable input speed function to the crank link servo system, and optimized the input by the 10th order B-spline curve designing (2000). Liu et al. [10] reduced the peak acceleration value of constant pitch screw by 20% by designing polynomial speed function, which is better than variable pitch screw (2001). Yao et al. [11] summarized the research of variable speed input mechanical system, introduced and exemplified the variable speed design steps (2004). Wen Hsiang Hsieh [12] proposed a new method to drive variable speed mechanism using planetary gear train (CCPGT) (2007). D. Mundo [13] used genetic algorithm to optimize the polynomial speed function, and designed the non-circular gear instead of the servo mechanism. Compared with the constant speed input mechanism, the peak acceleration of the lead screw decreased by 36.9% (2007). Heidari et al. [14] used the genetic algorithm to design the input speed function, optimized the kinematic characteristics of a geneva mechanism (2008). Yan et al. [15] proposed an integrated design method that takes dimensions, counterweights, input speed functions and controller parameters as design variables and designed the four-bar linkage, proved the feasibility of the method (2009). Hsu et al. [16] used cosine function to design the input speed function, reducing the peak acceleration of lead screw (2012). Saeidi et al. [17] compared the function design by the optimal control theory with the crank speed function design by the genetic algorithm (2012). Yang et al. [18] used B-spline curve to design input function, optimized a kind of screw drive system (2015).

The transmission system mentioned in this paper is a kind of ball screw drive mechanism driven by crank-linker. The input is the circular motion of crank, and the output is the reciprocating motion of ball screw. This system was originally a constant speed input system, but its output performance is poor. This mechanism is also widely used in industry, such as weft insertion mechanism [19] which is widely used in rapier loom of textile industry. The main purpose of optimization is to reduce the peak acceleration of ball screw to reduce the inertia load of lead screw and improve the output performance. For the traditional optimization method, the variable pitch lead screw design [20, 21] and the use of excellent materials are generally used [22]. In recent years, there are also many intelligent optimization algorithms for variable pitch design [23], but the design cost is relatively high compared with the variable speed method. In this paper, a cuckoo algorithm based on non-homogeneous search of quantum mechanism is used to optimize the 7-order polynomial input speed function, so as to optimize the kinematics of the transmission mechanism. Compared with the other optimal design results, this method is better.

2 Model of Ball-Screw Transmission System

The transmission system is composed of a crank-linker and a ball screw system in series, where the ball screw is driven by a crank-linker, the system input is the circular movement of the crank clockwise, and the output is the reciprocating rotation of the ball screw. As shown in Fig. 1, O-point is the driving point, 1 is the crank, 2 is the connecting rod, 3 is the slider, 4 is the ball screw.

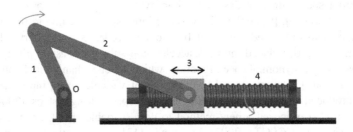

Fig. 1. Schematic diagram of ball-screw transmission mechanism

As shown in Fig. 2, suppose that the crank length R_1, the link length r_2, r_3 is the relative positions of the slider to **O**-point, θ_1 is the angle between the crank and the negative half of axis X, that is, the input crank angle displacement, θ_2 is the angle between the link and the negative half axis, s is the displacement of the slider to the left end of the ball screw. During the clockwise rotation of the crank, the link moves the slider to do a linear reciprocating motion, and the ball screw rotates to and under the push of the slider, and the system outputs is the angle of the ball screw.

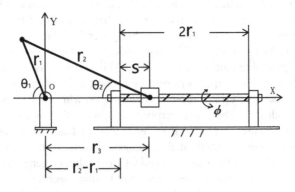

Fig. 2. Model of ball-screw transmission mechanism

From the position relationship shown in Fig. 2, the following equations can be obtained:

$$r_2\cos\theta_2 = r_1\cos\theta_1 + r_3 \tag{1}$$

$$r_2\sin\theta_2 = r_1\sin\theta_1 \tag{2}$$

It can be obtained by Eq. (1) and Eq. (2) joint operation:

$$\frac{\cos\theta_2}{\sin\theta_2} = \frac{\cos\theta_1}{\sin\theta_1} + \frac{r_3}{r_1\sin\theta_1} \tag{3}$$

After identity transformation, the following results are obtained:

$$r_1\sqrt{\frac{\sin^2\theta_1}{\sin^2\theta_2} - \sin^2\theta_1} = r_1\cos\theta_1 + r_3 \tag{4}$$

From Eq. (2):

$$\frac{\sin\theta_1}{\sin\theta_2} = \frac{r_2}{r_1}$$

Bring Eq. (5) into Eq. (4), the relative position of the slider to the **O**-point can be obtain as follows:

$$r_3 = r_1\left[\sqrt{(\frac{r_2}{r_1})^2 - \sin^2\theta_1} - \cos\theta_1\right] \tag{5}$$

From the position relation, the function relation between the slide displacement s and the rotation angle θ_1 is as follows:

$$s = r_3 - (r_2 - r_1) = r_1\left[\sqrt{(\frac{r_2}{r_1})^2 - \sin^2\theta_1} - \cos\theta_1 - (\frac{r_2}{r_1} - 1)\right] \tag{6}$$

In order to simplify the problem, only the moving process of the slider to the right is studied, and the parameters are dimensionless. Assuming that the time for the slider to move to the right side once is τ, the stroke of the slider is $2r_1$, the turning angle of the crank is π. The dimensionless parameters can be taken as follows:

$$T = \frac{t}{\tau}, t \in [0, \tau] \tag{8a}$$

$$S = \frac{s}{2r_1}, s \in [0, 2r_1] \tag{8b}$$

$$\Theta = \frac{\theta_1}{\pi}, \theta_1 \in [0, \pi] \tag{8c}$$

Where, $T, S, \Theta \in [0, 1]$. Take the above equation into Eq. (7) can obtain the dimensionless relationship between the input angle and the slider displacement as follows:

$$G(\Theta) = S = \frac{1}{2}\left[\sqrt{(2R)^2 - sin^2\pi\Theta} - cos\,\pi\Theta - (2R - 1)\right] \tag{9}$$

Where R is the ratio of the length of crank and connecting rod $R = \frac{r_1}{2r_2}$.

As a variable speed input mechanism, Θ is also a function of dimensionless time (T):

$$S = G(\Theta(T)) \tag{10}$$

For the lead screw with constant pitch P, the angle of the lead screw and the displacement of the slider are linear as follows:

$$\phi = \frac{2\pi}{p}s \tag{11}$$

Assuming that the dimensionless rotation angle of the ball screw is Φ, since the rotation angle of the ball screw is directly proportional to the displacement of the slider, within the time τ of the slider moving to the right, the constant relationship between the displacement and the rotation angle can be obtained as follows:

$$\Phi = \frac{\phi}{\phi_{max}} = \frac{s}{s_{max}} = \frac{s}{2r_1} = S \tag{12}$$

Because $\Phi = S$, the dimensionless speed V, acceleration A and jerk J of the ball screw can be obtained by differentiating S with respect to T:

$$V = \frac{d\Phi}{dT} = \frac{dS}{dT} = G'(\Theta)\dot{\Theta}(T) \tag{13a}$$

$$A = \frac{d^2\Phi}{dT^2} = \frac{d^2S}{dT^2} = G''(\Theta)\dot{\Theta}^2(T) + G'(\Theta)\ddot{\Theta}(T) \tag{13b}$$

$$J = \frac{d^3\Phi}{dT^3} = \frac{d^3S}{dT^3} = G'''(\Theta)\dot{\Theta}^3(T) + 3G''(\Theta)\ddot{\Theta}(T)\dot{\Theta}(T) + G'(\Theta)\dddot{\Theta}(T) \tag{13c}$$

Where $G'(\Theta)$, $G''(\Theta)$, $G'''(\Theta)$ is the first, second and third-order derivative of $G(\Theta)$ with respect to the input angle $G(\Theta)$. Where $\dot{\Theta}(T)$, $\ddot{\Theta}(T)$, $\dddot{\Theta}(T)$ is the first, second and third-order derivative of $\Theta(T)$ with respect to dimensionless time T, i.e. angular velocity, angular acceleration and angular jerk of the input.

The function of dimensionless speed (V), acceleration(A) and jerk (J) of the ball screw with respect to T can be obtained by bring $G(\Theta)$ (Eq. (9)) to the Eq. (13a–13c) as follows [11]:

$$V = \frac{d\Phi}{dT} = \frac{dS}{dT} = G'(\Theta)\dot{\Theta}(T) = \left[\frac{-\pi}{4}\frac{\sin 2\pi\Theta}{\sqrt{(2R)^2 - \sin^2\pi\Theta}} + \frac{\pi}{2}\sin\pi\right]\frac{d\Theta}{dT} \tag{14a}$$

$$A = \frac{d^2\Phi}{dT^2} = \frac{d^2S}{dT^2} = G''(\Theta)\dot{\Theta}^2(T) + G'(\Theta)\ddot{\Theta}(T)$$

$$= \left\{\frac{-\pi^2}{2}\frac{\cos 2\pi\Theta}{\sqrt{(2R)^2 - \sin^2\pi\Theta}} + \frac{\pi^2}{8}\sin^2(2\pi\Theta)\left[(2R)^2 - \sin^2\pi\Theta\right]^{-\frac{3}{2}} + \frac{-\pi^2}{2}\cos\pi\Theta\right\}\left(\frac{d\Theta}{dT}\right)^2$$

$$+ \left[\frac{-\pi}{4}\frac{\sin 2\pi\Theta}{\sqrt{(2R)^2 - \sin^2\pi\Theta}} + \frac{\pi}{2}\sin\Theta\right]\frac{d^2\Theta}{dT^2} \tag{14b}$$

$$J = \frac{d^2\Theta}{dT^3} = \frac{d^3S}{dT^3} = G'''(\Theta)\dot{\Theta}^3(T) + 3G''(\Theta)\ddot{\Theta}(T)\dot{\Theta}(T) + G'(\Theta)\dddot{\Theta}(T)$$

$$= \left[\frac{-\pi}{4}\frac{\sin 2\pi\Theta}{\sqrt{(2R)^2 - \sin^2\pi\Theta}} + \frac{\pi}{2}\sin\pi\right]\frac{d^3\Theta}{dT^3} + 3\left\{\frac{-\pi^2}{2}\frac{\cos 2\pi\Theta}{\sqrt{(2R)^2 - \sin^2\pi\Theta}}\right.$$

$$\left. - \frac{\pi^2}{8}\sin^2(2\pi\Theta)\left[(2R)^2 - \sin^2\pi\Theta\right]^{-\frac{3}{2}} + \frac{\pi^2}{2}\cos\pi\Theta\right\}\left(\frac{d\Theta}{dT}\right)\frac{d^2\Theta}{dT^2} \tag{14c}$$

$$+ \left\{\frac{\pi^3\cos 2\pi\Theta}{(2R)^2 - \sin^2\pi\Theta} - \frac{3\pi^3}{8}\sin(4\pi\Theta)\left[(2R)^2 - \sin^2\pi\Theta\right]^{-\frac{3}{2}}\right.$$

$$\left. - \frac{3\pi^3}{16}\sin^3(2\pi\Theta)\left[(2R)^2 - \sin^2\pi\Theta\right]^{-\frac{5}{2}} - \frac{\pi^3}{2}\sin\pi\Theta\right\}\left(\frac{d\Theta}{dT}\right)^3$$

Through the optimization design of the crank dimensionless speed function, the actual speed, acceleration and jerk of the ball screw can be obtained through the following variable conversion relations:

$$v = \frac{d\phi}{dt} = \frac{dS}{dt}\cdot\frac{2\pi}{p}\cdot 2r_1 = \frac{4\pi r_1}{p\tau}\cdot V \tag{15a}$$

$$a = \frac{dv}{dt} = \frac{dV}{dt}\cdot\frac{2\pi}{p}\cdot\frac{2r_1}{\tau} = \frac{4\pi r_1}{p\tau^2}\cdot A \tag{15b}$$

$$j = \frac{da}{dt} = \frac{dA}{dt}\cdot\frac{2\pi}{p}\cdot\frac{2r_1}{\tau^2} = \frac{4\pi r_1}{p\tau^3}\cdot J \tag{15c}$$

3 Optimization Design of Variable Speed Input Function

3.1 Description of the Optimization Problem

The dimensionless input speed function $\Theta(T)$ was choosed to the optimization. The optimization problem is to constantly change the parameters of the input speed function to find a set of optimal parameters c_i, so that the peak acceleration of the ball screw when the slider moves to the right is minimum, so as to reduce the inertia load of the system. We use polynomial function as the input angle function, and set seventh-order polynomial input speed function as follows:

$$\Theta(T) = c_0 + c_1 T + c_2 T^2 + c_3 T^3 + c_4 T^4 + c_5 T^5 + c_6 T^6 + c_7 \quad (16)$$

3.2 Restraint Conditions

When the slider moves from left end to the right end, there is dimensionless T, S, $\Theta \in [0, 1]$, and the following restraint condition can be obtained.

1. In the process of slider moving to the right, the crank keeps turning clockwise and the crank turns unchanged, that is $\Theta(T) > 0$.
2. In the process of continuous crank rotation, the input crank speed function $\Theta(T)$ must be continuous and the input crank speed must be at least the second-order continuous differentiable function to reduce the driver load.
3. According to the boundary condition of sliding block moving forward, it can be seen that $\Theta(0) = 0$, $\Theta(1) = 1$.

3.3 Objective Function

The main purpose of optimization is to reduce the peak acceleration of the ball-screw output, so as to reduce the inertia load and output vibration. However, in order to ensure that the function meets the constraints, it is necessary to ensure the rationality of the crank angle acceleration peak at the same time, so it is also necessary to reduce the input acceleration peak, that is, the peak value of the second derivative of the input function $|\ddot{\Theta}(T)|$. Therefore, the optimization problem is a typical multi-objective optimization problem which can be transformed into a single objective problem by weighting. Where, ω_1, ω_2 are the weight coefficients, for different optimization methods and values, C is the optimization parameter. The objective function $P(C)$ was shown as follows:

$$PC = \omega_1 |A|_{max} + \omega_2 |\ddot{\Theta}(T)|_{max}$$
$$= \omega_1 |G''(\Theta)\dot{\Theta}^2(T) + G'(\Theta)\ddot{\Theta}(T)|_{max} + \omega_2 |\ddot{\Theta}(T)|_{max} \quad (17)$$

Where $C = [c_1, c_2, c_3, c_4, c_5, c_6]$ were selected as the optimal parameters. From Eq. 3, the parameter c_0 and c_7 can be obtained as follows:

$$c_0 = 0 \quad (18)$$

$$c_7 = 1 - (c_1 + c_2 + c_3 + c_4 + c_5 + c_6) \quad (19)$$

Based on the above description the optimal polynomial parameter C_{best} can be obtained by solving the optimization problem via search algorithm.

4 Optimal Polynomial Parameterons Searching Based on NoCuSa

4.1 Nonhomogeneous Cuckoo Search Algorithm

NoCuSa (Nonhomogeneous cuckoo search algorithm) is an improved cuckoo algorithm proposed by Cheung [24] et al. in 2016 The algorithm introduces the idea of quantum strategy, and the nest has a nonhomogeneous update strategy. This new nonhomogeneous update law enables NoCuSa to have good ability of global search and local search when dealing with problems in different fitness environments.

As an improved cuckoo search algorithm, NoCuSa still satisfies the three idealization assumptions as follows [25]:

1. Each cuckoo lays one egg at one time and places its eggs in a randomly selected nest;
2. The best nest with high quality eggs will continue to the next generation;
3. The number of host nests is fixed, and the eggs laid by cuckoo will be found by the host birds with the probability of Pa $\in$ [0, 1]. In the case of being found by the host bird, the host bird will throw away its eggs or build a new nest. For simplicity, the nest found is replaced by a new random solution.

The steps of the NoCuSa are as follows:

Step 1. Initialize location of the nests. Sets the iteration number, difference factor α, discovery probability Pa, range of solution and other parameters. The initial nest is generated according to the following Eq. 20. Calculate the degree of each initial nest according to Eq. 16 and find out the best nest.

$$x_i^j = x_{min}^j + rand(0, 1) \cdot \left(x_{max}^j - x_{min}^j\right) \tag{20}$$

Where j is the dimension of solution. x_{max}^j and x_{minx}^j are the maximum and minimum values on the j dimension of solution set, i is the nest number. $rand(0, 1)$ is a random number evenly distributed in the interval [0, 1].

Step 2. Location update strategy. The update strategy of NoCuSa was shown in Eq. 21. The general cuckoo algorithm only has strategy ① But NoCuSa has a nonhomogeneous search laws with three kind of policies to update the location of the nests. ② ③ strategy was inspired by Schrödinger equation.

$$X_{i,k+1} = \begin{cases} ①X_{i,k} + \alpha \cdot s[X_{i,k} - g_k] \,, \text{sr} \in (\frac{2}{3}, 1] \\ ②\bar{X}_{i,k} + L[\bar{X}_k - X_{i,k}] \,, \text{sr} \in (\frac{1}{3}, \frac{2}{3}] \\ ③X_{i,k} + \varepsilon[g_k - X_{i,k}] \,, otherwise \end{cases} \tag{21}$$

Where $\bar{X}_{i,k}$ is the average value of $X_{i,k}$ in generation k. g_k is the best nest of generation k. α is a control constant to stepsize. $L = \delta \ln(1/\eta)$, $\varepsilon = \delta exp(\eta)$, η, sr is a random number evenly distributed in the interval [0, 1]. δ is control parameter, i.e., learning regulator. s is the random path of Levis flight by a simulate method (Mantegna 1994), defied as follows:

$$s = \frac{u}{|v|^{1/\beta}} \tag{22}$$

Where u, v is a orthodox distribution random number which meeting $u \sim N(0, \sigma_V^2), u \sim N(0, \sigma_V^2)$. σ_v, σ_u is defined as follows:

$$\sigma_u = \left[\frac{sin(\pi\beta/2) \cdot \Gamma(1+\beta)}{2^{(\beta-1)}\beta \cdot \Gamma((1+\beta)/2)} \right] \tag{23}$$

$$\sigma_v = 1 \tag{24}$$

Where $\Gamma(\cdot)$ is the gamma function. β is a constant to control Levis flight, usually set $\beta = 1.5$.

After updating, only the nest with better fitness value can replace the original nest.

Step 3. Updated randomly. Using the discovery probability Pa to randomly update the nest discovered by the Host bird, the updating rule is as follows:

$$X_{i,k+1} = \begin{cases} X_{i,k+1} + s', & \text{if } P > Pa \\ X_{i,k+1}, & \text{else} \end{cases} \tag{25}$$

Where $s' = rand(0, 1) \cdot (X_i - X_j)$. X_i, X_j are selected from the current generation randomly, $rand(0, 1)$ is a random numbers in the interval [0, 1]. Pa is the probability that the host bird finds the alien eggs, and the value of Pa affects the number of random updates, usually set Pa = 0.25.

In Eq. 25, only random updates that are better than the original position are accepted.

Step 4. Iteration. If the termination condition is met, the global optimal nest g i.e., optimal speed function parameters C_{best} will be output. Otherwise, step 2 will be returned to iterate until the termination condition is met.

4.2   Optimal Parameters Searching and Result Analysis

In this part, NoCuSa was used to optimize the seventh-order polynomial crank input function. Because the main purpose of the optimization of ball screw drive system is to reduce the peak acceleration of the ball screw, the weight coefficient can be taken $\omega_1 = 2$, $\omega_2 = 1$. It can be set that **Pa** = 0.25, the difference factor $\alpha = 0.1$, Levy factor $\beta = 1.5$, the learning regulator $\delta = 1.6$. Set the number of iterations as 1000. The convergence curve of the target value **p**, i.e., the optimization process of decreasing **p** with the increase of the iteration as shown in Figure 3. As can be seen from the

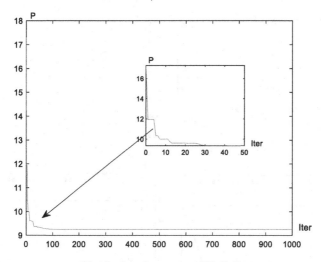

Fig. 3. Iterative curve of NoCuSa

figure, NoCuSa quickly converge to the optimal solution C_{best}. When running 60 generations on MATLAB®2019a, it converges to 99.9% of the optimal value.

The optimal parameters were obtained as follows:

$$C_{best} = [1.1621, -0.0058, -0.8304, 3.0000, -5.2108, 4.0000] \qquad (26)$$

The optimal input function can be obtained by taking the input speed function parameter C into Eq. (16), (18) and (19). And the result was obtained as follows:

$$\Theta(T) = 1.1621T - 0.0058T^2 - 0.8304T^3 + 3.0000T^4 - 5.2108T^5 + 4.0000T^6$$
$$- 1.1151T^7 \qquad (27)$$

The input speed function parameters obtained by different methods are compared in the Table 1.

Table 1. Comparison of input speed function parameters with four methods

	C_1	C_2	C_3	C_4	C_5	C_6	C_7
Constant input	1						
Liu [11]	0.9226	0.3376	0	1.5102	-5.7481	5.8519	-1.8742
GA [13]	1.1685	-0.2571	1.2e-5	2.3984	-5.9051	5.0918	-1.1151
NoCuSa	1.1621	-0.0058	-0.8304	3.0000	-5.2108	4.0000	-1.1151

The optimal angular velocity (Ω), angular acceleration (Ω'), and angular jerk (Ω'') curve of the crank can be obtained by solving the first, second, and third derivatives of the optimal input function $\Theta(T)$, as shown in Fig. 4.

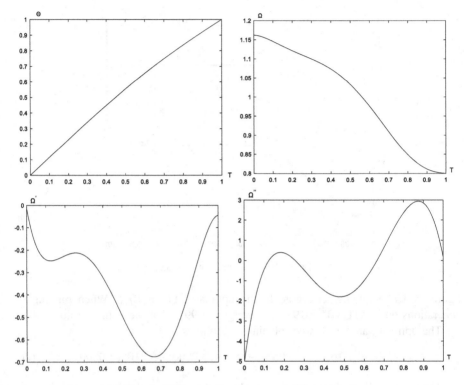

Fig. 4. Optimal movement curve of the crank

By taking Eq. (20) into Eq. (14a–14c), the optimal angular displacement (Φ), angular velocity(V), angular acceleration (A) and angular jerk (J) output curve of ball screw can be obtained as shown in Fig. 5. The other 3 curves in Fig. 5 were as for comparison.

Table 2. Comparison of kinematic characteristics of transmission mechanism with four methods

	Peak velocity	Peak acceleration	Peak jerk
Constant input	1.6692	**6.6972**	24.0429
Liu [11]	1.8142	**5.3061**	36.0797
GA [13]	1.7409	**4.3315**	29.7574
NoCuSa	1.7035	**4.2888**	26.1268

Where, the peak values in Table 2 refer to the maximum values after taking the absolute values of the curves.

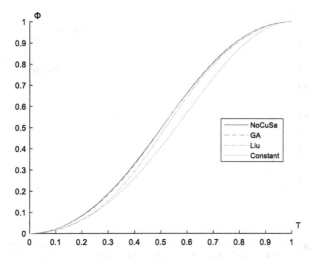

Fig. 5. Angular displacement curve of ball-screw in four methods

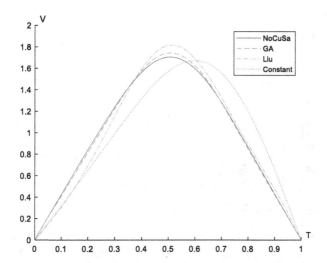

Fig. 6. Velocity curve of ball- screw in four methods

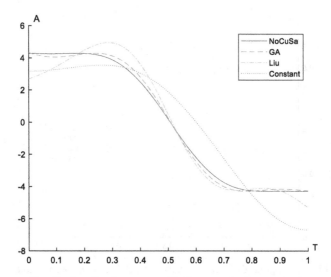

Fig. 7. Acceleration curve of ball- screw in four methods

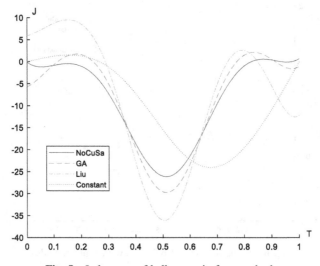

Fig. 8. Jerk curve of ball screw in four methods

It can be seen from Table 2 and Fig. 5, 6, 7 and 8 that the three kinds of ball screw motion characteristic parameters optimized by the cuckoo algorithm are superior to other optimization methods, especially the dimensionless peak acceleration is reduced by 35.96% compared with the constant rotation, 19.17% compared with the method proposed by Liu et al. and 0.99% compared with the genetic algorithm. It also can be seen from Fig. 7 and Fig. 8 that the output acceleration and jerk curve was more gentle. And the peak value of output velocity and jerk is lower than all of other optimization methods.

4.3   Actual Output Performance Simulation

In order to simulate the actual movement curve of the ball screw, this paper selects a set of actual parameters of the ball screw, as shown in the Table 3. The actual output of the ball screw can be obtained by taking the actual parameters of the ball screw into the Eq. (15a–15c). The simulation curves of the ball screw output are shown in the Fig. 9. Table 4 shows the simulation results of the actual output kinematic parameters of the four methods.

Table 3. Actual parameters of ball screw

Crank Length R1 (mm)	Connecting rod lengthR2 (mm)	Screw pitch (mm)	Lead screw length (mm)
100	280	40	300

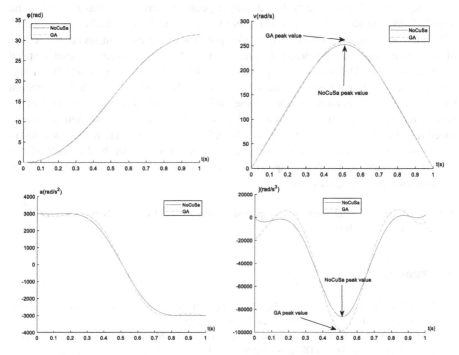

Fig. 9. Actual motion characteristic curve of ball-screw drive mechanism

Table 4. Comparison of actual motion parameters of ball screw in four methods

	Peak velocity	Peak acceleration	Peak jerk
Constant	247.5944	**4690.2**	79499
Liu [11]	269.0987	**3716.0**	119300
GA [13]	258.2302	**3033.5**	98394
NoCuSa	252.6764	**3003.6**	86389

Where, the peak values in Table 4 also refer to the maximum values after taking the absolute values of the curves. It can be seen from Table 4 that the peak acceleration of the ball screw obtained by the constant speed method is $4690.2\,\text{rad}/\text{s}^2$, while the peak acceleration obtained by this method is $3003.6\,\text{rad}/\text{s}^2$, which is 35.96% lower. Compared with the optimized design method proposed by Liu, the peak is reduced by 19.17%. Compared with the optimization of the genetic algorithm, the peak is reduced by 0.99%.

5 Conclusion

In this paper, a variable speed design method of ball-screw mechanism based on NoCuSa is proposed. By optimizing the input speed function of the crank, the output peak acceleration of the ball screw is reduced and its output performance is improved.

Compared with other methods, the output peak acceleration in this method is the lowest, the speed and jerk were lower than other optimization methods, and the curve of acceleration and jerk is more gentle. Compared with the constant speed, the reduce of peak acceleration is significant. The results proved the superiority of variable speed input once again. The lower peak acceleration obtained in this paper will be of significance for the transmission of heavy machinery. The fast convergence in iteration shows that NoCuSa is more suitable for the design of the screw speed function than the genetic algorithm, and this proves the applicability of NoCuSa in the design of optimal speed function. In addition, it shows that NoCuSa has great application space and broad application prospects in parameter optimization. In future research, this method can also be extended to reciprocating motion mechanism of other structures, such as reciprocating punching machine.

Acknowledgement. This work was supported in part by the Liaoning Natural Science Fund Project, China (No. 20170540725).

References

1. Mills, J.K., Notash, L., Fenton, R.G.: Optimal design and sensitivity analysis of flexible cam mechanisms. Mech. Mach. Theory **28**(4), 563–581 (1993)
2. Yu, Q., Lee, H.P.: Curvature optimization of a cam mechanism with a translating flat-faced follower. J. Mech. Des. **118**(3), 446–449 (1996)
3. Qiu, H., Lin, C.J., Yue, Y.: A universal optimal approach to cam curve design and its applications. Mech. Mach. Theory **40**(6), 669–692 (2005)
4. Chaudhary, K., Chaudhary, H.: Optimal design of planar slider-crank mechanism using teaching-learning-based optimization algorithm. J. Mech. Sci. Technol. **29**(12), 5189–5198 (2015). https://doi.org/10.1007/s12206-015-1119-5
5. Woelfel, M.: Introduction to electronic cam. Assembly Autom. **19**(1), 17–24 (1999)
6. Yan, H.S., Tsai, M.C., Hsu, M.H.: An experimental study of the effects of cam speeds on cam-follower systems. Mech. Mach. Theory **31**(4), 397–412 (1996)
7. Yao, Y., Zhang, C., Yan, H.S.: Motion control of cam mechanisms. Mech. Mach. Theory **35**(4), 593–607 (2000)

8. Yan, H.S., Chen, W.R.: A variable input speed approach for improving the output motion characteristics of Watt-type presses. Int. J. Mach. Tools Manuf. **40**(5), 675–690 (2000)

9. Yan, H.S., Chen, W.R.: On the output motion characteristics of variable input speed servo-controlled slider-crank mechanisms. Mech. Mach. Theory **35**(4), 541–561 (2000)

10. Liu, J.Y., Hsu, M.H., Chen, F.C.: On the design of rotating speed functions to improve the acceleration peak value of ball–screw transmission mechanism. Mech. Mach. Theory **36**(9), 1035–1049 (2001)

11. Yao, Y.A., Yan, H.S., Cha, J.Z.: Variable-input-speed mechanism design. In: 2004 Proceedings of the 11th World Congress in Mechanism and Machine Science, Tianjin, China, pp. 1733–1737 (2004)

12. Hsieh, W.H.: An experimental study on cam-controlled planetary gear trains. Mech. Mach. Theory **42**(5), 513–525 (2007)

13. Mundo, D., Yan, H.S.: Kinematic optimization of ball-screw transmission mechanisms. Mech. Mach. Theory **42**(1), 34–47 (2007)

14. Heidari, M., Zahiri, M., Zohoor, H.: Optimization of kinematic characteristic of geneva mechanism by genetic algorithm. World Acad. Sci. Eng. Technol. **20**, 387–395 (2008)

15. Yan, H.S., Yan, G.J.: Integrated control and mechanism design for the variable input-speed servo four-bar linkages. Mechatronics **19**(2), 274–285 (2009)

16. Hsu, M.H., Lin, S.C., Chang, J.C., Chen, C.Y.: Speed function design for improving acceleration characteristic of ball-screw systems. In: International Conference on Systems & Informatics. IEEE (2012)

17. Saeidi, H., Heidari, M., Zahiri, M.: Application of optimal control method to improve kinematic characteristics of crank-slider mechanisms. In: Design, Materials and Manufacturing, Parts A, B, and C, vol. 3 (2012)

18. Yang, J.H., Hsu, M.H., Yan, H.S.: Kinematic and dynamic characteristics design of a variable-speed machine with slider–crank and screw mechanisms. J. Mech. Robot. **8**(1), 014502 (2015)

19. Genini, G.: Mechanism for controlling the motion of the weft carrying grippers in looms. US (1977)

20. Yan, H.S., Cheng, H.Y.: Geometric design and machining of variable pitch lead screws with swinging and translating meshing rollers. JSME Int. J. **40**(1), 120–127 (1997)

21. Yan, H.S., Liu, J.Y.: Geometric design and machining of variable pitch lead screws with cylindrical meshing elements. J. Mech. Des. **115**(3), 490 (1993)

22. Su, Y.L., Lin, J.S.: Wear control in a variable-pitch lead screw with cylindrical meshing elements. Tribol. Int. **26**(3), 201–210 (1993)

23. Huang, H.Z., Bo, R., Chen, W.: An integrated computational intelligence approach to product concept generation and evaluation. Mech. Mach. Theory **41**(5), 567–583 (2006)

24. Cheung, N.J., Ding, X.M., Shen, H.B.: A nonhomogeneous cuckoo search algorithm based on quantum mechanism for real parameter optimization. IEEE Trans. Cybern. **47**(2), 1–12 (2016)

25. Yang, X.S., Deb, S.: Cuckoo search via Levy flights. Mathematics, pp. 210–214 (2009)

Robotic Vision and Machine Intelligence

Research on Short-Term Traffic Flow Forecast Based on Improved GA-Elman

Qi Li and Hongliang Wang[✉]

School of Computer and Communication Engineering,
Liaoning Petrochemical University, Fushun, China
895079750@qq.com

Abstract. Taking the short-term traffic flow of a single section as the research object, the dynamic Elman neural network is used to predict the short-term traffic flow. This paper proposes an improved short-term traffic flow prediction method based on the improved GA-Elman neural network. The replacement crossover operator optimizes the genetic algorithm, reduces the solution space size of the crossover process, and further improves the evolutionary efficiency of the prediction model. The experimental results show that the prediction effect of the proposed prediction model is better than the traditional model, and it is verified that the improved GA-Elman prediction model can reduce the error of traffic flow prediction.

Keywords: Short-term traffic flow prediction · Genetic algorithm · Crossover operator · Elman neural network

1 Introduction

In recent years, with the progress of society and the increase in the number of cars, the situation of traffic congestion in cities has become more and more serious, and it has therefore become a core issue restricting urban development. Road traffic control and induction can effectively relieve the pressure of urban road traffic system, and the prediction of short-term traffic flow is the key link.

At present, a lot of in-depth research has been carried out in the field of short-term traffic flow prediction at home and abroad, and prediction models covering various fields such as statistical algorithms and artificial intelligence algorithms have been established. The study of neural network-based prediction models is an important branch. Yanqiu W [1] applied the improved BP neural network and carried out simulation experiments with MATLAB software to establish a traffic flow prediction model to simulate the prediction algorithm of the dynamic timing of traffic flow; Ma W [2] et al. used the BP neural network algorithm to determine the weights and proposed a dynamic recurrent network model based on Elman, which can overcome the problem of low prediction accuracy when the network scale is simple or the number of neurons is small. Song XS [3] and other existing BP networks based on BP networks the highway traffic flow model was improved, and a more operable Elman dynamic neural

The original version of this chapter was revised: The affiliation of the authors has been corrected as "School of Computer and Communication Engineering, Liaoning Petrochemical University". The correction to this chapter is available at https://doi.org/10.1007/978-981-33-4932-2_31

J. Qian et al. (Eds.): ICRRI 2020, CCIS 1336, pp. 191–202, 2020.
https://doi.org/10.1007/978-981-33-4932-2_13

network model was proposed; He Ding [4] et al. applied the Elman dynamic neural network to the prediction model of short-term traffic flow, and compared with the traditional BP neural network After conducting a comparative study, it was found that the Elman network has a better effect than the BP neural network in predicting traffic flows with dynamic characteristics; Yang X [5] et al. Used genetic algorithms to optimize the weights and thresholds of the BP neural network and establish traffic Stream prediction model; Li Song [6] et al. combined the chaotic characteristics with the time series model and used GA to optimize the BP network in order to obtain the optimal solution when estimating the predicted value of traffic flow, which further improved the accuracy of the predicted value of the model. Zhang Qiuyu [7] et al. Optimized the weights and thresholds of Elman neural network through genetic algorithms, while improving the generalization ability and prediction accuracy of Elman neural network, it also overcomes the shortcoming of Elman neural network easily falling into the local minimum; The above research shows that the Elman neural network model is superior to the BP neural network model in the application of short-term traffic flow prediction. However, the above two improved models The type ignores the operation redundancy problem of genetic algorithms in the process of cross-operation. The reason is that it cannot effectively reduce the size of the solution space of the cross-process, which further leads to the low evolution efficiency of the prediction model.

This paper uses a crossover operator based on permutation to optimize the genetic algorithm [8]. This algorithm can not only use the mutation operator to avoid the network model from falling into a local optimum, but also reduce the solution space size of the crossover process, further improving the prediction model Evolutionary efficiency. In this paper, after using the neural network model to make short-term traffic flow prediction, a good prediction effect is obtained. The test results of the examples show that this method is effective and feasible for the traffic flow prediction model.

2   Elman Neural Network Prediction Model

Elman neural network [9–11] is a typical dynamic recurrent neural network. It is a feedback neural network model proposed by Elman in 1990. The difference between this model and BP neural network is that there is a receiving layer in the hidden layer. The layer can be used to memorize the output value of the previous layer of the hidden layer and return it to the input. It can be called a one-step delay operator to realize the memory function. The network can reflect the dynamic characteristics of the system and enhance the global stability of the network. The Elman neural network model is shown in Fig. 1.

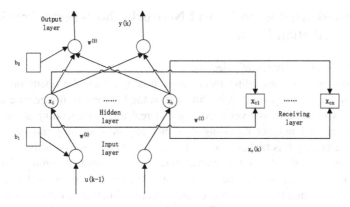

Fig. 1. Basic structure of Elman neural network

The non-linear state space expression of the Elman neural network [12, 13] is:

$$y(k) = g\left(w^{(3)}\right)x(k) + b_2 \tag{1}$$

$$x(k) = f\left(w^{(1)}x_c(k) + w^{(2)}(u(k-1)) + b_1\right) \tag{2}$$

$$x_c(k) = x(k-1) \tag{3}$$

Where k is the moment of the current state. y, x, u, x_c represents m-dimensional output node vector, n-dimensional intermediate layer node unit vector, r-dimensional input vector, and n-dimensional feedback state vector. $w^{(1)}$, $w^{(2)}$, $w^{(3)}$ represents the connection weights from the receiving layer to the intermediate layer, the input layer to the intermediate layer, and the intermediate layer to the output layer. $g\left(w^{(3)}x(k)\right)$ is the transfer function of the output neuron, which is a linear combination of the output of the middle layer. The purelin function is often used. $f\left(w^{(1)}x_c(k) + w^{(2)}(u(k-1))\right)$ is the transfer function of the middle layer neurons, often using the tansig function. b_1, b_2 is thresholds selected in the hidden and output layers, respectively.

Elman neural network uses the standard BP neural network algorithm to propagate the error back to correct the weights in the network so that the sum of squared errors at the output layer is minimized. The error function is defined as:

$$E = \frac{1}{2}\sum_{k-1}^{T} [y_d(k) - y(k)]^2 \tag{4}$$

In the formula. $y_d(k)$ represents the system output vector.

3 Improved GA-Elman Neural Network Short-Term Traffic Flow Prediction Model

1. Coding and fitness function selection

The selection of the encoding mechanism directly affects the operation efficiency and accuracy of the genetic algorithm, and is the key link of the genetic algorithm [14, 15]. Currently, two-level encoding and real number encoding are commonly used. Since real number encoding does not require encoding and decoding, the solution accuracy It is higher, and the neural network is more sensitive to the initial weight and threshold, so this paper chooses real number coding. The fitness function is the evolutionary rule of survival of the fittest in the natural world, which is obtained through the conversion of the objective function of the problem. In order to facilitate the selection of the fitness of the operator below, the standard fitness function is reciprocal, that is, the fitness function F:

$$F = \frac{1}{\sum_{i=1}^{n} abs(y_i - a_i)} \tag{5}$$

Where y_i is the actual output value of the i-th node, a_i is the predicted output value of the i-th node, and n is the number of Elman neural network output nodes.

2. Genetic operator
(1) Selection operator

The selection operation is the survival of the fittest of the individuals in the group. The main purpose is to improve the global convergence and calculation efficiency. This paper chooses the proportional selection operator, that is, the probability that the individual is selected is proportional to its fitness. The probability P_i of individual i being selected is:

$$P_i = \frac{F_i}{\sum_{i=1}^{N} F_i} \tag{6}$$

Where F_i is the fitness value of individual i and N is the number of individuals in the population.

(2) Crossover operator

The crossover operation of genetic algorithms is to generate new gene combinations and thus new individuals. Commonly used crossover operators are PMX, PBX, etc., and the time required for traditional crossover operations is $O(n^2)$, which leads to low evolutionary efficiency of genetic algorithms. Therefore, this paper uses a new type of crossover operator [8] to reduce the size of the crossover operation solution space by reducing the time required for the crossover operation. The specific operation steps are as follows:

1) Randomly select two tangent points on the parental chromosomes (p_1 and p_2): cp_1 and cp_2.
2) As shown in Fig. 2, swapping the positions of the two sets of chromosomal fragments between the two tangent points (cp_1 and cp_2) results in two sets of chromosomes without collision detection: o_1 and o_2.

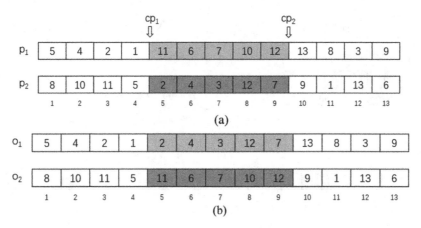

(a)

(b)

Fig. 2. (a) Parent chromosomes (b) Two sets of chromosomes without conflict detection

3) Define a swap list based on selected chromosome fragments.
4) As shown in Fig. 3, generate a directed graph of the exchange list and find all the different paths between the nodes in the graph.

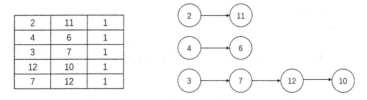

Fig. 3. Exchange list and directed graph

5) Create two lists L1 and L2 with the same length as the parent chromosome, use the first and second columns of the swap list as the indexes of L1 and L2, and set their corresponding values to 1. When one of the items is equal to 2, it indicates that it is an intermediate node, and the third column of the exchange list is set to 0.
6) For each path with more than two nodes, add an edge between the two endpoints and delete all intermediate nodes.
7) Update exchange list based on improved path.
8) Apply the updated exchange list to o_1 to generate op_1, as shown in Fig. 4

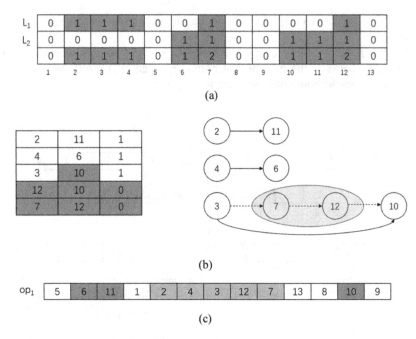

Fig. 4. (a) L1 and L2 (b) Updated exchange list (c) Newly generated op_1

9) Define a list F of the same length as the parent chromosome, initialized to zero.
10) Perform the following operations to produce op_2: $F(op_1[i]) = p_1[i]$; $op_2[i] = F(p_2[i])$ (i = 1, 2,..., n, n is the length of the parental chromosomes), as shown in Fig. 5.

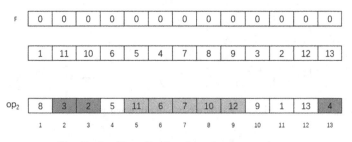

Fig. 5. Auxiliary list F and newly generated op_2

(3) Mutation operator

The mutation operation of the genetic algorithm is to randomly select several individuals from the population and randomly determine the mutation positions in the selected individuals. This paper uses the basic mutation operator, which adds a random number to the mutation gene with a uniform probability in a certain range [16].

The mutation probability adopts adaptive adjustment rules [17]:

$$P_m = \begin{cases} \frac{k_0(f_{max}-f_0)}{(f_{max}-f)}, & f_0 \geq -f \\ k_1, & f_0 < -f \end{cases} \tag{7}$$

Where f_0 is the fitness value of mutant individuals, and k_0, k_1 are mutation probability adjustment coefficients.

3. Forming a new generation.
4. When the maximum number of descendants is reached, the algorithm stop.
5. Generate optimized neural network initial weights and thresholds.
6. Train and learn the Elman neural network, update the weights, calculate the error of the objective function to determine whether it meets the requirements.
7. When the number of trainings is reached, the process ends.

4 Experimental Results and Performance Analysis

4.1 Experimental Setup

In order to verify the accuracy of the proposed theory, simulation experiments are performed on the traffic flow prediction model. The first five time periods are selected as the sample data to predict the current time period. Based on this, it can be determined that the Elman neural network has five input layer nodes and one output layer node. After testing, when the number of hidden layer nodes is 18, the model has the smallest prediction error and the best training effect. The learning rate of the network is set to 0.1, the maximum number of trainings is 1000, the initial population of the genetic algorithm is 100, the crossover probability is 0.6, and the number of genetic generations is 50 generations.

In order to evaluate and compare experimental results, the research uses performance indicators commonly used in traffic flow prediction, including:

Absolute average error:

$$MAE = \frac{1}{T}\sum_{t=1}^{T}|y(t) - y'(t)| \tag{8}$$

Absolute average error:

$$MSE = \frac{1}{T}\sqrt{\sum_{t=1}^{T}(y(t) - y'(t))^2} \tag{9}$$

Relative average error:

$$MAPE = \frac{1}{T}\sum_{t=1}^{T}\frac{|y(t) - y'(t)|}{y(t)} \tag{10}$$

Relative mean square error:

$$\text{MSPE} = \frac{1}{T}\sqrt{\sum\nolimits_{t=1}^{T}\left(\frac{|y(t) - y'(t)|}{y(t)}\right)^2} \tag{11}$$

Accuracy:

$$EC = 1 - \frac{\sqrt{\sum_{t=1}^{T}(y(t) - y'(t))^2}}{\sqrt{\sum_{t=1}^{T}y(t)^2} - \sqrt{\sum_{t=1}^{T}y'(t)^2}} \tag{12}$$

Where: $y(t)$ is the measured value, $y'(t)$ is the predicted value, and T is the total number of data samples.

In order to ensure that the magnitude difference between the dimensions is not large, and to avoid the network prediction error from increasing due to the large difference in the magnitude of the input and output data, the data pre-processing uses the L2 norm normalization [18].

$$\text{norm}(x) = \sqrt{x_1^2 + x_2^2 + \cdots + x_n^2} \tag{13}$$

$$X = \frac{x_i}{norm(x)} \qquad i = 1, 2 \cdots n \tag{14}$$

L2 norm normalized data can not only prevent overfitting, but also make the optimization solution stable and fast.

4.2   Analysis of Results

This article selects traffic data collected from the main arteries of the Interstate Highway in Los Angeles, California, USA [19]. The experimental sample is selected from November 2, 2015 to November 27, 2015. There are 1920 periods except Saturday As the data source, training with the data from the previous 19 days, and finally using the trained parameters to predict the traffic flow on the last day. Compare the measured data with the predicted results, verify the effectiveness of the improved GA-Elman algorithm, and compare the predicted results with the measured results. Part of the real traffic flow data is shown in Table 1.

Table 1. Part of the real traffic flow data

Time	Actual flow	Normalized	Date	Period
...	...	...	...	...
2015/11/27 7:00	133	0.50000	1	1
2015/11/27 7:15	166	0.62891	1	1

(*continued*)

Table 1. (*continued*)

Time	Actual flow	Normalized	Date	Period
2015/11/27 7:30	195	0.74219	1	1
2015/11/27 7:45	239	0.91406	1	1
2015/11/27 8:00	261	1.00000	1	1
2015/11/27 8:15	260	0.99609	1	1
2015/11/27 8:30	218	0.83203	1	1
2015/11/27 8:45	168	0.63672	1	1
2015/11/27 9:00	150	0.56641	1	1
...	...	...	...	...
2015/11/27 21:00	83	0.30469	1	0
2015/11/27 21:15	123	0.46093	1	0
2015/11/27 21:30	109	0.40625	1	0
2015/11/27 21:45	72	0.26172	1	0
2015/11/27 22:00	82	0.30078	1	0
2015/11/27 22:15	52	0.18359	1	0
2015/11/27 22:30	50	0.17578	1	0
2015/11/27 22:45	61	0.21875	1	0
2015/11/27 23:00	38	0.12891	1	0
...	...	...	...	...

The verification experiments used GA-BP neural network [5], GA-Elman neural network [7], and improved GA-Elman neural network to construct prediction models. The simulation results are shown in Fig. 6 and Fig. 7, and the predicted performance index pairs are shown in Table 2.

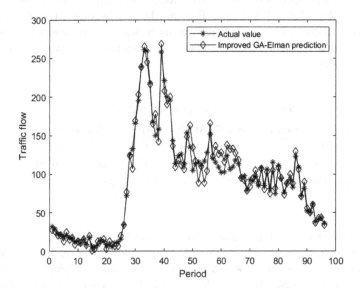

Fig. 6. Comparison of actual value and predicted output value

Figure 6 is a comparison curve between the actual value and the predicted value of the improved model. Through analysis, it can be seen that the model can better reflect the change trend of the traffic flow, and the deviation between the predicted value and the actual value is small, indicating that the model is used for The prediction of traffic flow is reasonable and effective.

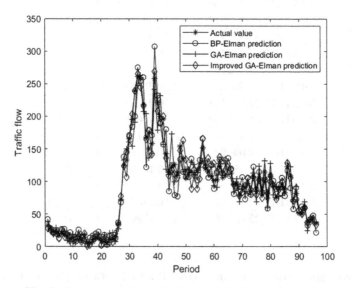

Fig. 7. Comparison of actual value and predicted output value

Figure 7 is a comparison curve between actual values and predicted values of three prediction models. Table 2 accurately lists the absolute average error, absolute mean square error, relative average error, relative mean square error, and accuracy coefficients of the three prediction models. Value, can better reflect the pros and cons of the prediction model.

Table 2. Evaluation index of percentage error

Predictive model	MAE%	MSE%	MAPE%	MSPE%	Accuracy
GA-BP	3.463	1.699	10.58	4.652	0.9628
GA-Elman	3.220	1.567	9.13	4.034	0.9732
Improve GA-Elman	2.759	1.402	8.848	3.847	0.9869

By analyzing Fig. 6, Fig. 7 and Table 2, it can be seen that all three prediction models can better reflect the change trend of traffic flow; overall, the improved GA-Elman has the smallest error, which is smaller than the traditional GA-Elman The model's error is smaller because the improved crossover operator reduces the evolutionary efficiency of the genetic algorithm and can obtain better weights and thresholds in shorter iterations; and the accuracy of the three prediction models is greater than 0.9 The improved GA-Elman model can even reach 0.985, indicating that the three models can fit the actual values well, especially the improved GA-Elman model has a better prediction effect. The above analysis shows that the improved GA-Elman prediction model has a good prediction effect. Based on GA-BP and traditional GA-Elman model, it can accurately and effectively predict traffic flow.

5 Conclusion

Based on the measured traffic flow dataset, the improved GA-Elman neural network model is tested. Simulation results show that the algorithm can be well applied to short-term traffic flow prediction. The traditional GA-Elman neural network model cannot effectively reduce the solution space size of the crossover process, which leads to low evolutionary efficiency of the prediction model and reduced prediction accuracy. This paper uses a new type of crossover operator to optimize the genetic algorithm, so that the evolutionary efficiency of the prediction model is improved, and the convergence speed and prediction accuracy are also improved.

The prediction model proposed in this article is to predict the traffic flow on working days, and the traffic flow on non-working days is significantly different from that on working days. In the next research work, time series can be introduced into the traffic flow prediction model, which overcomes the shortcomings of traditional traffic flow prediction models in the time series, and further improves the accuracy of traffic flow prediction.

Acknowledgments. This work was supported by a project supported by the Natural Science Foundation of Liaoning Province (Fund No. 20180550688).

References

1. Yanqiu, W., Qiang, L., Jian, Z., et al.: The city traffic flow prediction based on BP neural network. In: 2008 Chinese Control and Decision Conference, pp. 2550–2552. IEEE (2008)
2. Ma, W., Wang, R.: Traffic flow forecasting research based on Bayesian normalized Elman neural network. In: 2015 IEEE Signal Processing and Signal Processing Education Workshop (SP/SPE), pp. 426–430. IEEE (2015)
3. Song, X.S., Li, H., Wu, B.H., et al.: Elman neural network model of traffic flow predicting in mountain expressway tunnel. In: 2010 International Conference on Computational Intelligence and Software Engineering, pp. 1–4. IEEE (2010)
4. He, D., Xu, P.: Comparative study of Elman and Bp neural networks applied to traffic flow prediction. Ind. Eng. **13**(06), 97–100 (2010)
5. Yang, X., Zhang, L., Xie, W.: Forecasting model for urban traffic flow with BP neural network based on genetic algorithm. In: 2019 Chinese Control And Decision Conference (CCDC), pp. 4395–4399. IEEE (2019)
6. Li, S., Liu, L., Xie, Y.: Genetic algorithm optimized BP neural network for short-term traffic flow chaos prediction (2011)

7. Zhang, Q., Zhu, X.: Short-term traffic flow prediction method based on GA-Elman neural network. J. Lanzhou Univ. Technol. **39**(03), 94–98 (2013)
8. Koohestani, B.: A crossover operator for improving the efficiency of permutation-based genetic algorithms. Expert Syst. Appl. 113381 (2020)
9. Yang, Q., Zhang, B., Gao, P.: Short-term traffic prediction method based on improved dynamic recurrent neural network. J. Jilin Univ. (Eng. Technol. Ed.) **42**(04), 887–891 (2012)
10. Li, Q., Qin, Y., Wang, Z.Y., et al.: Prediction of urban rail transit sectional passenger flow based on Elman neural network. In: Applied Mechanics and Materials. Trans Tech Publications, vol. 505, pp. 1023–1027 (2014)
11. Liu, N., Chen, Y., Yu, Y., Fan, G.: Traffic flow forecasting method based on Elman neural network. J. East China Univ. Sci. Technol. (Nat. Sci. Ed.) **37**(02), 204–209 (2011)
12. Liu, Z., Guo, J., Cao, J., et al.: A hybrid short-term traffic flow forecasting method based on neural networks combined with K-nearest neighbor. Promet-Traff. Transp. **30**(4), 445–456 (2018)
13. Liu, Y., Zhang, Y.: Short-term traffic flow prediction based on wavelet denoising and GA-Elman neural network. J. Transp. Sci. Technol. **17**(6), 80–85 (2015)
14. Zhang, C., Zheng, J., Qian, J.: Comparison of genetic algorithm coding schemes. Appl. Res. Comput. **28**(3), 819822 (2011)
15. Yang, H., Hu, X.: Wavelet neural network with improved genetic algorithm for traffic flow time series prediction. Optik **127**(19), 8103–8110 (2016)
16. Sofronova, E.A., Belyakov, A.A., Khamadiyarov, D.B.: Optimal control for traffic flows in the urban road networks and its solution by variational genetic algorithm. Procedia Comput. Sci. **150**, 302–308 (2019)
17. Jiao, G., Wang, S.: Research on neural network parameter optimization based on genetic algorithm (2012)
18. Hoffer, E., Banner, R., Golan, I., et al.: Norm matters: efficient and accurate normalization schemes in deep networks. In: Advances in Neural Information Processing Systems, pp. 2160–2170 (2018)
19. Principal Arterial - Interstate [2015-11-02 to 2015-11-27]. https://www.transportation.gov/data

Tire Defect Detection Based on Faster R-CNN

Zeju Wu[1](✉) ⓘ, Cuijuan Jiao[1] ⓘ, Jianyuan Sun[2] ⓘ,
and Liang Chen[1] ⓘ

[1] The School of Information and Control Engineering,
Qingdao University of Technology, Qingdao 266525, China
wuzeju_qut@163.com
[2] National Centre for Computer Animation, Faculty of Media
and Communication, Bournemouth University, Bournemouth BH12 5BB, UK
sunj@bournemouth.ac.uk

Abstract. The tire defect detection method can help the rehabilitation robot to achieve autonomous positioning function and improve the accuracy of the robot system behavior. Defects such as foreign matter sidewall, foreign matter tread, and sidewall bubbles will appear in the process of tire production, which will directly or indirectly affect the service life of the tire. Therefore, a novel and efficient tire defect detection method was proposed based on Faster R-CNN. At preprocessing stage, the Laplace operator and the homomorphic filter were used to sharpen and enhance the data set, the gray values of the image target and the background were significantly different, which improved the detection accuracy. Moreover, data expansion was used to increase the number of images and improve the robustness of the algorithm. To promote the accuracy of the position detection and identification, the proposed method combined the convolution features of the third layer and the convolution features of the fifth layer in the ZF network (a kind of convolution neural network). Then, the improved ZF network was used to extract deep characteristics as inputs for Faster R-CNN. From the experiment, the proposed faster R-CNN defect detection method can accurately classify and locate the tire X-ray image defects, and the average test recognition rate is up to 95.4%. Moreover, if there are additional types of defects that need to be detected, then a new detection model can be obtained by fine-tuning the network.

Keywords: Rehabilitation robot · Faster R-CNN · Improved ZF convolutional neural networks · Recognition rate · Tire defect detection

1 Introduction

The tire defect detection method has a strong self-learning ability, which can help the robot achieve the function of automatic visual recognition [1, 2]. It meets the requirements of rehabilitation robots for smart devices, makes medical services more accurate, and provides patients with more advanced and effective rehabilitation treatment processes [3]. A radial tire is widely recognized because it has the advantages of high speed, energy saving, durability, safety, comfort, and driving performance. The radial tire manufacturing process is complex and requires high precision. Quality

© Springer Nature Singapore Pte Ltd. 2020
J. Qian et al. (Eds.): ICRRI 2020, CCIS 1336, pp. 203–218, 2020.
https://doi.org/10.1007/978-981-33-4932-2_14

problems often occur in the manufacturing process, which will directly or indirectly affect the service life of the tire, and even endanger personal safety [4, 5]. Therefore, it is necessary to carry out the non-destructive test on each tire before leaving the factory, which can help persons quickly adjust the machinery production equipment in the tire production process. This test can save production material costs and improve tire quality.

At present, the tire defect detection methods generally employ an X-ray machine to obtain the images of the tire, and then workers observe the obtained images to identify whether the tires have the defects and classify these defects according to the shape and gray features of the images [6]. However, artificial visual inspection is an objective method. The detection results are easily affected by the level of workers' professionalism. It is easy to misjudge when the workload and intensity of the workers are large [7, 8]. Therefore, in this paper, we propose a tire detection method based on Faster R-CNN. Using the convolutional neural network to obtain the feature point information of the image and analyze the upper semantics, and then replace the human eye to complete the task. It is applied to the research of rehabilitation robots, and it accomplishes a part of tasks that human eyes cannot judge competent in a special environment, and meets the needs of rehabilitation robots for high-precision positioning of smart devices. The tire detection system is used to realize the human eye guidance function of the rehabilitation robot, and the convolutional neural network and image processing technology are combined to complete the specific tasks of the rehabilitation robot in the given environment. The rapid development of artificial intelligence is of great significance to the development of rehabilitation robot automation.

In 2014, Ross Girshick et al. [9] designed the R-CNN target detection framework with a regional nomination strategy and CNN feature extraction algorithm, which made a great breakthrough in target classification and location tasks. To solve the problem of low efficiency and large training space of R-CNN, Girshick et al. [10] proposed the Fast R-CNN method. Firstly, Conv Character Map was obtained by carrying out the convolution operation of the image. At the same time, the region of interest (ROI) was obtained using the Selective Search method, and then the candidate region was mapped to the feature map of the CNN last layer. The image only needs to extract features once, which reduces the computational complexity. However, it still does not solve the problem of the slow computational speed of a selective search algorithm. The Faster R-CNN was proposed by Ren et al. [11], which used the region proposal networks (RPN) to select the proposal regions on the premise of absorbing the characteristics of Fast R-CNN. Moreover, most of the prediction is completed under the GPU, which greatly improves the detection speed and accuracy.

Organization of the paper as follows. Section 2 provides a review of the literatures regarding the development process of deep learning. In Sect. 3, we illustrate our algorithm elaborately. Section 4 provides the experimental results and Sect. 5 shows the conclusion of the paper.

2   Related Work

In recent years, the problem of tire defect detection has attracted a lot of attention for domestic researchers. For example, Q Liu et al. [12] of Shandong University of Finance and Economics proposed a tire defect detection algorithm based on Radon transformation in 2015; Bin Zhang et al. [13] of Qingdao University Soft Control Enterprise proposed a tire X-ray image impurity detection technology based on image processing in 2016. The technology has been completed in four steps: histogram column equalization, Fourier transform and low-pass filtering, binarization and closed operation, and based on which the tire bead impurity defects were identified. However, these methods can only detect and identify the tire bead impurity defects. Xuehong Cui et al. [14] of Qingdao University of Science and Technology proposed a tire X-ray image defect detection method based on inverse transformation of principal component residual information in 2017. The principal component analysis (PCA) is used to reconstruct the dominant texture of tire image, and then the defect can be found by subtracting the original image from the reconstructed dominant texture image (only using the small remaining eigenvalues and corresponding eigenvectors to restore the defect and noise). It can detect foreign matter sidewall, however, the extraction features are complex and require people to participate in the selection.

The above methods for tire defect detection are all based on traditional machine vision [15]. The features used for tire defect detection are artificially selected and designed [16], and the obtained features are jagged and poorly robust. Therefore, these methods are suitable for simple defect detection. The automatic recognition and localization of tire X-ray images with different defect areas, various shapes and complex background areas are powerless.

In 2006, Hinton et al. [17] proposed in-depth learning to solve this problem. In-depth learning can automatically learn the features of the target according to the training dataset, and abstract the high-level feature expression by integrating the transformation features of each layer from low to high, which makes classification and location easier. Since then, the emergence of excellent algorithms based on convolutional neural networks, and played a great role in target detection.

In this paper, in order to improve the recognition accuracy and the location accuracy, the tire defect location and recognition method is improved based on the Faster R-CNN. At preprocessing stage, the Laplace operator and the homomorphic filter are used to sharpen and enhance the data set due to the gray values of the image target and the background was significantly different. Moreover, we combine the convolution features of the third layer and the fifth layer of the ZF network as the input of the RPN layer. The convolution feature of the layer is combined as the input of the RPN layer; since the output of the third layer of the ZF network is 384 dimensions, and the output of the fifth layer is 256 dimensions, the convolution operation of is performed after the output of the third layer, so that its output is also 256 dimensions. The outputs of the layers are added.

Figure 1 shows the specific process. Faster R-CNN network structure mainly consists of three parts: feature extraction layer, RPN (Regional Proposal Network) layer, Fast R-CNN layer and so on. In the feature extraction layer, the ZF network is used to extract the feature map of the input image. The RPN layer is used to preliminarily extract regions of interest (ROI). Fast R-CNN layer is used to locate and identify tire defects.

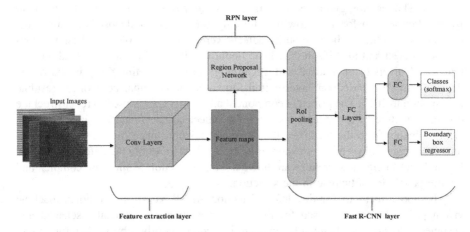

Fig. 1. Target detection framework based on Faster R-CNN (Color figure online)

In this paper, the faster R-CNN and ZF network [10] (a kind of convolution neural network) is introduced to solve the task of the tire defect recognition and location, which provides a more concise and efficient method for industrial tire defect detection.

3   Proposed Method

3.1   X-Ray Image Defect Detection of Tire

The X-ray images of tires mainly contain three types of defects: foreign matter sidewall, foreign matter tread and sidewall bubble, as indicated by the green arrow in Fig. 2. Figure 2 shows the flowchart of the proposed method. The red arrow is rectangular boxes. The training set is pre-processed before network training: homomorphic filtering is used to enhance the training set.

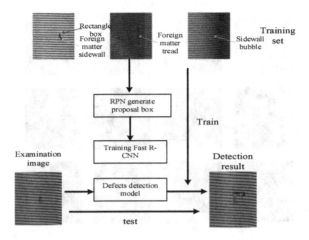

Fig. 2. Defect detection process based on Faster R-CNN

Using X-ray machine to make X-ray image of 360° tire irradiation, because the obtained X-ray image contains some noise. Target defects such as foreign matter sidewall and sidewall bubble, the gray difference between the target and background is very small in some areas. In this case, it is difficult for the human eye to find and recognize the target, which has negative for the labeling of the rectangular box and the training of the model, so it is necessary to do pre-processing to enhance the gray difference between these targets and background. This method is divided into two stages: training stage and testing stage.

Training stage: The input data of the training model was increased by the image geometric transformation method. And the rotation, the horizontal offset, the vertical offset, the scaling and horizontal flip were selected. The data augmentation method enables the convolutional neural network to learn more image invariant features and avoid overfitting. Defect detection model is obtained by determining the type of target to be detected (foreign matter sidewall, foreign matter tread, and sidewall bubble), selecting the network and training.

Testing stage: Testing the samples to be tested with the obtained model. The deep learning framework used in the experiment is MXNet, which is accelerated by GPU, and the feature extraction network is ZF. The image data used in this paper are provided by Soft Holding Company Limited.

Figure 3 shows the labeled dataset sample. The tire defect detection dataset is constructed according to PASCAL VOC dataset format standard. The rectangular box is labeled on the image by Labelme. Figure 3(a)–(d) show the foreign matter sidewall. Figure 3(e)–(h) show the foreign matter tread. Figure 3(i)–(l) show the sidewall bubble. An XML file is automatically generated for each image to record various information of the image, and the bounding box coordinate information.

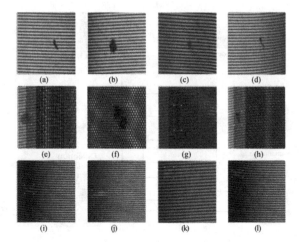

Fig. 3. Tire defect dataset sample

The training dataset contains 1022 pictures, including foreign matter sidewall (215), foreign matter tread (155), sidewall bubble (126) and normal tire X-ray images (including sidewall/ tread images) (526). The number of pictures in brackets represents the number of pictures. The proportion of each kind of defect sample is consistent with the frequency of occurrence of such defects. Defect image is the smallest rectangular image containing defects. Because of the uncertainty of defects in production, the length, width and size of defect image are different and distributed in 50 × 50–200 × 500 pixel. In order to unify the size of defective images to meet the requirements of the algorithm, and to maximize the representation of image defects, while reducing computational complexity, In this paper, each image is segmented into the sidewall and tread regions, and the long sidewall and tread regions are divided into 8 segments and then the training dataset is composed of these images.

Improved Convolutional Neural Network

Convolutional neural network is specially designed for image recognition. It benefits from its network structure similar to biological neural network, which reduces the complexity of network model and the number of weights [18–21]. In this paper, ZF network is used to extract image features.

Figure 4 shows the framework of the ZF network. The process of ZF network detection is:

- (1) The image of size is used as input, and the feature is extracted automatically by convolution layer;
- (2) Then the RPN is used to generate high quality proposal box, each image is about 300 proposal boxes;
- (3) The proposal boxes are mapped to the last convolution feature map of CNN;
- (4) The ROI pooling layer is used to fix the size of each proposal box;
- (5) Classification layer and boundary regression layer is used to make specific classification judgment and accurate border regression for the proposal areas.

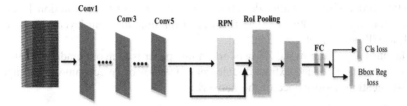

Fig. 4. Framework of the ZF network

Figure 5 shows the framework of the improved ZF network. From the above detection process, we can see that only the last layer of convolution is used in the recognition of ZF network. Since the image information can be extracted into the abstract feature of the target after it passes through the multi-layer convolution layer, Although these abstract features are helpful for judging the specific categories of targets, after multi-layer convolution feature extraction, the details of the target will also be lost, which makes the extracted features less sensitive to the size and location of the target. Therefore, if only the last layer of convolutional feature information is used to locate and identify in the detection process, there will be a great error for the smaller defect area, because the smaller defect target (if the defect area is 25×25 pixel) has only one or several convolution features after passing through five layers of convolution layer. Such little information not only has a negative impact on location precision, but also influences the recognition of the target.

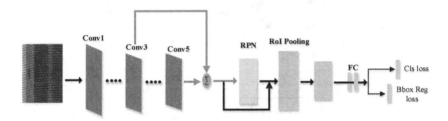

Fig. 5. Framework of the improved ZF network

According to the above situation, inspired by the Fully Convolutional Networks (FCN) on the success of the Semantic Segmentation task proposed by Jonathan Long et al. [22], to improve the recognition accuracy and location accuracy, the tire defect location and recognition method is improved based on Faster R-CNN. Based on the Faster R-CNN network, this paper combines the convolution features of the third layer and the fifth layer of ZF network as the input of RPN layer. The convolution feature of the layer is combined as the input of RPN layer; since the output of the third layer of ZF network is 384 dimensions, and the output of the fifth layer is 256 dimensions, the convolution operation of is done after the output of the third layer, so that the output of the third layer is 256 dimensions, and then the output of the fifth layer is added. Because the shallower convolution layer can extract local features in the feature extraction process of the target detection network, the deeper convolution layer can extract more abstract features, if the features of multiple different layers can be combined, it will help target detection [23].

The purpose of improving ZF network in this paper is to avoid gradient explosion, improve learning rate, and extract double depth features of defect samples to improve the classification performance of the network.

Region Proposal Network

Figure 6 shows the framework of the Region Proposal Network (RPN). The RPN is a full-convolution network. The RPN network can quickly generate anchor boxes of different sizes, and determine the probability of the target or background of the image in the frame for preliminary extraction of Region of Interest (ROI), which solves the speed problem of the Selective Search (SS) [24], and greatly improves the target detection speed. The specific process of constructing the RPN is to use a small sliding window (Convolution Kernel) to scan the feature map of the final convolution. After sliding convolution, a d-D vector is mapped. Finally, the d-D vector is sent to two fully connected layers, the regression layer and the classification layer. In each sliding window, region proposal boxes are predicted at the same time, so the regression layer has outputs to encode the coordinates of region proposal boxes, and the classification layer outputs scores to estimate the probability that each region proposal box is a target/non-target.

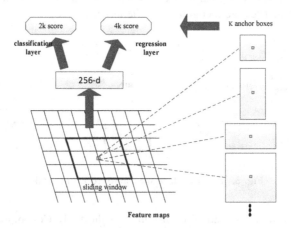

Fig. 6. Framework of the Region Proposal Network

The RPN network uses a Bounding Box Regression (BBox) to generate an ROI, which loss function is defined as (1).

$$L_{reg}\left(t_i, t_i^*\right) = \sum_{i=x,y,w,h} Smooth_{L_1}\left(t_i - t_i^*\right) \tag{1}$$

Where, x, y, w, h represents the central coordinates of ROI and their width and height, t_i represents four parametric coordinates of the predicted bounding box, t_i^* represents the coordinate vector of the Ground Truth (GT) bounding box corresponding to the positive anchor. For robust loss function in [10] is shown in (2) and (3).

When $|x| < 1$:

$$Smooth_{L_1}(x) = 0.5x^2 \tag{2}$$

$|x| \geq 1$:

$$Smooth_{L_1}(x) = |x| - 0.5 \tag{3}$$

The total loss function of the RPN network is defined as (4).

$$L(\{P_i\}\{P_i^*\}) = \frac{1}{N_{cls}} \sum_i L_{cls}(P_i, P_i^*) + \lambda \frac{1}{N_{reg}} P_i^* \sum_i L_{reg}(t_i, t_i^*) \tag{4}$$

Where, i represents the window index value generated by RPN in a single image sample feature graph, P_i represents probability indicating that the window is a target, P_i^* represents the predictive probability of the GT, N_{cls} is the normalized value of the classification term, i.e. the batch number 256; N_{reg} is normalized value of regression term, λ is weight value used to balance the two kinds of losses. In this paper, we set $\lambda = 1$ to make the two weights approximately equal.

In this paper, a total of 12000 tire defect images are selected to make tire defect samples for neural network training. The learning rate of the network is 0.1, and the total number of iterations is 80000.

Fast R-CNN Network

In practical calculation, 1000 ROIs are provided by RPN network, and 300 of them are randomly selected for Fast RCNN network training. The positive and negative samples are judged by Intersection over Union (IOU) [25]. In order to ensure the quality of positive samples, when IOU >0.6, the ROI is determined to be positive sample; when IOU <0.2, it is negative sample. Fast R-CNN uses pooling to process ROI of different sizes to ensure that input vector dimensions are the same when entering the full connection layer. ROI pooling input is divided into two parts, RPN network output proposals and CNN network output image feature map. Proposals correspond to the $M \times N$ scale, so firstly, it is mapped back to the $(M/16) \times (N/16)$ scale by using the spatial scale number of 1/16. Then the horizontal and vertical directions of each proposal are divided into seven equal parts, and each proposal is processed by Max pooling. After processing, proposals of different sizes are all 7×7 in size, which achieves fixed length output.

The proposal feature maps were obtained using the ROI pooling, and the full connection layer and softmax calculate the category of each proposal (foreign matter sidewall and foreign matter tread, sidewall bubble), and outputs the category probability vector. At the same time, the positional offset was obtained using BBox, which is used to return a more accurate target detection frame. On the premise of absorbing the characteristics of Fast R-CNN, Faster R-CNN uses RPN prediction suggestion frame to predict 300 high-quality candidate regions, and most of the prediction is completed in GPU, which greatly improves the speed of target detection.

4   Experimental Results and Analysis

4.1   Analysis of Preprocessed Results

The typical method of image denoising is filtering. Median filtering is a kind of nonlinear smoothing filter. Its core is to set the gray value of each pixel in the image to the median value of the gray value of all pixels in a neighborhood window of the point, so as to realize the gray value of each point within a certain range, and then remove the noise points in the image. Therefore, it can improve the image data, suppress unnecessary deformation or enhance the image details of defects in the tire, and improve the accuracy and accuracy of detection.

Figure 7(a) shows the original image, and Fig. 7(b) shows its corresponding histogram. It can be seen from the original image that the defect contains the foreign matter tread. Some defects are very close to the gray value of the background, and the gray distribution is relatively scattered. Figure 7(c) shows the image after preprocessing, and Fig. 7(d) shows the histogram after preprocessing. After processing, the gray value of the target and background of the image has obvious difference, and the gray distribution is more concentrated.

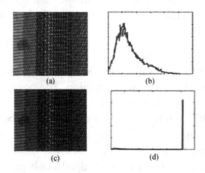

Fig. 7. The preprocessed results: (a) The original image, (b) The corresponding histogram, (c) The preprocessed image, and (d) the histogram of the preprocessed image.

4.2   Defect Detection Results

Figure 8 shows the results of image detection using an improved Faster R-CNN network model.

Figure 8(a) shows the detection results of the foreign matter sidewall, Fig. 8(b) shows the detection results of the foreign matter tread, and Fig. 8(c) shows the detection results of the sidewall bubble. It can be seen that for three kinds of defects of different sizes, the test result shows that the detection quality is high. Figure 8(d) shows the result of false detection. The yellow area in the picture is the real location of the bubble defect, and the normal area in the picture is mistake for as the bubble defect (such as the red box). This is mainly due to the fact that the weak edges of bubble defects in X-ray images can hardly be clearly displayed, while the detection accuracy of bubble defects with strong edges like Fig. 8(c) is higher.

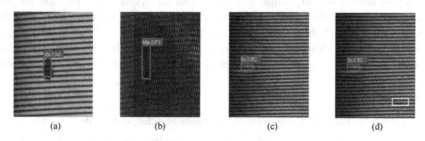

| (a) | (b) | (c) | (d) |

Fig. 8. The detection results (a) The detection results of the foreign matter sidewall, (b) The detection results of the foreign matter tread detection, (c) The detection results of the sidewall bubble, and (d) The result of false detection. (Color figure online)

The total number of pictures to be detected is 184. Table 1 shows the number statistics of test results, include foreign matter sidewall, foreign matter tread and sidewall bubble defects. And Table 2 shows the ratio statistics of test results.

Table 1. The number statistics of test results

Defects	Correct	Missed	False
Foreign matter sidewall	36	1	0
Foreign matter tread	55	2	0
Sidewall bubble	33	3	2
Total	124	6	2

Table 2. The ratio statistics of test results

Defects	Correct rate	Missed rate	False rate
Foreign matter sidewall	0.973	0.027	0
Foreign matter tread	0.965	0.035	0
Sidewall bubble	0.917	0.083	0.056
Total	0.954	0.046	0.015

From Table 1 and Table 2, it can be seen that the method has good detection results for all three kinds of defects. It has a higher detection rate for other defects except for some bubble defects.

4.3   Comparison of Test Results

In the experiment, mAP(mean average precision) [25] is used as the accuracy evaluation index of tire defect detection performance. The higher the value is, the higher the detection accuracy of the algorithm is.

Table 3 shows the comparison results of the different method. The detection accuracy of the original algorithm is 85.17%, and the speed is 17 fps. Add the preprocessing method to the original algorithm, the detection accuracy is 93.82% and the speed is 19 fps. The method in this paper, the model detection accuracy is 95.37% and the speed is 28 fps. Although the accuracy of the method in this paper is slightly improved from 93.82%, the detection speed is greatly improved, and the detection effect is better.

Table 3. Comparison of test results of different methods

Method	mAP/%	Speed/fps
Original algorithm	85.17	17
Original algorithm +Preprocessed	93.82	19
Method in this paper	95.37	28

The training parameters of RPN network are set in reference [10]. The values of other parameters in the training process are shown in Table 4.

Table 4. Parameter setting

Parameters	Values
TEST.SCALES	400
TEST.MAX_SIZE	600
TRAIN.BATCH_SIZE	64
TEST.RPN_MIN_SIZE	4
CONF_THRESH	0.6

The training results of in [10] algorithm and the results of direct detection after preprocessed are compared with the accuracy of the detection results of this method. Table 3 and 4 show the specific comparison results.

Figure 9 shows the detection results in [10] algorithm, Fig. 9(a) shows the detection results of the foreign matter sidewall, Fig. 9(b) shows the detection results of the foreign matter tread, and Fig. 9(c) shows the detection results of the sidewall bubble. The algorithm in [10] is poor for detection of foreign matter tread and sidewall bubble.

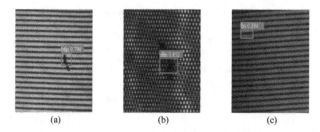

(a)   (b)   (c)

Fig. 9. The detection results in [10] algorithm (a) The detection results of the foreign matter sidewall, (b) The detection results of the foreign matter tread, (c) The detection results of the sidewall bubble.

Figure 10 shows the detection results of preprocessing tests, Fig. 10(a) shows the detection results of the foreign matter sidewall, Fig. 10(b) shows the detection results of the foreign matter tread, and Fig. 10(c) shows the detection results of the sidewall bubble. The pretreatment algorithm has greatly improved the sidewall bubble.

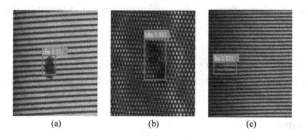

(a)   (b)   (c)

Fig. 10. The detection results of preprocessed tests (a) The results of the foreign matter sidewall detection, (b) The results of the foreign matter tread detection, and (c) The results of the sidewall bubble detection.

Figure 11 shows the comparison of the detection accuracy. The histogram of the detection accuracy obtained from the statistical results in Table 4 and 5. From Fig. 11, it can be seen that the three methods have high detection accuracy for foreign matter sidewall, but algorithm used in [10] is poor for detection of foreign matter tread and sidewall bubble. Because of the preprocessing of the training dataset and the improvement of the ZF network, the accuracy of detecting foreign matter tread and sidewall bubble has been greatly improved (Table 6).

Table 5. The number statistics of test results in [10]

Defects	Correct	Missed	False
Foreign matter sidewall	29	5	3
Foreign matter tread	11	3	1
Sidewall bubble	10	16	4
Total	50	24	8

Table 6. The number statistics of preprocessed test results

Defects	Correct	Missed	False
Foreign matter sidewall	62	5	1
Foreign matter tread	50	6	3
Sidewall bubble	32	7	2
Total	144	18	6

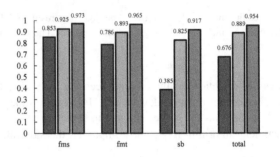

Fig. 11. Comparison of the detection accuracy of different methods

In Fig. 11, the blue part represents the accuracy of the original algorithm, the yellow part represents the accuracy of the method of add preprocessing to the original algorithm, and the red part represents the accuracy of the method in this paper.

5   Conclusions

This paper proposes a novel tire defect detection method based on Faster R-CNN, which is used to identify the type of tire defects and mark the location of the defects. In the pre-processing stage, the homomorphic filtering method is used to sharpen and enhance the training dataset, which avoids the generation of a large number of redundant windows and improves the detection speed and accuracy. By improving the framework of ZF network and referring to the structure of semantic segmentation network FCN, combining the third-level convolution feature with the fifth-level convolution feature, smaller defects can be detected. The target improves the accuracy of location detection and recognition in Faster R-CNN network. Compared with traditional methods, the detection process has been greatly improved in the degree of automation, and has strong robustness, which is an effective and feasible method. In the future, this method can be applied to the intelligent manufacturing of rehabilitation robots to improve the core competitiveness of enterprises.

The appearance of secondary tire or waste tire will waste a lot of raw materials and manpower, and if missed inspection enters the market, it will greatly increase the risk of traffic. Therefore, defect detection is the most important link in tire production. The development of a software for automatic detection and recording of tire defects will greatly improve the efficiency of tire production, and will also provide convenience for the future search and statistics of common tire defects, so as to improve the tire production process, liberate manpower, and truly realize the unmanned production of industry 4.0, which is of great practical significance to the modernization of China's industrial field.

At present, the method studied in this paper can accurately locate and identify defects, but the results of the bubble location are slightly poor. In the future, we will combine a variety of direct and non-direct detection methods to conduct in-depth research on bubble defects.

Acknowledgment. This work was supported by the Key Research and Development Program of Shandong Province (NO. 2018GGX101040), the Applied Basic Research Programs of Qingdao (NO. 18-2-2-62-jch), the National Natural Science Foundation of China (NSFC) (No. 61501278), the Natural Science Foundation of Shandong Province (No. ZR2015FQ013). There are no conflicts of interest.

References

1. Hou, Z.G., Zhao, X.G., Cheng, L., Wang, Q.N., Wang, W.Q.: Recent advances in rehabilitation robots and intelligent assistance systems. J. Autom. **42**(12), 1765–1779 (2016). https://doi.org/10.16383/j.aas.2016.y000006

2. Gu, T., Li, C.J., Zhan, Q.: Advances in application of rehabilitation robots for upper limb dysfunction in patients with stroke. J. Neurol. Neurorehabil. **13**(1), 44–50 (2017)

3. Zheng, X.Z.: Take the intelligent rehabilitation training robot as an example to analyze the influence of artificial intelligence on rehabilitation sports technology. Road Health (2), 9 (2018)

4. Wu, G.H., Xiong, H.J.: Current situation and research progress of radiographic testing technology in China. Instrum. J. **37**(8), 1683–1695 (2016). https://doi.org/10.3969/j.issn. 0254-3087.2016.08.001

5. Wu, Z., Lin, J., Liu, W.: Joint inspection in X-ray #0 belt tire based on periodic texture. Multimed. Tools Appl. **78**(7), 9299–9310 (2018). https://doi.org/10.1007/s11042-018-6507-2

6. Yan, W.X.: Research on deep learning and its application on the casting defects automatic detection. South China University of Technology, Guangzhou (2016). CNKI: CDMD:2.1016.770553

7. Zhang, Y.: Research on key problems of tire defect nondestructive testing based on computer vision. Qingdao University of Science and Technology (2014)

8. Zhang, Y., Lefebvre, D., Li, Q.: Automatic detection of defects in tire radiographic images. IEEE Trans. Autom. Sci. Eng. **14**(3), 1378–1386 (2017). https://doi.org/10.1109/tase.2015. 2469594

9. Girshick, R., Donahue, J., Darrell, T., et al.: Rich feature hierarchies for accurate object detection and semantic segmentation. In: Proceedings of 2014 IEEE Conference on Computer Vision and Pattern Recognition, pp. 580–587. IEEE, Columbus (2014)

10. Girshick, R.: Fast R-CNN. In: 2015 IEEE International Conference on Computer Vision, Santiago, pp. 1440–1448 (2015)

11. Ren, S., He, K., Girshick, R., et al.: Faster R-CNN: towards real-time object detection with region proposal networks. IEEE Trans. Pattern Anal. Mach. Intell. **39**(6), 1137–1146 (2017). https://doi.org/10.1109/tpami.2016.2577031

12. Liu, Q., Wang, G., Guo, Q., Liu, Y., Zhang, C.: Tire defect detection based on radon transform. Comput. Inf. Syst. **11**(21), 7841–7848 (2015)
13. Zhang, B., Lin, S., Gao, S.: Impurity detection technology of tire X-ray image based on image processing. Rubb. Technol. Equip. **42**(9), 50–54 (2016)
14. Cui, X.H., Liu, Y., Wang, C.X.: Defect automatic detection for tire X-ray images using inverse transformation of principal component residual. In: Artificial Intelligence & Pattern Recognition, pp. 1–8 (2016). https://doi.org/10.1109/icaipr.2016.7585205
15. Alaknanda, A.R.S., Kumar, P.: Flaw detection in radiographic weldment images using morphological watershed segmentation technique. NDT & E Int. **42**(1), 2–8 (2009). https://doi.org/10.1016/j.ndteint.2008.06.005
16. Bai, F., Zhang, M.L., Zhang, X.J., et al.: A survey of local binary feature description algorithms. Electron. Measur. Instrum. **30**(2), 165–178 (2016). https://doi.org/10.13382/j.jemi.2016.02.001
17. Hinton, G.E.: Reducing the dimensionality of data with neural networks. Science **313**(5786), 504–507 (2006). https://doi.org/10.1126/science.1127647
18. Zeiler, M.D., Fergus, R.: Visualizing and understanding convolutional networks. In: Fleet, D., Pajdla, T., Schiele, B., Tuytelaars, T. (eds.) ECCV 2014. LNCS, vol. 8689, pp. 818–833. Springer, Cham (2014). https://doi.org/10.1007/978-3-319-10590-1_53
19. Abou-Ali, M.G., Khamis, M.: TIREDDX: an integrated intelligent defects diagnostic system for tire production and service. Expert Syst. Appl. **24**(3), 247–259 (2003). https://doi.org/10.1016/s0957-4174(02)00153-7
20. Ren, L.L., An, D.F., Shen, Y.D.: Causes for defects of carcass ply cord in TBR tire and their counter measures. Tire Ind. **24**(9), 559–561 (2004)
21. Li, J.N., Zhang, B.H.: Face recognition by feature matching fusion combined with improved convolutional neural network. Laser Optoelectron. Progr. **55**(10), 246–253 (2018)
22. Long, J., Shelhamer, E., Darrell, T.: Fully convolutional networks for semantic segmentation. IEEE Trans. Pattern Anal. Mach. Intell. **39**(4), 640–651 (2014). https://doi.org/10.1109/tpami.2016.2572683
23. Yang, X.M., Wu, W., Qing, L.B., et al.: Image feature extraction and matching technology. Guangxue Jingmi Gongcheng Opt. Precis. Eng. **17**(9), 2276–2282 (2009). https://doi.org/10.1360/972009-1549
24. Xu, L.M., Lu, J.D.: Bayberry image segmentation based on homomorphic filtering and K-means clustering algorithm. Trans. Chin. Soc. Agric. Eng. **31**(14), 202–208 (2015). https://doi.org/10.11975/j.issn.1002-6819.2015.14.028
25. Feng, X.Y., Mei, W., Hu, D.S.: Aerial target detection based on improved faster R-CNN. Acta Optica Sinica **38**(6), 0615004 (2018)

Real-Time Target Tracking Based on Airborne Vision

Dongxu Zhang[1] and Hongwei Gao[1,2(✉)]

[1] College of Automation and Electrical Engineering,
Shenyang Ligong University, Shenyang 110159, China
30963915@qq.com
[2] State Key Laboratory of Robotics, Shenyang Institute of Automation, Chinese
Academy of Sciences, Shenyang 110016, China

Abstract. Aiming at the problem of occlusion that leads to the failure of the tracking task during the real-time tracking of the target by the drone, the tracking-by-detection strategy is used in this article, the system design based on intelligent algorithm of computer vision has been proposed. Yolov3-tiny detection algorithm is used to achieve target recognition, this algorithm combines accuracy and speed very well, and is easy to transplant. At the same time, in order to further improve the control performance, fuzzy adaptive Proportion Integration Differentiation (PID) control algorithm is used in this system to complete the UAV tracking of the car. Verified by simulation, the accuracy of the detection algorithm reaches more than 90%, and the detection speed reaches 30FPS, the tracking effect is good, and it can dynamically adapt to external changes.

Keywords: Uav · Object tracking · Fuzzy control · Ros · Yolov3-tiny

1 Introduction

In recent years, drones have begun to appear in all aspects of human society's production and life, and have been widely used in aerial photography, surveillance, security, disaster relief and other fields. However, the actual application of drones in various early scenarios is mostly based on human remote control or intervention, and the degree of automation is not high. The degree of automation of a drone is one of the decisive factors for its greater role in the future. As the demand for drone automation continues to expand, target tracking based on computer vision has become one of the hotspots in current research, it mainly includes target detection in image processing and flight control of UAV. The drone to carry the deep learning computing platform and camera is used to further realize the automation and intelligence of the drone.

Traditional target tracking algorithms usually need to manually select the initial frame, according to the information of the current frame and the information of the previous frame, a relevant filter is trained, then, the position of the target in the initial frame is predicted in the newly input frame, and the highest confidence is the most likely tracking result. However, in the process of target tracking, it is easy for the occlusion and temporary disappearance of the target. Once the target is lost, the

J. Qian et al. (Eds.): ICRRI 2020, CCIS 1336, pp. 219–230, 2020.
https://doi.org/10.1007/978-981-33-4932-2_15

tracking task will fail. Therefore, in the field of target tracking, most researchers use a tracking-by-detection strategy. The detection-based target tracking algorithm combines the detection and tracking process, identifies the target through the detector, and transmits the target information to the tracker, that is, while the tracker tracks the target, it continuously improves tracking process through information exchange with the detector, and enhances the robustness of the entire tracking process. For example, Milan et al. [1] proposed a continuous energy function minimization method, starting from the integrity of the moving target, a whole energy function that is more suitable for the movement characteristics was proposed. Then, by optimizing the energy function, a better tracking effect was obtained, and a better effect was obtained in dealing with the occlusion situation. Song et al. [2] proposed an online multi-target tracking algorithm based on the tracking by detection strategy, which used Gaussian Mixture Probability Hypothesis Density to handle error detection, and for the tracking debris phenomenon caused by the occlusion during the tracking process, a layered tracking frame was designed to link the trajectories of the target number exchange. This method has achieved good results when dealing with occlusion and fragmentation.

In recent years, deep learning has achieved good results in the field of detection and recognition, and more and more are applied to target tracking based on detection. Considering the excellent performance of deep learning in target detection, and the effect of tracking by detection strategy in solving occlusion and other problems, the tracking by detection strategy to detect targets with deep learning was used in this article, so as to obtain more accurate detection results for target tracking. The current mainstream detection algorithms can be divided into two-stage detection algorithms and one-stage detection algorithms according to structure. Two-stage detection algorithms are represented by R-CNN [3] and Faster R-CNN [4], etc., and one-stage detection algorithms are SSD (Single Shot Multibox Detector) [5] and YOLO [6–8], etc. to represent. The main difference between the two types of algorithms is that two-stage consists of two stages of detection and classification, and one-stage integrates detection and classification into one stage. The performance indicators of the detection algorithm are mainly accuracy and speed. The two-stage algorithm has advantages in accuracy, but the processing speed is too slow. The one-stage algorithm is slightly worse in accuracy, and the detection speed reaches more than 40fps, yolov3 has improved the performance of detecting small targets, and the accuracy has also been greatly improved. And this algorithm can use GPU to accelerate, the network structure is simple, the calculation amount is small, which improves the real-time detection and is easy to transplant, so this article uses yolov3-tiny as the detection algorithm.

2   Overall System Framework

This system is based on open source flight control software px4 and robot operating system (ROS), using the open source framework GAAS to encapsulate px4 and ROS. In this simulation platform (the simulator is Gazebo), it can quickly verify the algorithm, reducing the difficulty of upper-level algorithm developers to get started. The entire system adopts a modular design, which reduces the coupling of the system. The software architecture between px4 and the simulation environment and GAAS is shown in Fig. 1:

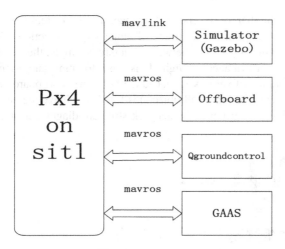

Fig. 1. Px4 on sitl

Sitl (software in the loop) simulates the operation of px4 flight control software on the computer. ROS integration with px4 follows this pattern: px4 communicates with the simulator (gazebo), receives sensor data through the simulator and sends it to the motor and actuator. You can also control px4 through the ground station and offboard mode.

ROS provides a series of operating system-level services designed for robot-related programming. With ROS, multiple independent programs running on one or more computers can easily implement interprocess communication (IPC), significantly reducing the difficulty of collaboration. In ROS, each running program is a ROS node, and all ROS nodes can establish communication with other programs through Roscore. The communication content between ROS programs is carried in the form of message. That is, the ROS program packages the data to be transferred into a message, and then transmits it. Each ROS node can publish data through a specified topic, and can also subscribe to a topic to receive corresponding data.

The communication between px4 and each module is connected by mavros. Mavros is the encapsulation of mavlink for ROS, which can encapsulate the mavlink data packets into ROS topics and messages, which are transmitted in the ROS network. ROS provides related APIs for Python and C++. Developers can simply write code to implement ROS node establishment, topic publishing and subscription.

3 Vision-Based Target Detection

3.1 Yolov3-Tiny Network Structure

Yolov3 algorithm [9–11] is a strong real-time deep learning target detection algorithm. On the basis of extracting image features through convolution, it adopts the method of residual network, and sets up a fast link connection between some layers, and adds the feature map of the upper layer of the image to the feature map of a relatively higher layer. When the number of layers is deepened, it can increase the mean average precision (MAP) value and the accuracy of small target detection. Yolov3-tiny is a lightweight version of yolov3. The network structure is simple, the calculation is small, and the real-time performance is high. It is easy to transplant to mobile devices. Through the integration of various advanced methods, the short board of the yolo series (fast but not good at detecting small objects) is filled to achieve the amazing effect and the speed of pulling out groups. The network structure diagram is shown in Fig. 2:

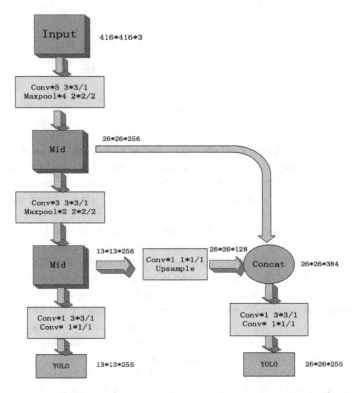

Fig. 2. Yolov3-tiny network structure

The image input of yolov3-tiny is 416 * 416 * 3. After multiple convolution and pooling operations, two detectors are finally obtained. One of the resolutions is 13 * 13, suitable for detecting larger targets; the other detector has a resolution of 26 * 26, suitable for detecting smaller targets. The detection process is to generate 3 anchors in each grid on the two feature maps of 13 * 13 and 26 * 26. The K-means clustering method is used to generate anchors on the labeled data set. The target bounding box is obtained by calculating the error between the prediction box and the groundtruth, and the probability that the target bounding box belongs to a certain category is calculated. In Fig. 3, x, y, a, and b are the offsets of the prediction box and groundtruth, respectively, and are used to regress the prediction target bounding box.

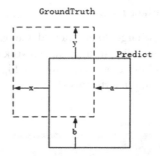

Fig. 3. Bounding-box regression

3.2 Simulation Production Data Set

The first step is to start the simulation environment. UAVs and cars appear on the ground. Then add some other models to expand the sample of the data set. The car is the target to be tracked. The drone is flying to the square of the car and using the visualization tool Rviz to subscribe to the camera information. At this time, Rosbag is used to record the video recorded by the camera, the aircraft is moved to obtain images of the car from different perspectives to expand the diversity of image data, or some obstacles are added to the accessories of the car. Then the script is used to convert the recorded video into a frame by frame image to get the image data we need, as shown in Fig. 4 below. The target to be detected is labeling in the picture with the LabelImg labeling tool to get the xml label.

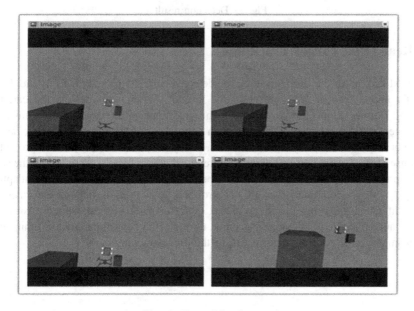

Fig. 4. Part of the data set

3.3   Detection and Recognition Result

According to the category of the target to be detected, the parameters of the yolov3-tiny network configuration file is modified to train the network model. The trained weight file is tested, and the recognition effect is shown in Fig. 5 under different angles, different heights, and surrounding interferences.

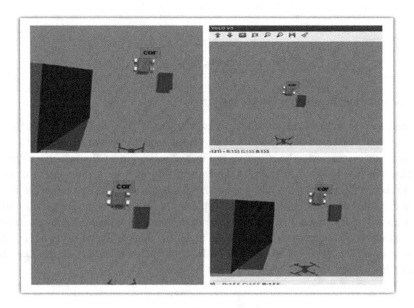

Fig. 5. Detection result

4   Real-Time Target Tracking

The next task is for the drone to track the detected car. The process is to fly the drone directly above the car by controlling the aircraft. When yolo detects the car, the tracking control algorithm is starting. The traditional PID controller requires an adjustment of its parameters to operate in an optimal for each change of variable, and it does not respond efficiently to disturbances. Fuzzy adaptive PID [12] can improve the response of the system and its parameters are tuned automatically according to the state of the process, thus enhancing the robust performance of the system. Fuzzy adaptive PID [12, 13] is a combination of PID control and fuzzy control, using fuzzy rules for fuzzy inference, querying the fuzzy matrix table for parameter adjustment, using the error E and error rate of change Ec as input, through PID control parameters tuning can better improve the control performance of the drone tracking car.

4.1 PID Control Model

The discrete-time PID controller model is shown in Eq. 1:

$$u(k) = k_p\{e(t) + \frac{1}{T_i}\sum_{j=0}^{1}Te(j) + T_d\frac{e(k) - e(k-1)}{T}\} \tag{1}$$

In the above formula, kp is the proportionality coefficient, Ti is the integration time, and Td is the derivative time. The role of kp is to increase the response speed of the system, the role of kp/Ti is to eliminate steady-state errors, and the role of $kp * Td$ is to improve the dynamic characteristics. For different control systems, PID parameter settings need to cooperate with each other to achieve the desired control performance. The error model of this system is shown in Fig. 6:

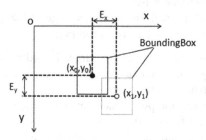

Fig. 6. Error model

Assuming that the car is detected at time t_0, the center pixel coordinate of the boundingbox is (x_0, y_0), and the center pixel coordinate of the boundingbox where the car moves to t_1 at the next time is (x_1, y_1). The errors on the pixel plane are:

$$E_x = y_1 - y_0 \quad E_y = x_1 - x_0 \tag{2}$$

The error E and error change rate Ec are taken in the system as the input of the controller, and the P, I, and D parameters Δkp, Δki, Δkd of the PID controller as the output. The fuzzy PID control structure is shown in Fig. 7:

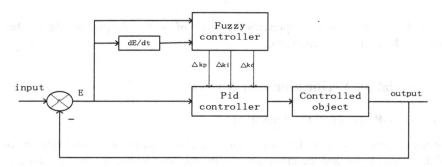

Fig. 7. Fuzzy PID control structure

4.2 Fuzzy PID System

Firstly, E and Ec are fuzzy processed, the fuzzy results are put into the knowledge base, and fuzzy reasoning is carried out according to the rules of the knowledge base. Finally, the results are clarified and defuzzified. The three parameters Δkp, Δki, Δkd of the PID controller are modified and used as the input of the PID controller.

4.2.1 Define Fuzziness and Membership

The error E and error change rate Ec and the fuzzy subsets of outputs Δkp, Δki, Δkd are taken as $\{NB, NM, NS, ZO, PS, PM, PB\}$, The domain of error E and error change rate Ec is $\{-3, 3\}$, and the quantization level is $\{-3, -2, -1, 0, 1, 2, 3\}$. In order to deal with the application of fuzzy controllers in different scenarios, by introducing quantization factors K_e K_{ec}, and scale factors K can be dealt with.

4.2.2 Building a Knowledge Base

Large, medium and small are defined as the language of fuzzy rules, and the control rules written according to experience are as follows:

Rule 1: When $|E|$ is large and $|Ec|$ is large, in order to improve the tracking performance of the system, larger kp and kd are taken, and at the same time, in order to avoid a large overshoot, ki takes 0;

Rule 2: When $|E|$ is large and $|Ec|$ is medium and small, the larger kp, medium and small kd are taken, and ki takes 0;

Rule 3: When $|E|$ is medium and $|Ec|$ is large, in order to make the system have a small overshoot, kp is medium, ki is 0, and kd is large;

Rule 4: When $|E|$ is medium and $|Ec|$ is medium and small, kp is medium, ki is 0, and kd is medium and small;

Rule 5: When $|E|$ is small and $|Ec|$ is large, kp is small, ki is large, and kd is large;

Rule 6: When $|E|$ is small, $|Ec|$ is medium and small, and kp, ki and kd are small.

4.2.3 Deblurring

Finally, the center of gravity method is used to blur, the formula is:

$$u = \frac{\sum x_i \mu_n(x_i)}{\sum \mu_n(x_i)} \tag{3}$$

In the above formula, u represents the center of gravity, i represents the number, and μ is the degree of membership.

5 Simulation Verification and Analysis

5.1 Filter-Based Tracking Algorithm

Kernel Correlation Filter (KCF) is to train a correlation filter based on the information of the current frame and the information of the previous frame, and then calculate the correlation with the newly input frame, and the confidence map obtained is the predicted tracking result, the point with the highest score (or block) is the most likely tracking result.

First start the simulation environment, run the KCF algorithm, manually select the initial rectangular box, transfer the initial position of the target to the tracking algorithm, start the tracking algorithm, and control the movement of the car, as shown in Fig. 8:

Fig. 8. KCF tracking results

It can be seen from the experimental results that when the target is occluded, the tracker loses the target, causing the mission to fail. The KCF algorithm has the following two shortcomings: First, because the KCF target frame has been set during the tracking process, the size of the target changes from beginning to end, but the target size in our tracking sequence changes from time to time. This It will cause the target frame to drift during the tracking process of the tracker, resulting in tracking failure. The second point is that KCF does not solve the problem of processing when the target is occluded during the tracking process.

5.2   Tracking Based on Target Detection

The implementation of yolo target detection under ROS is darknet-ros [14]. The weight file and configuration file of the trained yolov3-tiny network are placed in the yolo configuration folder corresponding to darnknet-ros. The ROS configuration file is modified to change the topic of the image subscribed by default to the topic published by the aircraft camera. Running yolo, when the target is detected, the control algorithm is starting to keep track of the target. The entire algorithm flow is shown in Fig. 9:

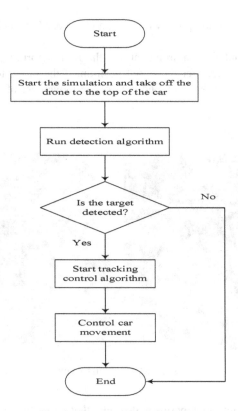

Fig. 9. Target tracking flowchart

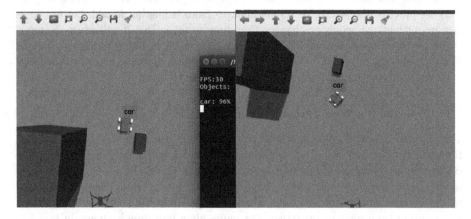

Fig. 10. Tracking result

The accuracy of the detection algorithm is shown in Table 1. At the same time, the detection rate of the algorithm reaches 30 Frames Per Second (FPS), which has good real-time performance (Fig. 10).

Table 1. Yolov3-tiny test results

Group	Object	Accuracy (%)
1	Car	96
2	Car	98
3	Car	100
4	Car	98

The tracking performance of the fuzzy adaptive PID control algorithm is shown in Table 2. In the table, x_error represents the difference between the center coordinate x of the entire pixel plane and the center coordinate x of the bounding box of the detection algorithm, and y_error represents the difference between the center coordinate y of the entire pixel plane and the center coordinate y of the bounding box of the detection algorithm. The standard to measure the tracking effect is that the smaller the x_error and y_error, the better the tracking effect of the control algorithm.

Table 2. Tracking algorithm control performance

Group	Distance between drone and car (m)	x_error (pixel)	y_error (pixel)
1	3	−13	21
2	4	−10	18
3	5	−7	10
4	6	−3	4

When the height of the plane from the ground car is different, the speed of the moving car changes on the pixel plane. Fuzzy adaptive PID control can use fuzzy sets and fuzzy inference to describe the dynamic characteristics of the system and make control decisions based on fuzzy rules. It has great potential in solving complex control problems and can dynamically adapt to changes in the outside world, further improving the tracking effect of the drone on the car.

6 Summary and Outlook

The simulation results show that the drone can complete the detection and tracking of the target autonomously through the offboard mode. A deep learning network is selected with a simple structure, a small amount of calculation, and strong real-time performance, and suitable computing platform resources as the host computer's autonomous control of the drone can better complete simple tracking tasks. For large-scale, wide-area detection and tracking problems, a single drone takes a long time. In future research, research on multi-machine collaborative sensing tasks can be conducted to rationally divide the operating area of each drone, improve work efficiency, and further achieve intelligence.

References

1. Milan, A., Roth, S.: Continuous energy minimization for multi-target tracking. IEEE Trans. Pattern Anal. Mach. Intell. **36**, 58–72 (2014). https://doi.org/10.1109/TPAMI.2013.103
2. Song, Y.-M., Jeon, M.: Online multiple object tracking with the hierarchically adopted GM-PHD filter using motion and appearance (2017). https://doi.org/10.1109/icce-asia.2016.7804800
3. Zhang, Y., Huang, Y., Wang, L.: What makes for good multiple object trackers? 467–471 (2016). https://doi.org/10.1109/icdsp.2016.7868601
4. Abbas, S., Singh, S.N.: Region-based object detection and classification using faster R-CNN, 1–6 (2018). https://doi.org/10.1109/ciact.2018.8480413
5. Wong, A., Shafiee, M.J., Li, F., Chwyl, B.: Tiny SSD: a tiny single-shot detection deep convolutional neural network for real-time embedded object detection (2018)
6. Lan, W., Dang, J., Wang, S.: Pedestrian detection based on YOLO network model, 1547–1551 (2018). https://doi.org/10.1109/icma.2018.8484698
7. Tan, L., Dong, X., Ma, Y., Yu, C.: A multiple object tracking algorithm based on YOLO detection, 1–5 (2018). https://doi.org/10.1109/cisp-bmei.2018.8633009
8. Mane, S., Mangale, S.: Moving object detection and tracking using convolutional neural networks, 1809–1813 (2018). https://doi.org/10.1109/iccons.2018.8662921
9. Redmon, J., Divvala, S., Girshick, R., Farhadi, A.: You only look once: unified, real-time object detection, 779–788 (2016). https://doi.org/10.1109/cvpr.2016.91
10. Redmon, J., Farhadi, A.: YOLO9000: better, faster, stronger (2016)
11. Redmon, J., Farhadi, A.: YOLOv3: an incremental improvement (2018)
12. Saeteros, M., Paucar, W., Molina, C., Caiza, G.: Development and analysis of a PID Controller and a fuzzy PID. In: Nummenmaa, J., Pérez-González, F., Domenech-Lega, B., Vaunat, J., Oscar Fernández-Peña, F. (eds.) CSEI 2019. AISC, vol. 1078, pp. 143–154. Springer, Cham (2020). https://doi.org/10.1007/978-3-030-33614-1_10
13. Lu, L., Zhao, H., Zhang, P.: The position control of machine tool based on fuzzy adaptive PID control, 842–847 (2019). https://doi.org/10.1109/fpm45753.2019.9035742
14. Bjelonic, M.: YOLO ROS: real-time object detection for ROS (2018). https://github.com/leggedrobotics/darknet_ros

Hyperspectral Image Classification Based on a Novel LDSLTSR Method

Yang Liu[1($\boxtimes$)] and Tiegang Bai[2]

[1] Shenyang Aerospace University, Shenyang, China
yang97_net@163.com
[2] Criminal Investigation Police, University of China, Shenyang, China

Abstract. Hyperspectral image classification is a research hotspot in remote sensing image analysis and application, and its high dimensionality and few samples make it a challenging problem. In this paper, we propose to use linear dynamic system model combined with linear transformation and sparse representation to achieve hyperspectral image classification. First, we perform image preprocessing based on the intuition that the neighboring pixel of a pixel likely share similar spectral characteristics. Second, we establish the linear dynamic system model of each sub-cube image extracted from the image, and at the same time, introduce linear transformation and sparse representation principles into the established model, and propose a novel linear dynamic system combining linear transformation and sparse representation (LDSLTSR). Finally, we use the error between the linear transformation of the model and the sparse representation to achieve classification. The spatial and spectral information of the image is fully considered in our method, which is conducive to the classification effect. The experiments are conducted on public AVIRIS data and Pavia University data to demonstrate the effectiveness of the proposed method. The results show that the proposed LDSLTSR method effectively outperforms other current state-of-the-art methods.

Keywords: Hyperspectral image classification · LDSLTSR · Linear dynamic system model

1 Introduction

With the development of remote sensing technology, the application range of hyperspectral images is getting wider and wider, which can be applied to fields such as precision agriculture, environmental management and social security [1]. In the application, hyperspectral image classification is one of the important technologies applied and has become a hot topic in research. However, the hyperspectral image has a narrow band and a large number of bands, resulting in strong correlation between pixels due to redundant information. The problem of high pixel dimension and correlation between pixels poses a huge challenge to the classification of hyperspectral images.

The original hyperspectral image classification method used only spectral information, without considering the spatial distribution of pixels. The main classification

© Springer Nature Singapore Pte Ltd. 2020
J. Qian et al. (Eds.): ICRRI 2020, CCIS 1336, pp. 231–242, 2020.
https://doi.org/10.1007/978-981-33-4932-2_16

methods include KNN method [2, 3], maximum likelihood estimation [4], artificial neural network [5], kernel based method [6] and so on, in which the support vector machine [7] demonstrates a superior performance. The hyperspectral images describe the feature information is spatially continuous, so ignoring the spatial information leads to undesirable classification results. With the in-depth study of remote sensing image classification, scholars agree that combining spectral information and spatial information can obtain better classification results. The sparse expression method is widely used in this field and has obtained some good results [8, 9]. In recent years, deep learning methods have become a hot spot in the field of pattern recognition and image analysis, and have also been applied to remote sensing image classification by many scholars and have obtained good classification results [10, 11]. However, the deep learning algorithm requires a lot of parameters, the model is complex, it is not easy to adjust the parameters, and the easier it is to overfit. In addition, deep learning requires a large number of training samples, while the number of samples for hyperspectral images is limited. Some traditional methods have simple models and less calculation, and satisfactory classification results can also be obtained.

In this paper, we apply linear dynamic system model and introduce linear transformation and sparse representation to propose LDSLTSR method to realize hyperspectral image classification. The proposed method fully describes the spatial and spectral characteristics of the image, and obtains satisfactory results. The overall flow chart of the algorithm is as follows (Fig. 1).

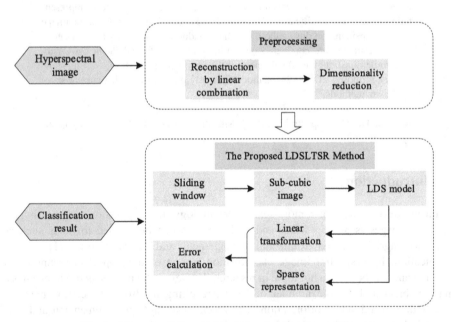

Fig. 1. The overall flow chart of the algorithm is as follows

2 Hyperspectral Image Preprocessing

For the specific task of hyperspectral image classification, not all spectral information is useful. Therefore, how to filter the spectral information useful for the classification task can not only improve the classification accuracy of the spectral image, but also simplify the calculation and reduce the storage space. Two layers of preprocessing are used in this article. The first layer of preprocessing is to consider the correlation between spatial pixels. Each pixel has a strong spatial correlation with the surrounding pixels in a certain neighborhood. Based on such correlation, each pixel can be reconstructed from a linear combination of surrounding pixels. The spectral dimension of the image remains unchanged after reconstruction and the reconstructed pixels have more distinct spatial characteristics. The purpose of the second layer of preprocessing is to reduce dimensionality. Data with hundreds of dimensions will cause the distance between classes to be small, reducing the accuracy of classification. We use the traditional PCA method for dimensionality reduction.

The ground objects have continuity in space, and each pixel in the hyperspectral image has similar spectral characteristics with the pixels in the neighborhood. Therefore, each pixel can be approximated to the vector reconstructed by linear combination of neighboring pixels. After the reconstruction is completed, the image has not only made obvious improvements in visual effects and enhanced robustness, but also the reconstructed image has more accurate structure and edge information, providing better distinguishable data for subsequent classification tasks.

$$y \approx \hat{y}, \hat{y} = \boldsymbol{D}\alpha \tag{1}$$

where $y \in \boldsymbol{R}^{m \times 1}$ is the center point pixel, $\hat{y}$ is the pixel reconstructed by $\boldsymbol{D}$, m is the number of spectral bands, $\boldsymbol{D} = [d_1, d_2, \ldots, d_n] \in \boldsymbol{R}^{m \times n}$ is a dictionary composed of neighborhood pixels, d_i is the i-th pixel in the neighborhood and α is the weight coefficient vector. The calculation process is to first get α, and then calculate the reconstructed pixel $\hat{y}$ by α. The calculation formula is as follows:

$$
\begin{aligned}
\hat{\alpha} &= \arg\min_{\alpha}\{\|y - \boldsymbol{D}\alpha\|_2^2 + \lambda\|\alpha\|_2^2\} \\
\hat{y} &= (\boldsymbol{D}^{\mathrm{T}}\boldsymbol{D} + \lambda \cdot I)^{-1}\boldsymbol{D}^{\mathrm{T}}y
\end{aligned}
\tag{2}
$$

where the regularization parameter λ in the formula is twofold, on the one hand, it can make the least square solution stable, on the other hand, a certain sparsity is imposed on the weight coefficient, and set $\lambda = 1$ in this article. In the reconstruction process, $\boldsymbol{D}$ is required to be a redundant dictionary $(m < n)$. And if the pixels with all spectral information are directly used for calculation, m is too large to meet the requirements and cannot achieve the desired effect. In our method, the spectral information is uniformly divided into subsets of hyperspectral data along the spectral dimension coordinates of the hyperspectral image, see Fig. 2.

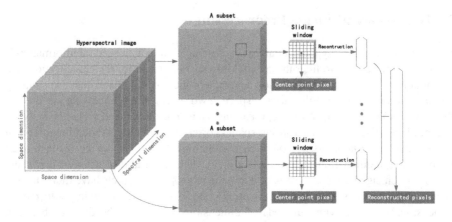

Fig. 2. The reconstruction process of a pixel

In Fig. 2, the hyperspectral images are divided into 5 subsets. Use the sliding window to slide from left to right and top to bottom on each subset. The center point pixel of the sliding window area is the pixel to be reconstructed, and other pixels in this area constitute a dictionary. The center point pixel is a linear combination of pixels in the dictionary. In this way, each pixel for each subset is reconstructed. For one pixel, 5 reconstructed subset pixels are generated. Then, these 5 reconstructed pixels are re-connected together in the original spectral order to obtain the final reconstructed pixel.

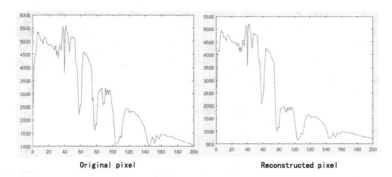

Fig. 3. The example of a pixel (center point pixel) and its reconstructed pixel

In Fig. 3, the horizontal axis represents the spectral bands, and the vertical axis represents the pixel value of each band. The center point pixel is reconstructed by linear combination of 48 pixels in the neighborhood. Since the pixels in the neighborhood have a strong correlation with the pixels at the center point, after linear combination reconstruction, the displayed information of the pixels is more refreshing, which is conducive to the task of subsequent image classification.

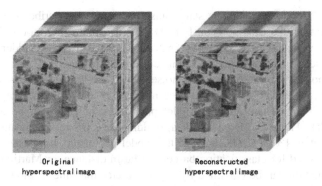

Original   Reconstructed
hyperspectral image   hyperspectral image

Fig. 4. Example of hyperspectral image and its reconstruction of AVIRIS data

The first layer processing considers both spatial and spectral information, which can be said to re-establish data with spectral spatial characteristics. As can be seen in Fig. 4, the work makes the edges of the image more discriminative. However, the dimensionality of the entire data is still very high, which is not conducive to image classification. Therefore, we use the second layer of preprocessing, namely dimensionality reduction. The PCA (Principal Components Analysis) dimensionality reduction algorithm is used to obtain spectral information with large amount of information, small correlation, and strong discriminativity between classes through mathematical transformation. One pixel of AVIRIS data is 200 dimension, we use PCA to reduce the dimension to 50 dimensions. One pixel of Pavia data is 103 dimension, we use PCA to reduce the dimension to 25 dimensions.

3   The Proposed LDSLTSR Method

As a mathematical model of dynamic texture, linear dynamic system (LDS) can effectively and accurately capture the appearance and motion information of texture, and can accurately describe the transfer characteristics of a 3D data in time and space domain [12]. We use the characteristics of LDS to describe the relationship between the spatial domain and the spectral domain of hyperspectral images to achieve the purpose of accurate classification.

3.1   LDS Model

Linear time-invariant dynamic system, referred to as linear dynamic system, as a linear Gaussian state space model, expressed by the following two equations:

$$\begin{cases} x_{t+1} = Ax_t + Bv(t) \\ y_t = Cx_t + w(t) \end{cases} \tag{3}$$

The two equations are called the state equation and the observation equation, where A, B, C are constant matrices that do not depend on the input. $y(t) \in R^m$, $x(t) \in R^n$,

$A \in R^{n \times n}$, $C \in R^{m \times n}$, $B \in R^n$. The state transition matrix A describes the behavior of the hidden state in the time domain. For 3D data, it captures the motion information. The observation matrix C describes the transition process from the hidden state to the output variable, and captures the appearance change information of the 3D data. Q is the process noise, R is the observation noise, which are the zero-mean-normally distributed random variables with covariance matrices Q and R, namely $v(t) \sim N(0, Q)$ and $w(t) \sim N(0, R)$. Therefore, the model parameters of LDS can be expressed as $\{A, B, C, Q, R\}$. Therefore, for a 3D data, building an LDS model means calculating model parameters $\{A, B, C, Q, R\}$. Once the model parameters have been determined, K-NN can be used for classification based on Martin distance, or Martin distance and the kernel function can be classified by support vector machine method.

In this study, the parameter $R = (A, B, C)$ is used as the model parameter. Given a linear dynamic system model denoted as $R = (A, B, C)$, it has an equivalent family member representation, which belongs to the knowledge of control theory. In this family, each member can be described as $P \cdot R = (P^{-1}AP, P^{-1}B, CP)$, where P is a linear non-singular transformation matrix, and $R = (A, B, C)$ and $P \cdot R = (P^{-1}AP, P^{-1}B, CP)$ are similar systems. Similar systems are characterized by different transition matrices and hidden states, but the same output variables. Similar systems in a family have the same output variables even though their model parameters are different, which means that family members can describe the same 3D data. The space composed of all systems obtained by non-singular matrix transformation is called linear transformation space.

3.2   The Proposed LDSSR Method

The problem to be solved is how to build the LDS model of one pixel of the hyperspectral image. Based on the strong correlation of the neighborhood space, we use the cubic data of a pixel neighborhood in the hyperspectral image to build the LDS model of the pixel, and combine the linear transformation theory to achieve the solution.

Using a moving sliding window to slide in the spatial domain to extract the neighborhood cubic data of each pixel, each obtained cubic data represents the center point pixel in the space. In this way we get a cube dataset of all categories of hyperspectral images. Randomly extract a certain proportion of data from each category in the cube data set to form a dictionary of sparse representation. Assuming the size of the dictionary is L, we use the method proposed in Reference [13] to obtain the LDS model parameters of each cube data, expressed as $M = \{(A_i, B_i, C_i)\}_{i=1}^{L}$, $A \in R^{n \times n}$, $C \in R^{m \times n}$, $B \in R^n$. The parameter m is the size of the observed feature vector, or the number of pixels in a frame of video image, and the parameter n is the size of the LDS model, which is equal to the size of the model parameters and the number of hidden variables. According to the sparse representation principle, each test data can be linearly represented by the data in the dictionary, see Eq. (6)

$$\begin{cases} x_{t+1} = \sum_{i=1}^{L} \alpha_i A_i x_t + \sum_{i=1}^{L} \alpha_i B_i w(t) \\ y_t = \sum_{i=1}^{L} \alpha_i C_i x_t + v(t) \end{cases} \tag{4}$$

Given a system, a model such as formula (1), a new state vector can be obtained by linear coordinate transformation as follows

$$\vec{x} = P^{-1}x \tag{5}$$

where P is the linear non-singular matrix of $n \times n$, so a new state space model is obtained as follows

$$\begin{cases} \vec{x}_{t+1} = \vec{A}\vec{x}_t + \vec{B}w(t) \\ y_t = \vec{C}\vec{x}_t + v(t) \end{cases} \tag{6}$$

The transformation relationship between the new model parameters and the original model parameters is $\vec{A} = P^{-1}AP$, $\vec{B} = P^{-1}B$, $\vec{C} = CP$. When looking at formulas (6) and (8) carefully, it is not difficult to find that they have the same form. In other words, the system reconstructed with sparse coding must be equivalent to a member of the similarity transformation space family. This relation is expressed as $\sum_{i=1}^{L} \alpha_i A \cong P^{-1}AP$, $\sum_{i=1}^{L} \alpha_i B_i \cong P^{-1}B$, $\sum_{i=1}^{L} \alpha_i C_i \cong CP$. This assumption is reasonable, that is, a linear transformation model can always be found in the linear transformation space, which is equal to or infinitely close to the sparse coding reconstruction system. The problem is formulated as follows

$$J = \min \left\| CP - \sum_{i=1}^{L} \alpha_i C_i \right\|_F^2 + \left\| P^{-1}AP - \sum_{i=1}^{L} \alpha_i A_i \right\|_F^2 + \left\| P^{-1}B - \sum_{i=1}^{L} \alpha_i B_i \right\|_F^2 \tag{7}$$

where $\{A, B, C\}$ is the model parameter of the test sample, $\{A_i, B_i, C_i\}$ is the model parameter of the system in the dictionary, and L is the size of the dictionary. In the modeling process, the coding coefficient x should be sparse, because only the items in the dictionary with the same class as the test sample should play an important role in the reconstruction and contribute a lot, while the corresponding coefficients of other types of samples should be zero or very small. Based on this assumption, formula (9) is updated to formula (10) as follows.

$$J = \lambda_C \left\| CP - \sum_{i=1}^{L} \alpha_i C_i \right\|_F^2 + \lambda_A \left\| P^{-1}AP - \sum_{i=1}^{L} \alpha_i A_i \right\|_F^2 + \lambda_B \left\| P^{-1}B - \sum_{i=1}^{L} \alpha_i B_i \right\|_F^2 + \lambda_\alpha \|\alpha\|_1 \tag{8}$$

where the parameters $\lambda_A, \lambda_B, \lambda_C, \lambda_\alpha$ are the constant coefficients of weight. However, the contribution of the data in the dictionary to the solution is different, a large coefficient means a large contribution, and vice versa. However, the contribution of the data in the dictionary to the solution is different, a large coefficient indicates a large contribution, and vice versa. Therefore, we added a coefficient weight term to the solution and proposed the LDSLTSR method as follows

$$J = \lambda_C \left\| CP - \sum_{i=1}^{L} \alpha_i C_i \right\|_F^2 + \lambda_A \left\| P^{-1}AP - \sum_{i=1}^{L} \alpha_i A_i \right\|_F^2 + \lambda_B \left\| P^{-1}B - \sum_{i=1}^{L} \alpha_i B_i \right\|_F^2 + \lambda_\alpha \|\alpha\|_1 + \|\alpha_i - \bar{\alpha}\|_2^2 \quad (9)$$

In our LDSLTSR method, there are three terms, the first term $\lambda_C \left\| CP - \sum_{i=1}^{L} \alpha_i C_i \right\|_F^2 + \lambda_A \left\| P^{-1}AP - \sum_{i=1}^{L} \alpha_i A_i \right\|_F^2 + \lambda_B \left\| P^{-1}B - \sum_{i=1}^{L} \alpha_i B_i \right\|_F^2$ represents the error between the linear representation of the dictionary and the linear space of the test data, and we hope this error is as small as possible; the second term $\lambda_\alpha \|\alpha\|_1$ indicates that the test data consists of a few basic vectors in the dictionary, and the coefficient of the base vector of the dictionary with the same label as the test data is large; the third term $\|\alpha_i - \bar{\alpha}\|_2^2$ represents the weight of the basis vector in the dictionary.

3.3   The Optimization of LDSLTSR Method

The task of optimization is to learn the parameters α^* and P^*, where $(\alpha^*, P^*) = \arg\min_{\alpha, P} J(\alpha, P)$. The optimal solution strategy in this paper is to fix one parameter and solve another parameter. It can be seen from the Eq. (11) that when the transformation matrix P is fixed, the equation is a convex optimization problem to optimize the coding coefficient α; when α is fixed, the formula is nonlinear with respect to P. The optimization process is that when these two variables are alternately fixed, the other parameter is obtained when the Eq. (11) is minimized. But the problem that needs to be raised is that the solution obtained by this kind of solution is a local optimum, and the result of optimization is closely related to the initial value of the selection.

When learning the parameter α, the linear non-singular matrix P needs to be fixed, and the optimization problem is transformed into a traditional sparse representation and solved as follows.

$$\min_{\alpha} \quad \left\{ \lambda_C \left\| CP - \sum_{i=1}^{L} \alpha_i C_i \right\|_F^2 + \lambda_A \left\| P^{-1}AP - \sum_{i=1}^{L} \alpha_i A_i \right\|_F^2 + \lambda_B \left\| P^{-1}B - \sum_{i=1}^{L} \alpha_i B_i \right\|_F^2 \right.$$
$$\left. + \lambda_\alpha \|\alpha\|_1 + \|\alpha_i - \bar{\alpha}\|_2^2 \right\}$$

$$(10)$$

Transform the above equation to get the following expression.

$$\min_{\alpha} \ \{\lambda_C \|C' - D_C\alpha\|_F^2 + \lambda_A \|A' - D_A\alpha\|_F^2 + \lambda_B \|B' - D_B\alpha\|_F^2 + +\lambda_\alpha \|\alpha\|_1 + \|\alpha_i - \bar{\alpha}\|_2^2\}$$

$$(11)$$

In the above expression, $D_c = [C_1, C_2, \cdots, C_L]$, $D_B = [B_1, B_2, \cdots, B_L]$, $D_A = [A_1, A_2, \cdots, A_L]$, $\alpha = [\alpha_1; \alpha_2; \cdots; \alpha_L]$, $C' = CP$, $A' = P^{-1}AP$, $B' = P^{-1}B$. In this paper, the feature-sign search method is used to solve the above optimization problem.

When learning the parameter P, the non-singular matrix α needs to be fixed, and the optimization problem is equivalent to the minimization problem as follows

$$\min_{P} \ \{\lambda_C \left\|CP - \sum_{i=1}^{L}\alpha_i C_i\right\|_F^2 + \lambda_A \left\|P^{-1}AP - \sum_{i=1}^{L}\alpha_i A_i\right\|_F^2 + \left\|P^{-1}B - \sum_{i=1}^{L}\alpha_i B_i\right\|_F^2\} \quad (12)$$

The space $P \in GL(n)$ formed by the linear non-singular matrix P is infinite. Considering the theoretical and computational difficulties, it is difficult to find the optimal P in the uncompressed $GL(n)$ space. The way to solve this problem is to abandon the space $GL(n)$, and replace it with the most compact subspace, that is the $n \times n$ matrix space $O(n)$, where. The optimized method applies the fast Jacobi-type algorithm proposed in [14].

3.4 The Classification of Hyperspectral Image

The parameters α^* and P^* are obtained by optimizing the proposed SS method. The representation error is used to classify the test sample, and the representation error calculation is as follows

$$e(j) = \sum_{t=1}^{F} \left\|y_t - \sum_{j=1}^{m_j}\alpha_j^* C_j Z_t^*\right\|_2^2 \quad (13)$$

where m_j is the number of samples of the j-th category in the dictionary. In this way, identify the category of the test sample: $identity(y) = \arg\min_{j}\{e_j\}$, where $Z_t^* = P^{-1} * x_t$.

4 Experiments and Analysis

To illustrate the effectiveness of the proposed method, the experiments are conducted on public AVIRIS data and Pavia University data. Some state-of-the-art classifiers are applied for comparison, including SVM, SOMP [15], NRS [16], JCR [17], KCRT-CK [18] and CNN [11]. The parameters set in the experiment are $\lambda_A = \lambda_B = \lambda_C = 1$. The number of iterations of optimization solution is 50, and In most cases, only 3 to 10 iterations are needed to get the optimal α^* and P^*. All methods use the same training data and test data. In all the experiments, we randomly chose 5% of the samples from each class as a function of training samples and the rest as test samples. Each classification experiment is repeated for 10 trials with different training and testing samples and overall classification accuracy is averaged over the 10 repeated trials.

(a) Classification analysis of AVIRIS data
AVIRIS data and Pavia data are applied to verify the validity of the algorithm. The hyperspectral image of AVIRIS dataset holds 220 bands with pixels in the 0.4- to 2.45-region of the visible and infrared spectrum with a spatial resolution of 20 m. The water-absorption bands and are removed, results in 200 spectral bands. The original Indian Pines dataset consists of 16 ground-truth land-cover classes. Some of the classes contain a small number of samples, which increased the difficulty of classification. The classification accuracy is illustrated in Table 1.

Table 1. Classification accuracy (%) for the AVIRIS on the test set (OA)

Class	SVM	SOMP	NRS	JCR	KCRTCK	CNN	LDSLTSR
Alfalfa	39.02	92.68	70.73	73.17	97.56	**100**	**100**
Corn-notill	76.73	92.22	91.44	93.23	96.58	98.36	**98.62**
Corn-min	81.53	93.44	93.71	95.98	95.18	97.8	**98.20**
Corn	70.89	91.08	69.01	72.30	100	97.20	**100**
Grass/Pasture	91.71	96.77	92.63	96.31	98.62	**99.30**	98.71
Grass/Tree	100	98.78	99.24	**100**	**100**	99.07	99.21
Grass/Pasture-mowed	88.00	**100**	52.00	36.00	92.00	**100**	**100**
Hay-windrowed	99.07	**100**	99.77	**100**	99.77	92.72	100
Oats	27.78	**100**	33.33	5.56	83.33	97.34	89.33
Soybeans-notill	83.98	91.08	90.39	93.82	**98.86**	98.23	97.89
Soybeans-min	94.70	96.70	99.55	99.68	99.46	97.66	**99.82**
Soybeans-clean	77.67	90.62	84.62	92.68	97.00	**99.32**	98.67
Wheat	99.46	98.91	97.83	98.91	99.46	99.01	**99.56**
Woods	93.23	99.30	98.77	99.38	**99.65**	98.60	99.58
Building-Grass-Trees	89.34	100	95.39	97.41	99.14	92.59	**100**
Stone-steel Towers	84.34	97.59	81.93	89.16	**98.80**	94.92	96.70
OA	88.24	95.6	94.34	96.06	98.48	98.33	**98.63**

(b) Classification analysis of Pavia data
Pavia data is acquired by the Reflective Optics System Imaging Spectrometer (ROSIS). The ROSIS sensor generates 115 spectral bands ranging from 0.43 to 0.86. The image has 103 spectral bands with pixels under the spatial resolution of 1.3 m. There are 9 classes in this image. The classification accuracy is illustrated in Table 2.

Table 2. Classification Accuracy (%) for the AVIRIS on the test set (using spectral–spatial information based on PCA and sparse coding)

Class	SVM	SOMP	NRS	JCR	KCRTCK	CNN	LDSLTSR
Asphalt	98.27	98.07	99.58	99.51	98.96	99.42	**99.50**
Meadows	99.91	99.51	99.97	99.98	99.97	99.93	**100**
Gravel	92.17	95.24	96.29	96.45	98.62	**98.69**	98.34
Trees	97.97	97.64	94.74	96.77	98.95	**99.88**	99.71
Painted metal sheets	99.42	**100**	99.92	**100**	**100**	99.97	**100**
Bare Soil	97.95	95.67	94.59	95.16	98.80	99.45	**100**
Bitumen	96.32	96.49	96.66	96.32	**99.75**	99.47	99.63
Self-Blocking Bricks	95.17	85.84	97.49	98.49	**99.03**	97.89	98.28
Shadows	97.18	97.07	99.65	99.53	97.54	99.96	**100**
OA	98.31	97.18	98.38	98.69	99.51	**99.54**	99.53

It can be seen from Table 1 and Table 2 that the LDSLTSR method proposed in this paper has achieved good classification results. The preprocessing process plays a role in removing noise, while considering the spatial correlation of pixels. The LDS model describes the texture and spectral characteristics of the image. At the same time, the introduction of linear transformation and sparse representation to optimize the model solution result in satisfactory classification results.

5 Conclusion

In this paper, we propose a linear dynamic system model combined with linear transformation and sparse representation method (LDSLTSR) for hyperspectral image classification. We use linear dynamic system model to describe the spectral and spatial-contextual information, and use linear transformation and sparse representation to explore the characteristics of the image. In contrast with some state-of-the-art methods, our method can better capture the similarity and distinctiveness among neighbouring pixels, resulting in better classification performance. The shortcoming of this paper is that there is no detailed analysis of the parameters such as the dimension of dimension reduction and the size of the pixel neighborhood. Undoubtedly these parameters will affect the classification, which is the next work plan.

Acknowledgement. This work is supported by the Liaoning Province Education Department Fund Project (No. JYT2020042).

References

1. Tarabalka, Y., Benediktsson, J.A., Chanussot, J.: Spectral-spatial classification of hyperspectral imagery based on partitional clustering techniques. IEEE Trans. Geosci. Remote Sens. **47**(8), 2973–2987 (2009)
2. Ma, L., Crawford, M.M., Tian, J.: Local manifold learning-based-nearestneighbor for hyperspectral image classification. IEEE Trans. Geosci. Remote Sens. **48**(11), 4099–4109 (2010)
3. Samaniego, L., Bárdossy, A., Schulz, K.: Supervised classification of recovery sensed image using a modified -NN technique. IEEE Trans. Geosci. Remote Sens. **46**(7), 2112–2125 (2008)
4. Zenzo, S.D., Bernstein, P., Degloria, S.D., Kolsky, H.G.: Gaussian maximum likelihood and contextual classification algorithms for multicrop classification. IEEE Trans. Geosci. Remote Sens. **25**(6), 805–814 (1987)
5. Atkinson, P.M., Tatnall, A.R.L.: Introduction neural networks in remote sensing. Int. J. Remote Sens. **18**(4), 699–709 (1997)
6. Guo, B., Gunn, S.R., Damper, R.I., Nelson, J.D.B.: Customizing kernel functions for SVM-based hyperspectral image classification. IEEE Trans. Image Process. **17**(4), 622–629 (2008)
7. Bazi, Y., Melgani, F.: Toward an optimal SVM classification system for hyperspectral remote sensing images. IEEE Trans. Geosci. Remote Sens. **44**(11), 3374–3385 (2006)
8. Xu, Y., Wu, Z., Li, J., et al.: Anomaly detection in hyperspectral images based on low-rank and sparse representation. IEEE Trans. Geosci. Remote Sens. **54**, 1990–2000 (2015)
9. Fang, L., Li, S., Kang, X., Benediktsson, J.A.: Spectral-spatial hyperspectral image classification via multiscale adaptive sparse representation. IEEE Trans. Geosci. Remote Sens. **52**(12), 7738–7749 (2014)
10. Maggiori, E., Tarabalka, Y., Charpiat, G., et al.: Convolutional neural networks for large-scale remote sensing image classification. IEEE Trans. Geosci. Remote Sens. **55**(2), 645–657 (2016)
11. Yu, C., Han, R., Song, M., et al.: A simplified 2D-3D CNN architecture for hyperspectral image classification based on spatial-spectral fusion. IEEE J. Select. Topics Appl. Earth Obser. Remote Sens. (99), 1 (2020)
12. Ghanem, B., Ahuja, N.: Sparse coding of linear dynamical systems with an application to dynamic texture recognition. In: Proceedings of the International Conference on Pattern Recognition (ICPR), pp. 987–990 (2010)
13. Soatto, S., Doretto, G., Wu, Y.N.: Dynamic textures. Int. J. Comput. Vis. **51**, 91–109 (2003). https://doi.org/10.1023/A:1021669406132
14. Jimenez, N.D., Afsari, B., Vidal, R.: Fast Jacobi-type algorithm for computing distances between linear dynamical systems. In: Control Conference IEEE (2013)
15. Chen, Y., Nasrabadi, N.M., Tran, T.D.: Hyperspectral image classification via kernel sparse representation. IEEE Trans. Geosci. Remote Sens. **51**(1), 217–231 (2013)
16. Li, W., Tramel, E.W., Prasa, S., Fowler, J.E.: Nearest regularized subspace for hyperspectral classification. IEEE Trans. Geosci. Remote Sens. **52**(1), 477–489 (2014)
17. Li, J., Zhang, H., Huang, Y., Zhang, L.: Hyperspectral image classification by nonlocal joint collaborative representation with a locally adaptive dictionary. IEEE Trans. Geosci. Remote Sens. **52**(6), 3707–3719 (2014)
18. Li, W., Du, Q., Xiong, M.: Kernel collaborative representation with Tikhonov regularization for hyperspectral image classification. IEEE Trans. Geosci. Remote Sens. Lett. **12**(1), 48–52 (2015)

Deep Convolutional Neural Network
for Remote Sensing Scene Classification

Tiegang Bai[✉] and Tao Lyu

Criminal Investigation Police, University of China, Shenyang, China
534964331@qq.com

Abstract. Remote sensing scene classification is an active issue in the field of remote sensing image analysis. However, due to the different shooting angles and scales, the characteristics of the images are changeable, which makes remote sensing scene classification challenging. Recently, convolutional neural network (CNN) has demonstrated excellent performance of classification in the field of remote sensing image analysis. In this paper, a deep learning based classification method using CNN is developed. Applying this deep learning framework for scene classification automatically integrates the traditional feature extraction and classification stages into a complete learning process without designing feature extraction and classifier respectively. The developed CNN can extract the depth characteristics of the images to reduce the impact of ground object deformation on the classification accuracy. Experiments are conducted on the well-known UCMLU and WHU19 datasets and the classification accuracy can achieve 98.10% and 96.84%. The experiment results compared with other advanced methods show the efficacy of the proposed method.

Keywords: Remote sensing scene · Classification · Convolutional neural network

1 Introduction

Remote sensing scene classification finds applications in a variety of domains including land use, urban planning, environment monitoring and has important significance for military application. However, the ground objects of remote sensing images are diverse and the viewing angle of the sensor and the size of the image are variable, which makes RSS classification a challenging problem. Many scholars have carried out research and made some progress. The advanced HOG-LBP feature combined with SVM is proposed in [1]. BOW method [2, 3] uses local and global features to improve classification performance. LBP method [4] uses multi-scale features for classification. The above methods can be classified as traditional methods. The research process of these methods is divided into two steps: feature extraction and classification. The classification results are limited to feature description and classifier performance. Recently, deep learning shows a powerful approach and widely used in Remote sensing scene classification [5–7]. It integrates the two steps of feature extraction and classification into an automatic learning process, which is an intelligent method closer to the human brain. Deep learning combined with metric learning [8, 9] has been applied to improve

© Springer Nature Singapore Pte Ltd. 2020
J. Qian et al. (Eds.): ICRRI 2020, CCIS 1336, pp. 243–254, 2020.
https://doi.org/10.1007/978-981-33-4932-2_17

the classification accuracy by forcing intra-class compactness and inter-class separability. Deep features [10–13] can be explored by deep learning methods and have obtained some progress.

In this paper, a deep convolutional neural network based on ZF-Net neural model is developed for remote sensing scene classification. The number of samples of remote sensing scene images is limited. The purpose of this paper is to apply the deep CNN learning method to achieve high-performance classification of these limited samples. The experiment results show that higher classification accuracy is achieved compared with other advanced methods.

2 Overview of the Proposed Method

The CNN network has been shown to provide better classification results than the SVM algorithm [14]. However, there are a few literatures on the application of CNN network to the classification of remote sensing scene image. In this paper, we have found that CNNs can be effectively employed to classify remote sensing scene classification.

The entire process of our proposed method is introduced in this section. The flowchart for remote sensing scene classification is shown in Fig. 1. For deep CNN networks, a large amount of training data is required to obtain ideal classification results. However, the amount of data in remote sensing scene images is not sufficient. Therefore, in the training phase, the data needs to be enhanced. The methods of enhancement are rotation and mirroring. After the training data is enhanced, its amount will increase exponentially. Data normalization and enhancement belong to the data preprocessing process. When the preprocessing of the data is completed, the deep CNN network is trained. After training the network, use the trained weights and bias parameters to classify the test data.

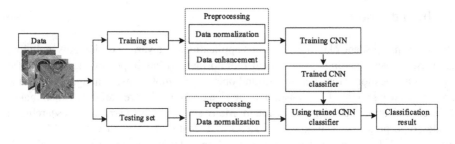

Fig. 1. Flowchart for remote sensing scene classification

3 Convolutional Neural Network Architecture

A typical CNN network usually consists of several convolutional layers, max pooling layers, and fully connected layers. Convolutional layer and max pooling layer generally appear in pairs. The network finally ends with a fully connected neural to output classification result. A CNN network with only one convolutional layer, one max pooling layer and one fully connected layer is shown in Fig. 2.

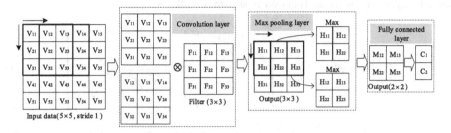

Fig. 2. A CNN with a convolutional layer, a max pooling layer and a fully connected layer

Convolution layer: For an input image, using a sliding window to move from left to right on the image, the local area of the image covered by the sliding window is performed convolution calculation with a filter. In Fig. 2, the size of input image is 5 × 5 and stride is set as 1, where the symbol $\otimes$ represents the convolution operation. And the filter size of the convolutional layer is 3 × 3, and we obtain the output of convolution layer, where $H_{ij} = \sum\limits_{i=1,j=1}^{i=3,j=3} F_{ij} \times V_{ij}$ (Take the sliding window moving to the first local area as an example) with size 3 × 3. The stride defines the step length of each step of the sliding window. The larger the stride, the smaller the output data size of the convolutional layer, which can result in a smaller network calculation, but the more information may be lost in the image. The convolution operation can describe the texture features of the image as much as possible, while helping to reduce the overall number of network training parameters and reduce computational cost and can help to obtain more effective training and more effective models. Usually, the convolutional layer is followed by a max pooling layer.

Max pooling layer: The pooling layer also has the effect of reducing the image size. In Fig. 2, the size of the output layer of the convolutional layer is reduced from 3 × 3 to 2 × 2 after passing through the max pooling layer. In fact, the max pooling layer operation is also a convolution filtering operation, and a filter is also required. At this time, the value of each position in the filter is 1, and the convolution operation is expressed as $M_{ij} = \max\{ H_{ij} \times 1\}_{i,j=1}^{i,j=2}$ (Take the sliding window moving to the first local area as an example). In other words, maximum pooling is to extract the maximum value in each sliding window field to replace the whole of sliding window field. Another important function of the pooling operation is to keep the features of the image invariant to the location. Pooling operation can also be mean pooling expressed as $M_{ij} = \text{mean}\{H_{ij} \times 1\}_{i,j=1}^{i,j=2}$. Maximum pooling is more widely used in CNNs than mean pooling. Therefore, it can be seen that the pooling operation reduces the amount of calculation of the higher layer, while providing a form that is invariant to the transformation.

Fully connected layer: A CNN ends with the fully connected layer for classification. Each output of the fully connected layer is connected to each neuron in the previous layer. In order to achieve classification, it is necessary to add softmax layer to the last layer. The softmax function is shown as follows.

$$P(y^{(i)} = n|x^{(i)}; F) = \begin{bmatrix} y^{(i)} = 1|x^{(i)}; F \\ y^{(i)} = 2|x^{(i)}; F \\ \vdots \\ y^{(i)} = N|x^{(i)}; F \end{bmatrix} = \frac{1}{\sum\limits_{j=1}^{n} e^{F_j^T x^{(i)}}} \begin{bmatrix} e^{F_1^T x^{(i)}} \\ e^{F_2^T x^{(i)}} \\ \vdots \\ e^{F_n^T x^{(i)}} \end{bmatrix} \tag{1}$$

In Eq. (1), $P(y^{(i)} = n|x^{(i)}; F)$ means the probabilistic for ith training data out of n number of training data, the jth class out of number of classes for $i = 1, 2, \ldots, m$, where F presents filter and $F_n^T x^{(i)}$ are inputs of the software layer.

4 Deep CNN for Remote Sensing Scene Classification

In many researches on image recognition, the number of samples of training images is limited, and the training process of large-scale convolutional neural networks can easily lead to overfitting. Therefore, most of CNN used in image recognition have lower layers. The number of remote sensing scene images is also very limited, so we choose to use fewer layers of CNN for image recognition.

4.1 Neural Network Architecture

The CNN architecture adopted in this paper is illustrated in Fig. 3. A ZF-Net neural model [15] consisting of five convolution layers, three max pooling layers, two fully connected layers and one software layer is developed to realize classification task. The following describes the network structure in detail.

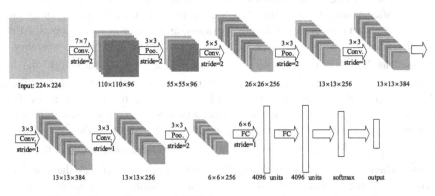

Fig. 3. The architecture of CNN for remote sensing scene classification

In this architecture, we consider gray images as inputs to the CNN model and all images are cropped to images with size of 224 × 224. Therefore, the size of input layer is 224 × 224. The filter size of the first convolutional layer is 7 × 7, and the number is 96, using a stride of 2. Since the input data size of the first few layers is relatively large, the number of filters should not be too large, and the input data size of the following base layer is relatively small, so you can use a larger number of filters to extract the feature map, such as the third time in the network structure In the convolution operation, the number of filters reaches 384. In this way, the output of the first convolutional layer is activated through a rectified linear function $ReLU(x) = max(x, 0)$ and contains 110 × 110 × 96 nodes. The calculation formula of the spatial size of output data is $(size\_of\_input - size\_of\_filter)/stride + 1$. There are 4800 parameters that need to be trained in this layer. The first convolution layer is followed by the first max pooling layer with size 3 × 3 and stride 2. The max pooling layer is the second hidden layer of the entire network, and the output has 55 × 55 × 96 nodes. The function of this max pooling layer is to reduce the amount of data, thereby reducing the amount of calculation, and to make the data invariant to location. There are no parameters to be trained in this layer. The output of the first pooling layer is used as the input of the second convolutional layer. The size of the filter of the second convolutional layer is 5 × 5 with stride 2. There are 6656 parameters that need to be trained in this layer and the output size is 26 × 26 × 256. Since the second max pooling layer size is 3 × 3 with stride 2, its output contains 13 × 13 × 256 nodes. The second max pooling layer is followed by three consecutive convolutional layers, the size of the filter is all 3 × 3, and the stride is all 1. The spatial size of the input and output data of these three layers has not changed, the number of filters are 384, 384, 256 and the number of parameters that need to be trained are 3840, 3840, 2560 respectively. The output of the fifth convolution layer has 13 × 13 × 256 nodes. Next, there is a max pooling layer, namely the third max pooling layer whose output data size is 6 × 6 × 256. The last two layers are fully connected layers. Each output node of the fully connected layer is connected to each neuron of the input data. Therefore, there are $(6 \times 6 \times 256 + 1) \times 4096$ parameters that need to be trained in the first fully connected layer, and 4097×4096 parameters that need to be trained in the second fully connected layer.

4.2 Training Process

The ZF-Net learning algorithm is developed for the training and classification. The training process is an iterative process, including forward propagation and backward propagation. The forward propagation channel starts from the input data and performs operations such as convolution, pooling and full connection to obtain classification results and network parameters. It can be said that it is a process from input to output. The back propagation channel is to calculate the error between the result and the expected result after the result obtained in the forward direction to update the network parameters, see Fig. 4. The arrows indicate the sequential direction of operations and processes. Any operation and process can only be carried out in the direction of the arrow, and cannot be carried out against the direction of the arrow. The symbol ⊖ represents the subtraction operation between the data input into the symbol.

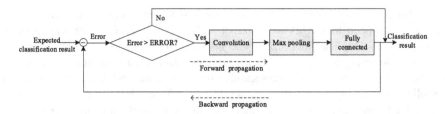

Fig. 4. Block diagram of remote sensing scene classification

The forward propagation channel is the path from the input of the block diagram to the output. It can output classification results through calculations of convolution, max pooling and fully connected layers. The backward propagation channel is the path from the output of the block diagram back to the input. It is to calculate the error between the classification result obtained by forward propagation and the expected classification result to update the network parameters. By continuously iterating the forward propagation and back propagation processes, the error is continuously reduced, and finally it is less than the set value or the number of iterations reaches the maximum, and the iteration process is stopped, and the network parameters are saved for the classification of the test data. As explained in Fig. 4, the forward propagation channel and the back propagation channel form a closed loop, based on the idea of negative feedback, which makes the error between the currently obtained classification results and the expected classification results smaller and smaller, when the error is small enough to complete the training process.

In the process of forward propagation, the output of the previous layer L_{i-1} is the input of the current layer L_i, denoted as x_i. Then the output of the current layer, x_{i+1} is calculated as follows.

$$x_{i+1} = f_i(u_i), \quad u_i = W_i^T x_i + b_i \tag{2}$$

where W_i is the weight matrix, b_i is the bias vector and $f_i(\cdot)$ is the activation function of the applied for L_i layer. A ReLU activation function is used in this architecture. Since remote sensing scene classification is a multi-classification problem, a softmax operation is required to be defined as follows.

$$y = \frac{1}{\sum\limits_{k=1}^{C} e^{W_{L,K}^T x_L + b_{L,K}}} \begin{bmatrix} W_{L,1}^T x_L + b_{L,1} \\ W_{L,2}^T x_L + b_{L,2} \\ \vdots \\ W_{L,C}^T x_L + b_{L,C} \end{bmatrix} \tag{3}$$

C is the number of output categories, L is the number of CNN layers.

In the process of forward propagation, the main task is to update the trainable parameters, including the weight W_i and the bias b_i, which are updated as follows.

$$W_i(updated) = W_i + \Delta W_i, \quad b_i(updated) = b_i + \Delta b_i \tag{4}$$

The gradient descent method is used to update the parameters. The specific method is to calculate the loss function as follows.

$$J(\theta) = -\frac{1}{m} \sum_{i=1}^{m} \sum_{j=1}^{C} 1\left\{ j = Y^{(i)} \right\} \log(y_j^{(i)}) \tag{5}$$

where m is the number of training samples. The variable Y is the desired output. The variable $y_j^{(i)}$ represents the jth entry of the actual output $y^{(i)}$ of ith training sample. The variable θ_i contains W_i and b_i. Then, θ is updated based on the following formula.

$$\theta = \theta - \eta \cdot \nabla_\theta J(\theta) \tag{6}$$

where η is the learning rate, and

$$\nabla_\theta J(\theta) = \left\{ \frac{\partial J}{\partial \theta_1}, \frac{\partial J}{\partial \theta_2}, \cdots, \frac{\partial J}{\partial \theta_L} \right\} \tag{7}$$

where

$$\begin{aligned}
\frac{\partial J}{\partial \theta_i} &= \left\{ \frac{\partial J}{\partial W_i}, \frac{\partial J}{\partial b_i} \right\} \\
\frac{\partial J}{\partial W_i} &= \frac{\partial J}{\partial u_i} \circ \frac{\partial u_i}{\partial W_i} = \frac{\partial J}{\partial u_i} \circ x_i \\
\frac{\partial J}{\partial b_i} &= \frac{\partial J}{\partial u_i} \circ \frac{\partial u_i}{\partial b_i} = \frac{\partial J}{\partial u_i}
\end{aligned} \tag{8}$$

where the symbol $\circ$ indicates that the elements are multiplied correspondingly. The flowchart of training the CNN model is shown in Fig. 5.

In the training process, as the number of iterations increases, the error between the output and the expected output becomes smaller and smaller. During each iteration, the weight and bias parameters are updated according to the direction of the gradient of the loss function. When the error is small enough, the iterative process stops, and the obtained CNN parameters are saved as following classification parameters.

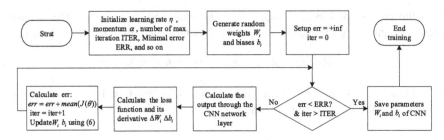

Fig. 5. Flow chart of CNN training process

4.3 Classification

After the training process is over, all the parameters of the CNN network are ready, and the classification can be performed. The working process of classification is just like the process of forward propagation stage as shown in Fig. 6, which is an open-loop process. The test data is sent to the CNN network, which is processed by the convolutional layer, the pooling layer and the fully connected layer to output the classification results. The weights of the filters and the biases in the convolution all come from the parameters completed in the training process.

Fig. 6. Flow chart of CNN classification

5 Experiments and Analysis

The experiment is carried out on the UCMLU data and WHU19 data to evaluate the developed deep CNN for remote sensing scene classification.

5.1 The Data Sets

The UCMLU data set is a 21-level land use image database for research purposes. Each of the following categories has 100 images as shown in Fig. 7. The pixel resolution of the public domain images of this database is 1 foot, the pixel size of most images is 256×256, and there are 2100 scene images in 21 categories. Each image is a 3-dimensional RGB image.

Fig. 7. Example images of UCMLU data

The WHU19 data set is collected from Google Earth, which provides high-resolution satellite images up to 0.5 m. It is a brand new public data set, composed of 950 pictures with a size of 600 × 600 pixels, evenly distributed in 19 scenes, as shown in Fig. 8. Each image is a 3-dimensional RGB image.

Fig. 8. Example images of WHU19 data

5.2   Experimental Results and Analysis

The whole experiment process is divided into 4 steps, including data separation, pre-processing, training, and classification.

Data separation: 80% of the randomly selected data in each category is used as training data, and the rest as test data.

Preprocessing: First, in UCMLU data, some image pixel sizes are not uniform, so normalized to 256 × 256. When the size of original image is larger than 256 × 256, it is cropped; when the size of original image is smaller than 256 × 256, the edge pixels are supplemented by edge interpolation algorithm. Next, in order to obtain better classification results, the training data is enhanced. A window with a size of 224 × 224 is used to randomly select sub-image as training data. Thus, five images with a size of 224 × 224 pixels are extracted from one image. And for WHU19, all images are compressed to a pixel size of 256 × 256 and then performed the same operation as UCMLU data. Each sub-image is rotated 90° gradually and their mirror image, see Fig. 9. The image in the red box represents the image used for training data extracted from the original image, the three images side by side are rotated images, and the second row of images represents the mirror image of the first row of images. In this way, an original image is enhanced to become 40 images. Therefore, for UCMLU we have 67200 samples for training, and for WHU19 we have 30400 samples for training.

Fig. 9. Example images of an enhanced image from UCMLU data (Color figure online)

Training: In the experiment, the training parameters include the learning rate, momentum, bias and the number of iterations. The settings of these parameters are shown in Table 1. All the weights and biases are initialized to be a random value between −1 to 1.

Table 1. Settings of the training parameters.

Training parameters	Values
Learning rate	0.001
Momentum	0.9
Weight	−1 to 1
Bias	−1 to 1
Number of iterations	2000

Classification: To evaluate the effectiveness of the CNN model, the results of comparison with other advanced methods are reported. 20% of the samples in the two data sets are used to test the CNN. The classification accuracy is shown in Table 2.

Table 2. Classification accuracy compared with different methods.

Method	UCMLU data	WHU19 data
MS-CLBP [4]	90.60%	93.4%
MS-CLBP-FV [4]	93.00%	94.32%
CBRCN [7]	94.53%	—
SICNN [6]	96.00%	—
CaffeNet+FV [10]	95.71%	93.68%
TL-MFF [13]	96.67%	95.47%
Deep CNN	98.10%	96.84%

From Table 2, we can see that our deep CNN achieves the highest classification accuracy, which demonstrates the effectiveness of the developed method. There are two main factors that restrict classification accuracy. One factor is the number of training samples, which is limited in the number of training samples in remote sensing scene classification problems; the other factor is the number of CNN layers, and deeper CNN can achieve better classification results accompanied by the demand for more training samples.

6   Conclusion

A CNN-based approach for remote sensing scene classification is developed using a deep learning method. Compared with MS-CLBP, MS-CLBP-FV, CBRCN, SICNN, CaffeNet+FV, TL-MFF, the developed method could achieve competitive accuracy using public UCMLU data and WHU19 data. In this study, the architecture of our CNN, which is constructed with ZF-Net neural model to perform classification task, contains five convolutional layers, three pooling layers and two fully connected layers. A total of 97600 images are used for training, and 610images are used for testing. In fact, CNN-based deep learning method needs a large number of labeled training samples to achieve higher classification accuracy. In the future, under the constraints of a limited number of samples, the research direction is to improve the robustness and efficiency of the method through parameter setting and network structure adjustment.

References

1. Konstantinidis, D., Stathaki, T., Argyriou, V., et al.: Building detection using enhanced HOG–LBP features and region refinement processes. IEEE Sel. Top. Appl. Earth Observ. Remote Sensing **10**(3), 888–905 (2017)
2. Zhu, Q., Zhong, Y., Zhao, B., et al.: Bag-of-visual-words scene classifier with local and global features for high spatial resolution remote sensing imagery. IEEE Geoence Remote Sensing Lett. **13**(6), 747–751 (2017)
3. Cheng, G., Li, Z., Yao, X., et al.: Remote sensing image scene classification using bag of convolutional features. IEEE Geoence Remote Sensing Lett. **14**(10), 1735–1739 (2017)

4. Longhui, H., Chen, C., et al.: Remote sensing image scene classification using multi-scale completed local binary patterns and fisher vectors. Remote Sensing **8**(6), 483 (2016)
5. Cheng, G., Xie, X., Han, J., et al.: Remote sensing image scene classification meets deep learning: challenges, methods, benchmarks, and opportunities. IEEE J.Sel. Top. Appl. Earth Observ. Remote Sensing (99), 1 (2020)
6. Liu, Y., Zhong, Y., Zhao, J., et al.: Scene semantic classification based on scale invariance convolutional neural networks. In: 2017 IEEE International Geoscience and Remote Sensing Symposium (2017)
7. Zhang, F., Du, B., Zhang, L.: Scene classification via a gradient boosting random convolutional network framework. IEEE Trans. Geoence Remote Sensing **54**(3), 1793–1802 (2016)
8. Cheng, G., Yang, C., Yao, X., et al.: When deep learning meets metric learning: remote sensing image scene classification via learning discriminative CNNs. IEEE Trans. Geoence Remote Sensing **56**(5), 2811–2821 (2018)
9. Lihua, Y.E., Lei, W., Wenwen, Z., et al.: Deep metric learning method for high resolution remote sensing image scene classification. Acta Geodaetica Cartogr. Sin. **48**(6), 698 (2019)
10. Yi, L., Ziqi, Z., Li, Y., et al.: Combined fisher Kernel coding framework with convolutional neural network for remote sensing scene classification. Remote Sensing Inf. (2018)
11. Zhou, Y., Xiao-Dong, M., Feng-An, Z.: Scene classification of remote sensing image based on deep network and multi-scale features fusion. Optik **171**, 287–293 (2018)
12. Liu, Q., Hang, R., Song, H., et al.: Learning multiscale deep features for high-resolution satellite image scene classification. IEEE Trans. Geosci. Remote Sensing **56**(1), 117–126 (2018)
13. Jun, Z., Peng, X., Min, Z., et al.: Land use classification algorithm based on multi-scale feature fusion. Comput. Eng. Des. **4**(41), 1099–1104 (2020)
14. Sutskever, I., Hinton, G.: Deep, narrow sigmoid belief networks are universal approximators. Neural Comput. **11**(20), 2629–2636 (2008)
15. Zeiler, M.D., Fergus, R.: Visualizing and understanding convolutional networks. In: Proceedings of the Computer vision–European Conference on Computer Vision, Amsterdam, The Netherlands, 8–16 October, pp. 818–833. IEEE, New York (2014)

Optimization Method in Monitoring

Localization Method of Multiple Leaks in Fluid Pipeline Based on Compressed Sensing and Time-Frequency Analysis

Xianming Lang[1,2(✉)], Jiangtao Cao[1], and Ping Li[1]

[1] School of Information and Control Engineering, Liaoning Shihua University,
Fushun 113001, Liaoning, China
lnpulxm@163.com

[2] Shenyang Academy of Instrumentation Science CO., LTD., Shenyang 110043,
Liaoning, China

Abstract. Considering the problems that large number of acoustic sampling data and the multiple leaks localization errors are largre, multiple leaks localization method based on compressed sensing and time-frequency analysis is presented. In this scheme, the wavelet analysis is denosied acoustic signal collected at the ends of pipeline, and then CS is used to reconstruct the denoised signal accurately to reduce the amount of collected acoustic signal. Then, the cross-correlation function of multiple leaks is analyzed with the smooth Affine-Wigner distribution. The multiple peak time is extracted simultaneously, and the multiple peak time is the time delay. The multiple leaks positions can be estimated by the multiple time delay and the distance between upstream acoustic sensor and downstream acoustic sensor. Field experiment results that the proposed method can accurately locate the multiple leaks.

Keywords: Multiple leaks · Compressed sensing · Time-frequency analysis · Fluid pipeline

1 Introduction

Fluid pipeline leak is a common phenomenon in fluid transportation [1, 2]. However, the existing leak detection and localization methods are generally aimed at a certain leak position, when the pressure of the local section of the fluid pipeline is too high or affected by corrosion, wear, vibration and so on, sometimes there are two or more leak points in the fluid pipeline. The traditional detection and localization methods for one leak localization often fails [3–6]. Therefore, it is important significance that studied multiple leaks localization method of fluid pipeline to locate leaks.

It is a common used method of pipeline leakage detection to eliminate noise and obtain effective leakage signal by signal processing method. Wavelet analysis (WA) can effectively denoise leakage acoustic signal. Zhao et al. [7] used wavelet analysis to denoise the collected acoustic signal, so as to realize the accurate locate the leak. Additionally, compressed sensing (CS) can simultaneously samples and compresses the acoustic signal [8, 9], overcomes the shortcomings of Nyquist sampling theorem. Moreover, it can reconstructs the original signal with high precision by a

J. Qian et al. (Eds.): ICRRI 2020, CCIS 1336, pp. 257–267, 2020.
https://doi.org/10.1007/978-981-33-4932-2_18

small number of signal sampling points. Thus, CS saves the storage space and computing resources of sampling data. Moreover, it can reduce the signal sampling rate. Most leak detection and localization systems use high-speed sampling for leak localization. Because of the large amount of processing and analysis of the collected acoustic signal, they increase the burden of computer calculation and storage.

Most of the studies of pipeline leak localization were studied by the assumption of a single leak. Time delay estimation is a common localization method by calculating time delay estimation of the leakage acoustic signal arriving at ends of the pipeline. Moreover, the peak value of the cross-correlation function is obtained, and localization is calculated by the time delay estimation. Li et al. [10] designed a leakage positioning method based on the pipeline model by CS algorithm. Xiao et al. [11] proposed a estimate sound velocity method by the dispersion effect of the leakage signal, and the leak localization according to the pipeline length and the cross-correlation time delay. It is difficult to distinguish the peak information of the cross-correlation function when the multiple leaks wave propagates upstream and downstream from each leak. Therefore, the leak localization method based on single leak is not applicable to multiple leaks. However, time-frequency analysis (TFA) can analyze the non-stationary leak acoustic signals from the time-domain and frequency-domain, so as to locate multiple leaks by time-frequency analysis.

In this paper, a method of multiple leak localization by CS and TFA is proposed. Firstly, multiple leaks acoustic signal is denoised by WA and reconstructed by CS. Secondly, the multiple leaks time delay of cross-correlation function calculation with the smooth Affine-Wigner distribution are obtained. Finally, multiple leaks localization is calculated by CS and TFA.

2 Multiple Leaks Localization Method

2.1 Multiple Leak Localization Analysis

Suppose the length of the pipeline is L and a leak occurs at position x from the upstream of the pipeline. Then, the time delay of transmission of the leakage acoustic signal to the ends of pipeline is as follows:

$$\Delta t = (L - 2x)/a \tag{1}$$

The leak position x is calculated as follows:

$$x = (L - a\Delta t)/2 \tag{2}$$

When multiple leakage occurs, the time delay of multiple leaks is estimated by the cross-correlation function of multiple leak acoustic signal. The time delay estimation generated by the single leak is Δt. Because multiple leaks will produce multiple peaks, the corresponding time delay can locate leaks with Eq. (2). Due to the interaction between multiple leak signals, the superimposed signals from multiple leaks are obtained at the ends of pipeline.

$$x_a(t) = \sum_{i=1}^{n} s_i(t) + n_1(t) \tag{3}$$

$$x_b(t) = \sum_{i=1}^{n} s_i(t - \tau_i) + n_2(t) \tag{4}$$

where x_a and x_b are superimposed upstream and downstream signals, respectively. n_1 and n_2 are noise of the outside pipeline, and τ_i is the delay time between each leakage acoustic signal, and the cross-correlation function is obtained as follows:

$$\begin{aligned} R_{x_a x_b}(\tau) &= E\{x_a(t)x_b(t - \tau)\} \\ &= \sum_{i=1}^{n} R_{s_i s_i}(t - \tau_i) + \sum_{i=j=1,i \neq j}^{n} R_{s_i s_j}(t - \tau_j) \\ &+ \sum_{i=1}^{n} R_{s_i n_2}(t - \tau_i) + \sum_{j=1}^{n} R_{s_j n_1}(t - \tau) + R_{n_1 n_2}(\tau) \end{aligned} \tag{5}$$

Suppose $n_1(t)$ and $n_2(t)$ are random noises and independent of each other.

$$R_{x_a x_b}(\tau) = \sum_{i=1}^{n} R_{s_i s_i}(t - \tau_i) + \sum_{i=j=1,i \neq j}^{n} R_{s_i s_j}(t - \tau_j) \tag{6}$$

where $R_{x_a x_b}(\tau)$ is the sum of cross-correlation function of each leak acoustic signal, and $\sum_{i=j=1,i \neq j}^{n} R_{s_i s_j}(t - \tau_j)$ is the cross-correlation function of the acoustic signal at each leak.

According to Eq. (6), when multiple leaks occurs, the correlation function is the sum of cross-correlation function of each leak acoustic signal, and there is multiple peaks of correlation function. Thus, it divided into two cases:

(1) There is single leak in pipeline, then $R_{x_a x_b}(\tau) = R_{s_i s_i}(t - \tau_1)$, that is, the cross-correlation function can obtain the peak value at τ_1. According to Eq. (3), we can calculate the leak localization with $\Delta t = \tau_1$.

(2) there are multiple leaks, and two leakage signals are orthogonal to each other. Thus, $R_{s_i s_j}(t - \tau_j)$ is equal to 0 for i, j. Therefore, Eq. (6) is the sum of the auto-correlation functions of each leak. There are multiple values with the cross-correlation function in $\tau = \tau_i$ $(i = 1, 2, \ldots, n)$. If the leak detection and localization system increases the sampling frequency and has sufficient resolution, multiple leaks can be located by the cross-correlation method. However, a large number of data is collected to increase the burden of computer storage and analysis.

2.2   Wavelet Analysis and Compressed Sensing

Wavelet analysis decomposed [12, 13] the leakage acoustic signal into components with different frequency bands and time by wavelet transformation. The noise can eliminate by the wavelet coefficient of the threshold value. Then, the leakage acoustic signal is reconstructed to achieve the purpose of denoising.

CS is a signal processing method that the leakage acoustic signal is sampled and compressed at the same time, and the original signal is reconstructed with less leak acoustic signal sampling. Because the signal sampling rate is reduced, the signal transmission cost is reduced.

The process of compressed sensing algorithm is as follows:

The measurement matrix is $\Phi \in R^{M \times N}$ ($M \leq N$), the measured sparse signal is $X \in R^N$, and the field measured signal value $Y \in R^M$, the compressed sensing model of the signal is as follows:

$$Y = \Phi X \tag{7}$$

According to the definition of compression ratio (Cr), it can be written as follows:

$$Cr = \frac{S_a - S_b}{S_a} \times 100\% \tag{8}$$

where Cr is compression ratio of the multiple leaks signal, S_a is the sample data length of the original multiple leaks signal, and S_b is the estimated sample data length after the measurement matrix observation.

The larger Cr, the higher the compression rate of multiple leaks. Therefore, according to the design of different Cr, multiple leaks can be compressed and collected.

2.3   Time-Frequency Analysis

Cross-correlation function of multiple leaks acoustic signal of pipeline is analyzed by the smooth Affine-Wigner distribution.

$$S_{x_a x_b}(\tau, \omega) = ASPW(R_{x_a x_b}(\tau)) \tag{9}$$

where $S_{x_a x_b}(\tau, \omega)$ is the time-frequency distribution of cross-correlation function multiple leaks acoustic signal.

The smooth Affine-Wigner distribution is described as follows:

$$ASPW(t, \omega) = \frac{1}{\omega} \int_{-\infty}^{+\infty} \int_{-\infty}^{+\infty} h\left(\frac{\tau}{\omega}\right) g\left(\frac{u - \tau}{\omega}\right) x_a\left(u + \frac{\tau}{2}\right) x_b\left(u - \frac{\tau}{2}\right) du d\tau \tag{10}$$

where $g(t)$ and $h(t)$ are smooth functions in time domain and frequency domain of multiple leaks acoustic signal, respectively.

The peak time of the time-frequency distribution $S_{x_a x_b}(\tau, \omega)$ of the cross-correlation function from multiple leaks signals corresponds to the delay time between multiple leaks signal.

$$\left[\omega_{1,...,n}, \hat{t}_{1,...,n}\right] = \arg \max S_{x_a x_b}(\tau, \omega) \tag{11}$$

where $\omega_{1,...,n}$ is the peak frequency corresponding to the peak frequency distribution peak of the multiple leaks acoustic signal, and $\hat{t}_{1,...,n}$ is the time delay of the multiple leaks acoustic signal.

2.4   Multiple Leak Localization Method

According to the characteristics of cross-correlation function of multiple leaks acoustic signal, multiple leaks localization equation is calculated as follows:

$$\hat{l}_{1,...,n} = \frac{L + \omega_{1,...,n} t_{1,...,n}}{2} \tag{12}$$

where $\hat{l}_{1,...,n}$ is the calculated localization of multiple leaks, and L is the length of the pipeline.

The multiple leaks acoustic signal at the ends of the pipeline are collected. The schematic diagram of localization method of multiple leaks in pipeline based on compressed sensing and time-frequency analysis is shown in Fig. 1.

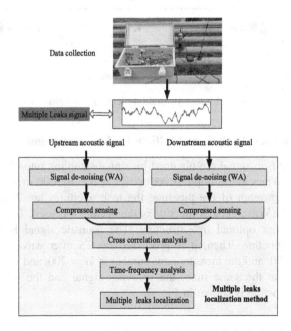

Fig. 1. The schematic diagram of localization method of multiple leaks in pipeline based on CS and TFA

3 Experiment of Multiple Leaks Localization of Pipeline

3.1 Experimental System of Multiple Leaks Localization

The length of pipeline is 2000 m, and the pipeline inner diameter is 50 mm. The upstream pressure of pipeline is 0.8 MPa, and the flow rate is 120 L/min. The acoustic signals of multiple leaks at the ends of the pipeline are obtained by acoustic sensor (PCB 106B). The multiple leaks experimental system of loop pipeline is shown in Fig. 2. The loop pipeline experimental system consists of a loop pipeline and multiple leaks localization system. In the loop pipeline system, centrifugal pumps are used to extract the fluid from the storage tank and pressurize the fluid, and the fluid returns to the storage tank through the loop pipeline. The multiple leaks localization system mainly consists of PCB 106B, industrial controlling computer and data collector (NI cDAQ-9181). According to the actual pipeline fluid transportation process, the sampling frequency is set to 1000 Hz. Multiple leaks position simulate multiple leaks. Moreover, multiple leaks acoustic signal is obtained by PCB 106B at the ends of pipeline, and collected by NI 9253. Then, the collected acoustic signal data is stored in the industrial controlling computer for multiple leaks localization.

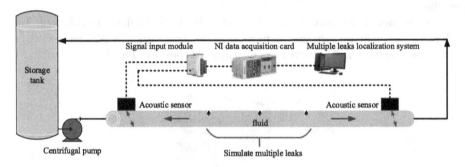

Fig. 2. The experimental setup for multiple leaks localization in loop pipeline

3.2 Time-Frequency Analysis of Multiple Leaks Localization

Multiple leaks acoustic signal are obtained by signal sampling rate of 1000 Hz at the ends of pipeline. When the multiple leaks occurs at 500 m, 600 m and 800 m at the same time from upstream of the pipeline, the leak apertures are 2 mm, 3 mm and 4 mm, respectively. Wavelet basis function of daubechies 4 is used. The acoustic signal is reconstructed with optimal tree structure. The acoustic signal is obtained from upstream of the pipeline. Then, it is processed by CS after wavelet analysis. The sparsity is set to 100, and the measurement number is $M = 200$, and $Cr = 80\%$. When three leaks occur at the same time, the recovery signal and the original signal of upstream of the pipeline is shown in Fig. 3.

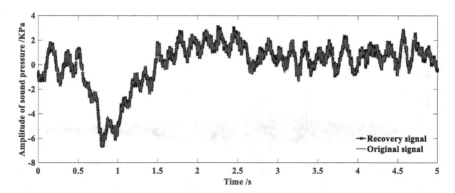

Fig. 3. The recovery signal and the original signal of upstream of the pipeline

It can be seen from Fig. 3, the time-domain signal is a relatively obvious signal inflection point. If the obtained acoustic signal upstream and downstream is calculated according to the cross-correlation function, there will be a relatively obvious peak value, so it is impossible to obtain the multiple leaks delay time and locate the leaks.

The original acoustic signal is estimated by CS, so the number of signal sampling points is increased by calculating the sampling rate of 200 Hz with CS. Moreover, the number of time-domain sampling points are increased by CS, and CS reduce sample of signal acquisition. The error between the original signal and CS recovery signal is shown in Fig. 4.

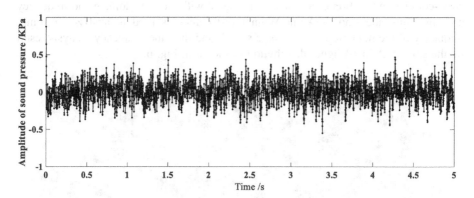

Fig. 4. The error between the recovery signal and the collected acoustic signal

It can be seen from Fig. 4, the error between the recovery signal and the original signal is ±0.2, which can better recover the original multiple leak acoustic signal. Then, the cross-correlation function is used to calculate the multiple leak recovery signal upstream and downstream.

The time delay estimation of the three leaks of cross-correlation function is shown in Fig. 5.

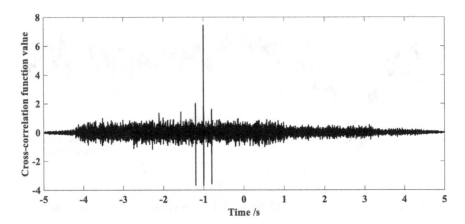

Fig. 5. Time delay estimation with the cross-correlation in 3 leaks

It can be seen from Fig. 5, because the multiple leaks sound waves are mixed together, the amplitude attenuation characteristics of multiple leaks sound wave are more complex when it propagates from each leak to the upstream and downstream. When calculating time delay estimation, there is a relatively obvious peak, which is difficult to distinguish the peak value of the cross-correlation function of each leak signal, so it is unable to obtain the time delay estimation of multiple leaks.

Because the cross-correlation analysis can not obtain accurate delay time of multiple leaks, the time-frequency analysis method is used to analyze the peak value of cross-correlation function of multiple leaks signal with the time domain and frequency domain. Thus, the smooth Affine-Wigner distribution is used to analyze the time-frequency of the multiple leaks acoustic signal, and the time-frequency analysis result by the smooth Affine-Wigner distribution is shown in Fig. 6.

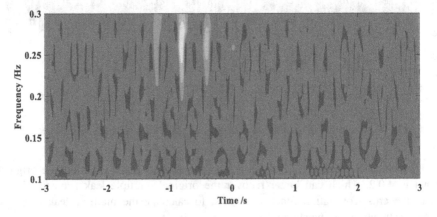

Fig. 6. Time-frequency analysis of cross-correlation of three leaks signals

It can be seen from Fig. 6, the spectrum of multiple leakage signal has three obvious spectrum peaks in the time range of 0–3 s in the frequency range of 0.1–0.3 Hz, and the time delay corresponding to the spectrum peak is estimated, so it can be calculated multiple leaks delay time.

Time-frequency distribution of three dimensional is used to highlight the cross-correlation function time spectrum of multiple leaks, as shown in Fig. 7.

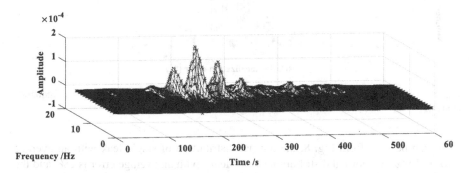

Fig. 7. Cross-correlation in time-frequency analysis in 3 dimensional display

It can be seen from Fig. 7, the peak value of the three leaks signal is the time delay of the corresponding leaks. Three leaks localization can be calculated by Eq. (12).

The multiple leaks acoustic data are collected in the loop pipeline, and 25 groups of experimental data are carried out for single leak, two leaks and three leaks, respectively. Then, the collected multiple leaks acoustic data are processed according to CS and TFA. Some of the experimental data and the results of multiple leaks localization by the proposed method is shown in Table 1.

Table 1. Localization results of time-frequency of correlation function with multiple leaks

Leak points	Leak position (m)	Localization (m)	Relative error (%)
1	800	797	0.15
2	600	804	0.2
	800	780	1
3	600	620	1
	800	777	1.15
	1000	960	2

In the experiment of single leak, two leaks and three leaks, the average value of each leak localization error is calculated, and the statistical distribution diagram of relative error of time-frequency location of multiple leaks is established, it is shown in Fig. 8.

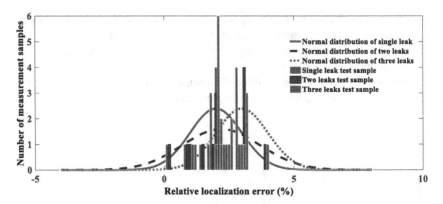

Fig. 8. The statistical distribution of multiple leaks localization errors by the time-frequency analysis

It can be seen from Fig. 8, the normal distribution of single leak with an average error is 1.9%, the normal distribution of two leaks with an average error is 2%, and the normal distribution of three leaks with an average error is 2.2%. Because the time-frequency analysis can highlight the time delay of multiple leaks more clearly, and then the localization of multiple leaks can be calculated. Thus, the proposed method can locate multiple leaks from the above experimental results.

4   Conclusion

A method of multiple leaks localization based on compression sensing and time-frequency analysis is proposed. The multiple leaks acoustic signal is obtained at the ends of the pipeline, and it is de-noised by wavelet analysis, and then the signal is reconstructed by compression sensing after de-noising. Cross-correlation function is applied to calculate time delay estimation of multiple leaks acoustic signal with the smooth Affine-Wigner distribution. In the multiple leaks experiment of loop pipeline, the minimum relative error of single leak is 0.15%, the minimum relative error of two leaks is 0.2%, and the minimum relative error of three leaks is 1%. Field experiments show that the proposed method based on CS and TFA can effectively estimate the localization of multiple leaks.

Acknowledgment. This work was supported in part by the National Natural Science Foundation of China under Grant 61673199, the Natural Science Foundation of Liaoning Province under Grant 2019-BS-158, the Scientific Research Funds of Liaoning Provincial Department of Education under Grant L2020017, in part by Funded by China Postdoctoral Science Foundation under Grant 2020M670796, and the Supported by Talent Scientific Research Fund of LSHU (No. 2019XJJL-008) of Liaoning Shihua University.

References

1. Verde, C., Torres, L.: Modeling and Monitoring of Pipelines and Networks. Springer, Cham (2017). https://doi.org/10.1007/978-3-319-55944-5
2. Wang, X., Ghidaoui, M.S.: Identification of multiple leaks in pipeline II: iterative beamforming and leak number estimation. Mech. Syst. Signal Process. **119**, 346–362 (2019)
3. Dina, Z., Manoj, K.T., Gupta, A.K., Sen, D.: A review of leakage detection strategies for pressurised pipeline in steady-state. Eng. Fail. Anal. **109**, 104264–104282 (2020)
4. Lang, X., Li, P., Guo, Y., Cao, J., Lu, S.: A multiple leaks localization method in a pipeline based on change in the sound velocity. IEEE Trans. Instrument. Meas. **69**(7), 5010–5017 (2020)
5. Kim, S.H.: Multiple leak detection algorithm for pipe network. Mech. Syst. Signal Process. **139**, 1–18 (2020)
6. Lang, X., et al.: Localization of multiple leaks in a fluid pipeline based on ultrasound velocity and improved GWO. Process Saf. Environ. Protect. **137**, 1–7 (2020)
7. Zhao, Y., Zhuang, X., Min, S.: A new method of leak location for the natural gas pipeline based on wavelet analysis. Energy **35**(9), 3814–3820 (2020)
8. Zhang, P., Wang, J., Li, W.: A learning based joint compressive sensing for wireless sensing networks. Comput. Netw. **168**, 1–10 (2020)
9. Park, C., Lee, B.: Online compressive covariance sensing. Signal Process. **162**, 1–9 (2019)
10. Li, J., Wang, C., Zheng, Q., Qian, Z.: Leakage localization for long distance pipeline based on compressive sensing. IEEE Sensors J. **19**(16), 6795–6801 (2019)
11. Xiao, Q., Li, J., Sun, J., Feng, H., Jin, S.: Natural-gas pipeline leak location using variational mode decomposition analysis and cross-time-frequency spectrum. Measurement **124**, 163–172 (2018)
12. Kalra, M., Kumar, S., Das, B.: Seismic signal analysis using empirical wavelet transform for moving ground target detection and classification. IEEE Sensors J. **20**(14), 7886–7895 (2020)
13. Li, F., Wu, B., Liu, N., Hu, Y., Wu, H.: Seismic time–frequency analysis via adaptive mode separation-based wavelet transform. IEEE Geosci. Remote Sensing Lett. **17**(4), 696–700 (2020)

Research on the Reconstruction Method
of Furnace Flame Temperature Field
with Small Singular Value Based
on Regularization

Zhe Kan[1]([⊠]), Zhen Sun[2], Jingyu Han[3], and Qin Lv[1]

[1] Liaoning Shihua University, Fushun 113001, China
kanzhe_kz@163.com
[2] Sinopec Baling Branch, Yueyang 414000, China
[3] State Grid Shandong Electric Power Dong'a Power Supply Company,
Liaocheng 252200, China

Abstract. In view of the ill-posed problems in the reconstruction of the temperature field in the furnace, on the basis of based on the Markov radial basis function and the Tikhonov regularization algorithm (MTR), propose a based on the Markov radial basis function and the correction Tikhonov regularization algorithm (MCTR). The algorithm performs singular value decomposition on the coefficient matrix of the temperature field measurement system, through the proportion of the sum of standard deviations of small singular values to standard deviations, determine the cutoff value of small singular values, select eigenvectors corresponding to small singular values to construct a new regularization matrix. Compared with the MTR algorithm, the unit matrix is used as the indifference correction of the regularization matrix, MCTR algorithm selectively corrects only small singular values, avoid the introduction of deviation when correcting larger singular values, get more reliable parameter solutions. The simulation experiment proves that the MCTR algorithm has higher reconstruction accuracy and stronger anti-noise ability in the reconstruction of acoustic temperature field.

Keywords: Singular value · Regularization matrix · Temperature field

1 Introduction

The distribution of the temperature field in a large boiler can quickly reflect the combustion situation in the furnace, in the furnace directly affects the ignition and burnout of pulverized coal and the safety of the boiler, the measurement of temperature field is extremely important for boiler combustion control and diagnosis [1]. The detection technology of the temperature field in the acoustic furnace is to use the sound wave flight time to reverse the temperature distribution of the area to be measured, it belongs to the inverse problem research of inferring the cause from the result. Compared with the traditional contact temperature measurement, the acoustic temperature measurement has the advantages of non-contact, large measurement object range and visualization [2, 3], and has attracted much attention.

© Springer Nature Singapore Pte Ltd. 2020
J. Qian et al. (Eds.): ICRRI 2020, CCIS 1336, pp. 268–279, 2020.
https://doi.org/10.1007/978-981-33-4932-2_19

The typical acoustic two-dimensional temperature field reconstruction algorithms include least squares method [4–6], Fourier regularization method [7], Gaussian function and regularization method [8], algebraic reconstruction method [9, 10], RBF neural network method [11], Landweber iteration method [12–15], genetic method [16], etc. These algorithms all require that the number of grid divisions of the area to be measured is less than or equal to the number of effective acoustic wave paths, which limits the accuracy of temperature field reconstruction due to too few original temperature sampling points. Literature [17] proposed Tikhonov regularization method, literature [18] proposed a method based on radial basis function and singular value decomposition, the number of grid divisions is not limited by the number of effective acoustic paths, and more original temperature sampling points can be obtained, which improves the reconstruction accuracy of the temperature field.

Acoustic temperature field reconstruction is an ill-posed inverse problem, and very sensitive to noise changes during flight time measurement, practical temperature field reconstruction algorithms need to have a certain degree of anti-noise ability to be better applicable to complex industrial environments.

The Markov radial basis function uses the method of function approximation to simplify the complicated actual situation into easy-to-understand mathematical relations and establish a positive problem model. The Tikhonov regularization improves the ill-conditionedness of the matrix by introducing regularization parameters [26] and regularization matrices [19–23]. The regularization matrix is a correction to the ill-conditioned matrix, this paper uses the eigenvectors corresponding to small singular values to construct a new regularization matrix to replace the standard Tikhonov identity matrix, the differential correction of different singular values, that is, it avoids the introduction of bias when correcting larger singular values, and can effectively reduce the variance of the parameter estimation, thereby obtaining a more stable parameter solution. In the reconstruction process of the temperature field, higher reconstruction accuracy and stronger anti-noise ability can be obtained.

2 Acoustic Method Temperature Field Reconstruction Principle

The principle of acoustic temperature measurement is to obtain the medium temperature indirectly according to the propagation speed of sound wave in the medium, the two have the following relationship [24–26]:

$$c = \sqrt{\frac{rRT}{M}} = Z\sqrt{T} \tag{1}$$

c : the propagation speed of sound wave in the medium, m/s
r: the ratio of specific heat at constant pressure to specific heat at constant volume for gas medium;
R: the ideal gas constant, worth 8.314 $J/mol \cdot K$;

T: absolute temperature of a gas, K;

M: as a molar mass of gas, kg/mol;

Z: as a constant determined by the gas composition, the flue gas is 19.08;

Acoustic method temperature field reconstruction needs to arrange several acoustic wave transceivers around a certain typical plane, and the acoustic wave emitted by any transceiver can be received by all other transceivers. After measuring the flight time on the sound wave path, the sound velocity distribution of the area to be measured can be reconstructed by using an appropriate reconstruction algorithm on the premise that the position of the transceiver is known, and then formula (1) to find the temperature distribution of the surface to be measured. Figure 1 is a schematic diagram of 8 acoustic wave transceivers around the square plane and the 24 effective acoustic wave flight paths generated by them.

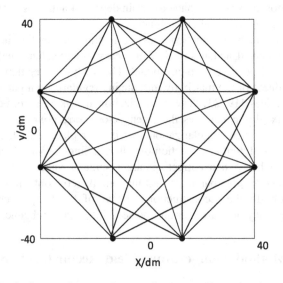

Fig. 1. 8 acoustic transceivers and corresponding acoustic paths

3   Image Reconstruction Linear Model

The acoustic method can reconstruct the temperature field on a typical level in two steps: the first step is to establish a positive problem model, that is, to establish an analytical formula for the propagation time of sound waves along the multipath and the sound velocity distribution in the area to be measured. The second step is to solve the inverse problem on the established forward problem model, that is, the reconstruction of the temperature field.

Suppose the reciprocal of the sound velocity distribution on the plane to be measured is $f(x, y)$, then the flight time of the sound wave along any path is:

$$g_k = \int_{l_k} f(x, y)ds; k = 1, \cdots m \tag{2}$$

l_k—the length of the k-th acoustic path;
ds—the integral derivative of the k-th acoustic path;
m—number of effective acoustic paths.

The area to be reconstructed is divided into n small blocks of uniform size. At this time, the distribution $f(x, y)$ of the reciprocal of sound velocity is expressed as a linear combination of n radial basis functions centered on the fast geometric center of the divided cell, that is:

$$f(x, y) = \sum_{i=1}^{n} \varepsilon_i \varphi_i(x, y) \tag{3}$$

Suppose the coordinates of the center point of the i-th cell block is (x_i, y_i), which is also the center of the radial basis function. In this paper, the Markov radial basis function is used, and the expression is as follows:

$$\varphi_i(x, y) = \exp\left(-\theta\sqrt{(x - x_i)^2 + (y - y_i)^2}\right) \tag{4}$$

Among them, ε_i to describe the undecided parameters of the reciprocal distribution of sound speed, θ is the shape parameter of radial basis function.

Combine (2), (3), and (4) and obtain:

$$g_k = \sum_{i=1}^{n} a_{ki}\varepsilon_i \tag{5}$$

a_{ki} in the formula:

$$a_{ki} = \int_{l_k} \varphi_i(x, y)ds \tag{6}$$

For the parameters to be determined ε_i, following are defined:

$$\begin{cases} A = (a_{ki})_{k=1,\cdots m, i=1,\cdots n} \\ g = (g_1, \cdots, g_m)^T \\ \varepsilon = (\varepsilon_1, \cdots, \varepsilon_n)^T \end{cases} \tag{7}$$

Then, the linear model of the positive problem is as follows:

$$g = A\varepsilon \tag{8}$$

4 Solving the Inverse Problem and Constructing a New Regularization Matrix

4.1 Regularization Solution of Acoustic CT Inverse Problem

The inverse problem of CT, that is, image reconstruction, is based on the measured sound wave flight time to reverse the temperature distribution image of the layer to be measured. Singular value decomposition of matrix A can be obtained:

$$A_{m \times n} = U_{m \times m} S_{m \times n} G_{n \times n}^T = \sum_{i=1}^{n} \alpha_i u_i v_i^T \tag{9}$$

Formula: α_i for the singular value of the Matrix A, $\alpha_1 \geq \alpha_2 \geq \cdots \alpha_n$, u_i and v_i is the Matrix left and right singular value vector. If the equation is ill conditioned, α_1 is much larger than α_n, and α_n is a smaller value close to zero.

The least square solution of formula (8) can be expressed as:

$$\varepsilon_{LS} = \sum_{i=1}^{n} \frac{u_i^T g}{\alpha_i} v_i \tag{10}$$

Since the inverse problem is ill-posed, and matrix A is an ill-conditioned matrix, there are many singular values close to zero. In the presence of noise, the noise will be amplified and accurate estimation of the parameters will not be obtained. And Tikhonov regularization [27–34] is an effective method to solve ill-posed problems. Tikhonov regularization is to use prior knowledge (such as the norm is bounded, the image is smooth, etc.) to convert ill-conditioned problems into well-conditioned problems. CT image reconstruction is to find the following optimal function.

$$\arg \min_{\varepsilon} = \left(\|A\varepsilon - g\|^2 + \lambda \|\varepsilon\|^2 \right) \tag{11}$$

Formula: λ is the regularization parameter, its role is to balance the weight of the measured data and the solution norm.

Under normal circumstances, the temperature measurement in the furnace takes the stability functional as the parameter's second norm constraint, that is, the unit matrix is used as the regularization matrix. Then the adjustment result of Eq. (11) is:

$$\varepsilon_t = \left(A^T A + \lambda I \right)^{-1} A^T g \tag{12}$$

Eigenvalue decomposition of the square matrix $A^T A$ can be obtained:

$$A^T A = G \Lambda G^T = \sum_{i=1}^{n} G_i \Lambda_i G_i^T \tag{13}$$

Where the eigen value is $\Lambda_i = \alpha_i^2$, that is, the eigenvalue of the normal equation coefficient matrix is the square of the singular value of the observation equation coefficient matrix, and the eigenvalue Λ_i is the right singular value vector of α_i.

Then the standard regularization approximate solution can be obtained:

$$\varepsilon_t = \sum_{i=1}^{n} \frac{\alpha_i^2}{\alpha_i^2 + \lambda} \frac{u_i^T g}{\alpha_i} v_i \tag{14}$$

It can be seen from formula (14) that in the standard Tikhonov that uses the unit matrix as the regularization matrix, the regularization parameter corrects each singular value, and the degree of correction is the regularization parameter λ. The model is ill-conditioned. The condition number of the observation matrix is large. The maximum singular value differs from the minimum singular value by several orders of magnitude. The regularization parameter is a smaller value much smaller than the maximum singular value. Therefore, the correction of the larger singular value is not only can not effectively reduce the variance of parameter estimation, but introduce more bias. Selectively correct only small singular values, which can reduce the variance of parameter estimates and reduce the introduction of bias [35].

4.2 Construct a New Regularization Matrix

The influence of matrix ill-conditionedness on parameter estimation is mainly reflected in the amplification of small singular value variance. It can be seen that the influence of eigenvalues on variance can be embodied as the influence of singular values on standard deviation. The smaller the singular value, the larger the proportion of its standard deviation component in the entire set. Regarding the selection criteria for small singular values, Lin Dongfang of Central South University gave the following formula [23]:

$$\sum_{i=k}^{n} \frac{\sigma_0}{\alpha_i} \geq 95\% \sum_{i=1}^{n} \frac{\sigma_0}{\alpha_i} \tag{15}$$

In the equation: σ_0 is standard deviation.

In the singular value matrix S, $\alpha_1 \geq \alpha_2 \geq \cdots \geq \alpha_k \geq \cdots \geq \alpha_n$, to determine the boundary value of the small singular value, a new regularization matrix is constructed using the small singular value and its corresponding eigenvector as follows:

$$R = \sum_{i=k}^{n} G_i G_i^T \tag{16}$$

Then CT image reconstruction is to find the following optimal function:

$$\arg\min_{\varepsilon} = \left(\|A\varepsilon - g\|^2 + \lambda\|R\varepsilon\|^2 \right) \tag{17}$$

Taking the derivative of Eq. (17) and making the derivative equal to zero, the optimized solution is:

$$\varepsilon_s = \left(A^T A + \lambda \sum_{i=k}^{n} G_i G_i^T \right)^{-1} A^T g \tag{18}$$

Then the small singular value regularization approximate solution corresponding to the regularization parameter is:

$$\varepsilon_s = \sum_{i=1}^{n} \frac{\alpha_i^2}{\alpha_i^2 + \lambda R} \frac{u_i^T g}{\alpha_i} v_i \tag{19}$$

Correcting small singular values can improve the system's ability to suppress noise while retaining the edge information of the image. Compared with the standard Tikhonov's indifference correction, it reduces the introduction of deviation when correcting larger singular values.

5   Temperature Field Reconstruction Simulation Analysis

In order to verify the effectiveness of the algorithm, a typical single peak symmetrical temperature field is simulated. The quality of temperature field reconstruction is evaluated by root mean square error, maximum error, maximum relative error, and average error. The root mean square error is defined as:

$$E = \frac{\sqrt{\frac{1}{n}\sum_{i=1}^{n} \left[T(i) - \hat{T}(i) \right]^2}}{T_{mean}} \times 100\%; i = 1, \cdots, n \tag{20}$$

In the formula: $T(i)$ is the model temperature of the i-th pixel; $\hat{T}(i)$ is reconstruction temperature of the i-th pixel; n is the number of pixels divided; T_{mean} is the average temperature of the model temperature field.

As shown in Fig. 2, (a) is the model temperature field, (b) is the least square method to reconstruct the temperature field (LSM), (c) is based on the Markov radial basis function and Tikhonov reconstruction (MTR), and (d) is based on the Markov radial basis function and the correction Tikhonov reconstruction (MCTR).

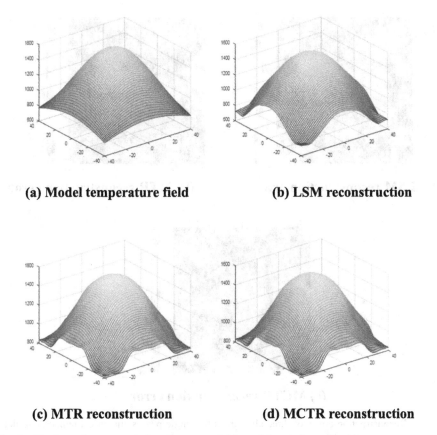

(a) Model temperature field   **(b) LSM reconstruction**

(c) MTR reconstruction   **(d) MCTR reconstruction**

Fig. 2. Single peak symmetrical temperature field model and three algorithms to reconstruct temperature field

As shown in Fig. 3, (a) is the reconstruction error of the least squares method, (b) is the Tikhonov reconstruction error based on the Markov radial basis function, and (c) is the correction Tikhonov reconstruction error based on the Markov radial basis function.

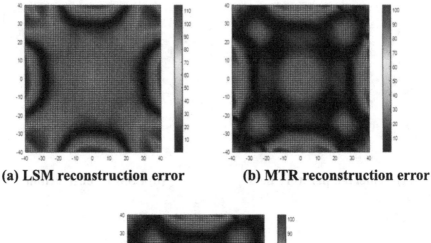

(a) LSM reconstruction error   (b) MTR reconstruction error

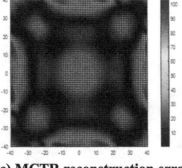

(c) MCTR reconstruction error

Fig. 3. Reconstruction errors of three algorithms for single peak symmetrical temperature field

The reconstruction errors of the three algorithms of single-peak symmetric temperature field in the absence of noise are shown in Table 1. It can be seen from Fig. 2 that the three reconstruction algorithms can reflect the temperature distribution of the model temperature field to a certain extent, and all have certain applicability. It can be seen from Fig. 3 that the two regularization algorithms, because the number of grid divisions is not limited by the number of effective acoustic paths, the reconstruction of the four corners and edge temperatures of the image is more reasonable and closer to the model temperature.

It can be seen from Table 1 that the reconstruction accuracy of the two regularized reconstruction algorithms is much higher than that of the least square method without noise, and the reconstruction evaluation indexes of MCTR are better than that of MTR algorithm, which has higher reconstruction accuracy and can reconstruct the temperature distribution closer to the model.

Table 1. Error table of three reconstruction algorithms

Temperature model	Reconstruction algorithm	Root mean square error /%	Maximum relative error /%	Maximum error / K	Mean error / K
Single peak	LSM	4.53	13.32	113.14	69.27
symmetrical	MTR	2.72	3.87	55.80	12.54
	MCTR	2.71	3.86	55.45	12.16

Acoustic temperature field reconstruction is a ill conditioned inverse problem, which is very sensitive to the change of noise in time of flight measurement. Practical temperature field reconstruction algorithm needs to have a certain degree of anti noise ability. In this paper, we add Gaussian white noise with mean value of 0 and standard deviation of 1e−4, 5e−4 and 1e−3 to flight time. The reconstruction errors of the three reconstruction algorithms at different noise levels are shown in Table 2.

Table 2. Reconstruction error under different noise levels

Noise standard deviation	Reconstruction algorithm	Root mean square error /%	Maximum relative error /%	Maximum error / K	Mean error / K
	LSM	4.68	13.90	118.11	72.89
1E-4	MTR	2.72	4.23	57.49	15.84
	MCTR	2.71	4.18	57.13	15.45
	LSM	5.35	16.20	137.61	86.76
5E-4	MTR	2.87	6.49	64.18	28.81
	MCTR	2.86	6.44	63.83	28.41
	LSM	6.25	18.99	161.19	103.65
1E-3	MTR	3.37	9.23	77.59	44.59
	MCTR	3.36	9.17	77.13	44.19

It can be seen from Table 2 that as the noise level increases, the reconstruction error of the model field gradually increases, but the error change rate of the MCTR and MTR reconstruction algorithms is much smaller than that of the LSM reconstruction algorithm, indicating that these two algorithms are at the same noise level The anti-noise ability of the MCTR algorithm is better than that of the LSM algorithm. At the same time, the reconstruction error of the MCTR algorithm is smaller than that of the MTR algorithm at any noise level, indicating that the MCTR algorithm has stronger anti-noise ability and is more suitable for reconstructing the temperature field in a complex industrial environment.

The quantitative analysis of experimental data shows that the MCTR algorithm based on small singular values to construct a regularization matrix has higher reconstruction accuracy while taking into account a certain degree of anti-noise ability.

6 Conclusion

The MCTR algorithm proposed in this paper based on small singular values to construct a regularization matrix, the simulation analysis results show that it has a good reconstruction effect under the condition of less acoustic data, and has a stronger anti-noise ability.

However, in the MCTR algorithm, the reconstruction of the edge temperature of the area outside the center point of the outermost division grid of the reconstruction area is mainly obtained by obtaining the temperature of the sensor point and using an interpolation algorithm, which causes a certain reconstruction error. In the next research, we will find a more suitable way to obtain the edge temperature, and then improve the overall image reconstruction accuracy.

Acknowledgments. This work was supported by the General Project of the Liaoning Province Education Department (L2020019), China.

References

1. Li, Y., Liu, S.: Acoustic three-dimensional temperature field reconstruction with dynamic evolution information. J. Electron. Meas. Instrument. **31**(11), 1711–1718 (2017)
2. Zhiqiang, S., Tiehua, L., Haitao, L.I.: Application of furnace temperature field detection using acoustic pyrometer. Guangdong Electr. Power **17**(6), 10–14 (2004)
3. Tong, L., Xin, J., Liping, P.: Developing measurement of the temperature field of the furnace. Mod. Electr. Power **19**(4), 14–20 (2002)
4. Fumio, I., Masayasu, S.: Fundamental studies of acoustic measurement and reconstruction combustion temperature in large boilers. Trans. Jpn. Soc. Mech. Eng. **53**(489), 1610–1614 (1985)
5. Mao, J., Wu, Y.F., Fan, W., et al.: Acoustic temperature field measurement in deep sea hydrothermal vents and re—construction algorithm. Chin. J. Sci. Instrument **31**(10), 2339–2344 (2010)
6. Wei, F., Chen, Y., Pan, H.C., et al.: Experimental study on under water acoustic imaging of 2 -D temperature distribution around hot springs on floor of Lake Qiezishan. Exp. Thermal Fluid Sci. **34**(8), 1334–1345 (2010)
7. Bramanti, M., Emanuele, A., Salerno, A., et al.: An acoustic pyrometer system for tomographic thermal imaging in power plant boilers. IEEE Trans. Instrument. Meas. **45**(1), 159–167 (1996)
8. Tian, F., Sun, X.P., Shao, F.Q., et al.: A study of complex temperature field reconstruction algorithm based on combination of Gaussian functions with regularization method. Proc. CSEE **24**(5), 212–220 (2004)
9. Shen, G.Q., An, L.S., Jiang, S.H., et al.: Simulation of two-dimensional temperature field in furnace based on acoustic computer tomography. Proc. CSEE **27**(2), 11–14 (2007)
10. An, L.S., Wang, R., Shen, G.Q., et al.: Simulation study on reconstruction of 3D temperature field in boiler furnace by acoustic CT algorithm. J. Chin. Soc. Power Eng. **35**(1), 13–18 (2015)
11. Tian, F., Liu, Z.S.H., Sun, X.P., et al.: Temperature field reconstruction algorithm based on RBF neural network. Chin. J. Sci. Instrument **27**(11), 460–1464 (2006)
12. Zhilan, L., Hua, Y., Guannan, C.: Acoustic temperature field reconstruction algorithm based on modified Land weber iteration. J. Shenyang Univ. Technol. **30**(1), 90–93 (2008)

13. Jang, J.D., Lee, S.H., Kim, K.Y., et al.: Modified iterative Landweber method in electrical capacitance tomography. Meas. Sci. Technol. **17**, 1909–1917 (2006)
14. Landweber, L.: An iterative formula for Fredholm integral equations of the first kind. Am. J. Math. **73**(3), 615–624 (1951)
15. Tautenhahn, U.: On the asymptotical regularization method for nonlinear ill-posed problem. Inverse Prob. **10**(6), 1405–1418 (1994)
16. Kan, Z.H., Meng, G.Y., Wang, X.L., et al.: Research of boiler temperature field reconstruction algorithm based on genetic algorithm. J. Electron. Meas. Instrument. **28** (10), 1149–1154 (2014)
17. Wang, R., An, L.S., Shen, G.Q., et al.: Three-dimensional temperature field reconstruction with acoustics based on regularized SVD algorithm. Chin. J. Comput. Phys. **32**(2), 195–201 (2015)
18. Yan, H., Cui, K.X., Xu, Y.: Temperature field reconstruction based on a few sound travel-time data. Chin. J. Sci. Instrument **31**(2), 470–475 (2010)
19. Hang, L., Ge, J., Zhang, Y.: Correction method of regularized matrix based on ratio of singular value. Mod. Radar **41**(4), 54–58 (2019)
20. Zhe, Z., Tingying, Z., Yingchun, S.: A method for solving Ill-posed problem with fuzzy prior information. J. Geodesy. Geodyn. **38**(5), 524–526 (2018)
21. Metzler, B., Pail, R.: GOCE data processing: the spherical cap regularization approach. Stud. Geophys. Geod. **49**(4), 441–462 (2005)
22. Xiaoniu, Z., Daizhi, L., Xihai, L., et al.: An improved singular value modification method for ill-posed problems. Geom. Inf. Sci. Wuhan Univ. **40**(10), 1349–1354 (2015)
23. Lin, D., Zhu, J., Song, Y., et al.: Construction method of regularization by singular value decomposition of design matrix. Acta Geodaetica et Cartographica Sinica **45**(8), 883–889 (2016). https://doi.org/10.11947/j.AGCS.2016.20150134
24. Lu, J., Takahashi, S., et al.: Acoustic computer tomographic pyrometry for two-dimensional measurement of gases taking into account the effect of refraction of sound wave paths. Meas. Sci. Technol. **11**(6), 692–697 (2000)
25. Hostein, P., Raabe, A., Muller, R., et al.: Acoustic tomography on the basis of travel-time measurement. Meas. Sci. Technol. **15**(7), 420–1428 (2004)
26. Hua, Y., Shanhui, W., Yinggang, Z.: Acoustic CT temperature field reconstruction based on adaptive regularization parameter selection. Chin. J. Sci. Instrument **33**(6), 103–109 (2012)
27. Li, Z.C., Huang, H.T., Wei, Y.: Ill-conditioning of the truncated singular value decomposition, Tikhonov regularization and their applications to numerical partial differential equations. Numeric. Linear Algebra Appl. **18**(2), 205–221 (2011)
28. Jiang, P., Peng, L.H., Xiao, D.Y.: Tikhonov regularization based on the second order derivative matrix for electrical capacitance tomography image reconstruction. J. Chem. Ind. Eng. (China) **59**(2), 409–419 (2008)
29. Liu, C.H., Diao, X.F., Wang, Y.M.: Regularization method study in ultrasound inverse scattering imaging. J. Zhejiang Univ. **39**(3), 195–199, 210 (2005)
30. Liu, S.H., Lei, J., Li, Z.H.H.: Image reconstruction a logarithm for ECT based on modified regularization method. Chin. J. Sci. Instrument **28**(11), 1977–1981 (2007)
31. Hansen, P.C.: Analysis of discrete ill-posed problem by means of the L-curve. SLAM Rev. **34**(4), 561–580 (1992)
32. Wu, L.M.: A parameter choice method for Tikhonov regularization. Electron. Trans. Numer. Anal. **16**, 107–128 (2003)
33. Wang, H.X., Tang, L., Yan, Y.: Total variation regularization algorithm for electrical capacitance tomography. Chin. J. Sci. Instrument **28**(11), 2014–2018 (2007)
34. Krawczyk-Stando, D., Rudnicki, M.: Regularization parameter selection in discrete ill-posed problems-the use of the U-curve. Int. J. Appl. Math. Comput. Sci. **17**(2), 157–164 (2007)
35. Yunzhong, S., Peiliang, X., Bofeng, L.: Bias-corrected regularized solution to inverse Ill-posed models. J. Geodesy **86**(8), 597–608 (2012)

Research and Development of the Detection Robot in Pipeline Based on Electromagnetic Holography

Guo Xiaoting[⊠], Song Huadong, Yang Ruixuan, Lang Xianming,
Tang Yinlong, Song Yunpeng, Zhu Haibo, Xu Chunfeng,
and Liu Guanlin

Shenyang Academy of Insrumentation Science Co., Ltd.,
Shenyang 110043, China
tingting0924@163.com

Abstract. It is important of the detection equipment and technology in the pipeline which means to prevent pipeline accidents before they occur. In this paper, an electromagnetic holography detection robot for nondestructive detection of oil and gas pipelines is designed. The mechanical structure and function, electronic system and upper computer analysis system of the robot are introduced. The integration of defect detection and defect identification can realize the identification and evaluation of high-risk factors for safe operation of oil and gas pipelines, such as pipeline corrosion, weld defects, mechanical damage, etc. The experimental results show that the robot can identify, locate and quantify the defects in the process of pipeline defect detection, which is an important means to guide the reasonable maintenance and integrity management of pipeline.

Keywords: Detection in the pipeline · Magnetic flux leakage · Eddy current · Pipeline corrosion · Weld defects

1 Introduction

Oil and gas pipeline network are the main arteries of energy transmission and an important "lifeline" of national economic and social development [1]. In recent years, China has accelerated the construction of oil and gas pipeline network. According to the 14th five-years plan, the total mileage of long-distance pipeline is expected to reach 240000 km by the end of 2025 [2]. The safety of oil and gas pipelines has been widely concerned by the society. Therefore, effective pipeline integrity management must be adopted to prevent pipeline damage. Specific measures include reasonable cleaning and maintenance of pipeline, periodic nondestructive testing to evaluate the status of internal and external pipelines, etc. Defects and possible hazards shall be found in time before causing troubles, so as to maintain the normal operation of the pipeline [3]. The detection equipment and technology in the pipeline are the most important measures we need to prevent pipeline accidents.

J. Qian et al. (Eds.): ICRRI 2020, CCIS 1336, pp. 280–288, 2020.
https://doi.org/10.1007/978-981-33-4932-2_20

Magnetic flux leakage detection is a relatively mature and most widely used industrial detection technology in the pipeline, which can detect the corrosion, mechanical damage and other metal loss defects inside and outside the pipeline [4]. At present, the domestic pipeline detection technology is not mature. Due to the complex application conditions and corrosion factors of oil and gas pipelines, the requirements of detector's reliability and environmental adaptability are strict. Even though this technology is highly valued and tracked in China for more than 20 years, no substantive breakthrough has been made in product realization. Due to the gap between the domestic detection technology and the actual demand, "foreign detection" has long monopolized the detection market of China's oil and gas pipelines [5]. In order to break the monopoly of foreign technology, improve the technical level of detection equipment in the pipeline and ensure the safety of energy channel, it is imperative to develop the detection robot technology in the pipeline. In 2018, our company successfully developed a new type of electromagnetic holography inline detector with independent intellectual property rights.

2   Structure and Function of Electromagnetic Holography Detection Robot

2.1   Structural Composition

The electromagnetic holography detection robot is mainly composed of magnetization system, defect sensing probe, geometry sensing probe, mileage sensing probe, data acquisition and storage unit, power management system, universal joint assembly and other devices as well as driving unit, Structural diagram of electromagnetic holography detection robot as shown in Fig. 1.

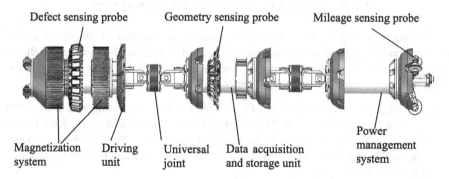

Fig. 1. Structure diagram of electromagnetic holography detection robot

As shown in Fig. 1, the electromagnetic holography detection robot is divided into magnetic leakage section, electronic section, battery section and mileage wheel recording device. The battery compartment provides energy for the whole system, the magnetic flux leakage section magnetizes the pipeline and collects magnetic flux leakage information of the defect, the mileage wheel records the specific location of the defect, and all the information is stored in the electronic compartment for subsequent analysis.

2.2   Major Function

(1) The magnetic flux leakage joint is composed of the power cup, magnetization system and defect sensing sensor unit. The power cup seals the pipeline and provides the driving force for the detector to move forward based on the pressure difference between the front and rear media. The magnetization system is composed of high magnetic permeability iron core, magnet and steel brush. The steel brush is in full contact with the inner wall of the pipeline 360°, thus forming a closed magnetic circuit. When the pipe wall is defective, the magnetic permeability is abnormal, the magnetic field line leaks out from the defect, resulting in magnetic leakage [6, 7]; The defect sensing sensor unit is composed of multiple groups of sensor probes, and the installation diagram of a single defect sensing sensor is shown in Fig. 2.

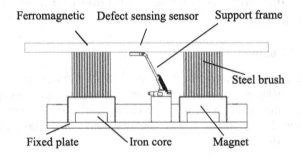

Fig. 2. Installation diagram of single defect sensing sensor

As shown in Fig. 2, the defect sensing sensor is installed in the center of the magnetic circuit by plate spring suitable for small-diameter pipeline, ensuring reliable contact between the magnetic circuit and the pipeline wall and flexibly adapting to the change of the internal diameter of the pipeline.

(2) The electronic section consists of a geometry sensing probe together with a data acquisition and storage unit. This section provides the installation platform for the data recording circuit, provides the enclosed and pressure bearing space, provides the signal input and output structure, and ensures the safe and reliable operation of the data recording circuit.

(3) The battery compartment consists of a power management system. The battery compartment provides sealed and pressure bearing structure for the battery, and provides voltage output interface to ensure that the battery can normally provide power for other devices in the working environment.

(4) The mileage recording device mainly provides the installation platform for the mileage sensor, and provides the accurate measurement pulse reference for the data acquisition circuit through the gear train.

(5) The universal joint assembly connects all the devices and realizes relative rotation among them, so that each device can smoothly pass the pipe elbow with specified radius.

(6) The buffer device is mainly used to prevent the detection robot in the oil and gas pipeline from colliding with the pipeline rigidly during the pipeline operation, which could cause damages to the equipment itself or the pipeline.

(7) First of all, the power cup device provides power for the oil and gas pipeline nondestructive testing device to move in the pipeline, ensures the smooth movement of the testing device in the pipeline under normal conditions; secondly, it serves as the support and buffer device of the battery device. The supporting cup is mainly used to provide support and buffer for each corresponding device to avoid rigid collision with the pipeline.

3 Electronic System of Electromagnetic Holography Detection Robot

The data acquisition and storage unit are located in the electronic section. The defect information signals, geometric deformation signals and mileage information signals collected by the defect sensing probe, geometric sensing probe and mileage sensing probe, as well as the temperature signals and attitude signals integrated in the acquisition system, are collected and stored in the internal memory through the FPGA + ARM architecture. After the detector exits the pipeline, download the data through the debugging interface and upload it to the upper computer data analysis system for analysis [8]. The structure diagram of the electronic system is shown in Fig. 3.

As shown in Fig. 3, the online electronic system based on FPGA + ARM mainly realizes power supply, initial signal processing, A/D conversion, multichannel signal parallel storage, etc. The power supply system can continuously supply power to the whole electronic system for up to 60 h. The data storage capacity is 500 g, and larger storage medium can be selected as required.

There are two modes in the electronic system, one is the normal mode driven by the mileage wheel, which is the ideal state if the mileage wheel does not slip. All defect sensing probes and geometry sensing probes are instructed to collect signals every 2.5 mm; The other is the fast mode driven by time. All main probes collect data at the frequency of 2 kHz, even if the mileage wheel slips or does not rotate [9, 10], the sensing defect detection probes' data will not be lost.

Considering the energy consumption of the detector at the state of rest, the sleeping function is loaded in the electronic system. When the detector stops for more than 30 min, the electronic system will sleep and stop collecting and storing data at the same time; when the detector moves forward again, the electronic system will stop sleeping after receiving the start signal and enter the collection state to continue collecting and storing all the data. The sleep function greatly saves the energy consumption of the electronic system and ensures the working time of the power supply system. The field application proves the effectiveness of the function.

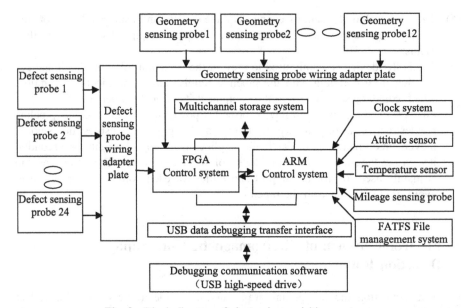

Fig. 3. Block diagram of electronic acquisition system

4 Analysis Software of Upper Computer of Detection Robot

The upper computer software of the electromagnetic holography detection robot is programmed in Python language as a whole. In the specific implementation, python language is used to pretreated the original data and then output the data file after pretreatment. At the same time, C++ language and QT are used to form the upper computer interface to call and extract the data after pretreatment. The upper computer software structure diagram is shown in Fig. 4.

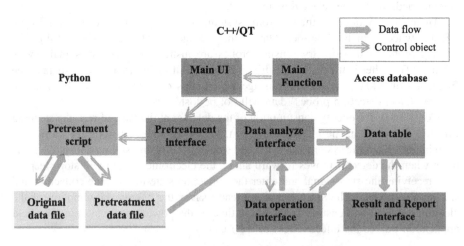

Fig. 4. Upper computer software structure block diagram

After analysis in the data analysis window, the database is formed as output. If there is any need to modify the database, just go back to the data analysis interface through the data operation interface, and then modify and save the new data.

The self-developed upper computer analysis system is called pipeline detection data analysis software, which has four parts: data of pretreatment function, data analysis function, chart and report, instructions and help. The function interface of pipeline detection data analysis software is shown in Fig. 5.

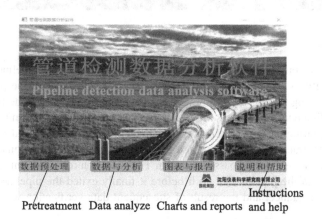

Pretreatment Data analyze Charts and reports Instructions and help

Fig. 5. The function interface of pipeline detection data analysis software

As shown in Fig. 5, the upper computer analysis system receives the data collected by the electromagnetic holography detection robot (defect signal, geometric signal, mileage signal, etc.). In the software background, it relies on the powerful data calculation function of the computer to carry out pipeline defect inversion and calculate the distribution and characteristics of pipeline defects. According to the calculation results, the corresponding detection drawings, tables and relevant data of the pipeline are output to provide support for the generation of subsequent pipeline analysis reports. At the same time, the data processing software also displays the corresponding detection results in the form of visual interface in front of the analysts, which provides convenient conditions for pipeline detection and analysis.

5   Application and Data Analysis of Electromagnetic Holography Detection Robot

5.1   Field Application

The electromagnetic holography detection robot is a significant means to insure the normal operation of the pipeline. In August 2019, the in-service detection was carried out in a diesel pipeline, and the field application of the detector is shown in Fig. 6.

Fig. 6. Field application of detector

As shown in Fig. 6, the pictures of the robot in and out of the pipeline are given. The total length of a diesel pipeline is 160 km, and the diameter of the pipeline was measured by using the pig with aluminum diameter measuring plate. The results show that the pipeline deformation is serious and reaches the deformation limit required by the detector. During the detection process, the detection robot was tracked. The receiver located at each tracking point monitored the detector, the receiver's signal was transmitted out when the detector passed through. The detection robot's running speed was normal, it run in the pipeline for 26 h before it finally exited the pipeline with intact mechanical appearance, and the result of subsequent data verification was normal.

5.2 Data Analysis

The data was downloaded through the debugging interface and uploaded to the upper computer data analysis system for analysis. Through the upper computer display, the signal curve of the pipeline component, which is part of the processed detection data, is shown in Fig. 7.

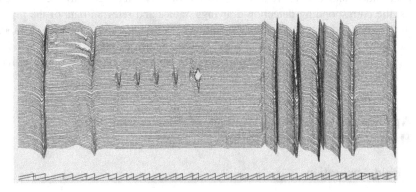

Fig. 7. Signal curve of pipe assembly

As shown in Fig. 7, the electromagnetic holography detection robot detects that the signal of pipeline components such as valve, tee and instrument branch pipe is obvious, which can clearly distinguish pipeline components and provide useful information for defect location.

The electromagnetic holography detection robot mainly detects and identifies the pipeline defects, and locates the serious defects. The signal curve of pipeline weld and defect is shown in Fig. 8.

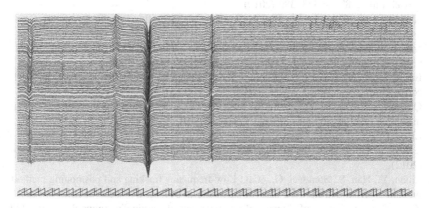

Fig. 8. Signal curve of pipeline weld and defect

As shown in Fig. 8, the electromagnetic holography detection robot can detect pipeline welds and defect signals sensitively, can lay a solid foundation for accurate description of defects, is able to pinpoint the position of pipeline defects, and can provide accurate detection reports for pipeline health maintenance.

6   Conclusion

In this paper, the mechanical structure and function, electronic system and upper computer analysis system of the electromagnetic holography detection robot developed for the normal operation and maintenance of the pipeline are systematically described, and the in-service detection application of the detection robot in a diesel pipeline has been carried out successfully. The results show that the detection device has the characteristics of accurate identification, safety and reliability, cost saving, and can accurately identify the pipeline components and defects. Furthermore, it can provide detailed and accurate detection report for pipeline operators, realizes the integration of defect sensing, defect location and defect identification, improves the technical level of internal corrosion detector in China, and lays an important foundation for the domestic development of detector series.

Fund projects. Funded by China Postdoctoral Science Foundation under Grant 2020M670796.

References

1. Ouyang, X., Hu, T., Di, Q.: Development of newly-constructed pipelines internal geometry inspection tool mechanical system. Dev. Innov. Mach. Electr. Products **27**(2), 89–91 (2014)
2. Wang, F., Feng, Q., Zhang, H., Song, H., Chen, J.: Pipeline feature recognition based on triaxial magnetic flux leakage internal detection technology. Nondestr. Test. **33**(01), 79–84 (2011)
3. Kishawy, H., Gabbar, H.: Review of pipeline integrity management practices. Int. J. Press. Vessel. Piping **87**(7), 373–380 (2010)
4. Xianghong, F., Shaohua, W., Jing, N.: Pipeline magnetic flux leakage detection technology and achievements in China. Petrol. Sci. Technol. Forum **8**, 55–57 (2007)
5. Jian, L., Jianlei, C., Dong, L.: Design of the main control system of high definition internal detector. Sensors Microsyst. **31**(5), 129–131 (2012)
6. Bai, G.S., Xu, Z., Wu, N.X.: Development of triaxial high definition magnetic flux leakage corrosion detector. Petrol. Mach. **42**(10), 103–106 (2014)
7. Qingshan, F.: Triaxial high definition magnetic flux leakage internal detection technology of in-service pipeline. Oil Gas Storage Transp. **28**(10), 72–75 (2009)
8. Li, J., Cui, J., Liu, D.: Design of master control system of high resolution magnetic-flux leakage detector inside the pipeline. Transducer Microsyst. Technol. **31**(05), 129–131-138 (2012)
9. Lijian, Y., Han, S., Songwei, G., et al.: Low-frequency tracking and positioning method for pipeline internal detector. Oil Gas Storage Transp. **37**(06), 613–619 (2018)
10. Zang, Y., et al.: Causes of slippage of odometer wheels in pipeline inline detectors. Oil Gas Storage Transp. **35**, 1–5 (2016)

Data-Driven Optimal Tracking Control for Linear Systems Based on Output Feedback Approach

Shikang Chen, Zhenfei Xiao, and Jinna Li[✉]

School of Information and Control Engineering, Liaoning Shihua University,
Fushun 113001, Liaoning, China
lijinna_721@126.com

Abstract. In this paper, an off-policy Q-learning method is proposed to solve the linear quadratic tracking problem of discrete-time system based on the output feedback of the system when the system model parameters are unknown. First, a linear discrete-time system with unknown parameters in the system matrix is given. Then, based on the Q-learning method and dynamic programming, an off-policy Q-learning algorithm without knowing system model parameters is proposed, such that the optimal controller is designed to obtain the control strategy which uses the system output data to learn the output feedback data driven optimal tracking control for linear discrete time systems with output feedback. Finally, the simulation results verify the effectiveness of the method.

Keywords: Off-policy Q-learning · LQT problem · Output feedback · Optimal control

1 Introduction

Linear quadratic tracking (LQT) of discrete time systems (DT) is a very important problem in the field of control. The basic idea is that the performance index is a quadratic function defined by the accumulation of deviation between the reference signal and the system output and the accumulation of control input. By designing an optimal controller, the performance index is minimized so that the output of the system can follow the track of the reference signal by an optimal approach. The traditional method to solve LQT problem is to solve a Riccati algebraic equation [1–3]. The traditional controller design methods all require the system model information to be known.

Reinforcement learning (RL) is a one kind of machine learning methods, which was born in 1950s and 1960s [4–6]. During the dynamic interaction with unknown dynamic environment, performance is evaluated and action is updated, such that the optimal performance together with the optimal action can be learned [7–11]. Reinforcement learning has many advantages and strong adaptability. At present, it has become an important learning method for solving optimization problems. It is widely used in robot, artificial intelligence, intelligent systems and other fields, and it is one of the research hot spots in recent years [12–15].

© Springer Nature Singapore Pte Ltd. 2020
J. Qian et al. (Eds.): ICRRI 2020, CCIS 1336, pp. 289–304, 2020.
https://doi.org/10.1007/978-981-33-4932-2_21

In the existing researches on the optimal tracking control results which do not depend on the system model by reinforcement learning method, most of them use the system state data to learn the state feedback control strategy and track the reference input of the system in the optimal or nearly optimal way, such as the optimal tracking control [16–22], etc. In [19], a Q-learning method based on off policy iteration was proposed to solve the optimal tracking control problem of networked control system. In [20], when the system parameter model is unknown, an optimal control method of linear network control was proposed. According to the state feedback information of the system, a novel off-policy Q-learning algorithm was proposed in [21], which solved the problem of linear quadratic tracking in discrete time under the condition of unknown system parameters. In [22], an optimal tracking control scheme was proposed.

In this paper, a Q-learning algorithm is developed to design the output feedback optimal tracking control strategy, such that the optimal quadratic tracking problem can be solved without the knowledge of system dynamics.

The innovation of this paper lies in (a): different from the traditional research which needs the traditional model information [1–3], the research of this paper is to learn the optimal tracking control strategy when the system model parameters are unknown. (b): compared with other model-free research on the state feedback of the system [20–22], this paper adopts a fully data-driven off policy Q-learning method to solve the linear quadratic tracking control problem of the discrete-time system based on the output feedback of the system and independent of the system model parameters. Finally, simulation experiments and practical application examples are given to verify the effectiveness of the algorithm.

2   On the Optimal Control Problem

This section will introduce the optimal control of linear quadratic tracking problem for discrete-time systems. The state equations of the following linear discrete systems are considered below:

$$\begin{cases} x_{k+1} = Ax_k + Bu_k \\ \quad y_k = Cx_k \end{cases} \tag{1}$$

where x_k is the state of the controlled object, u_k is the input of the controlled object, and y_k is the output of the controlled object. A, B, and C are matrices of appropriate dimensions, respectively. The reference signal of our interest is as follows:

$$r_{k+1} = Fr_k \tag{2}$$

where r_k is the input of the reference object, F is also a matrix with appropriate dimensions. For the linear quadratic tracking problem of discrete-time system, we need to control the output signal y_k in the system (1), and gradually track and catch up with our reference input signal r_k as time goes on. According to the actual problems, we design and select the output feedback controller as follows:

$$u_k = -K_1 y_k - K_2 r_k \tag{3}$$

The purpose of our controller is to optimize performance index J. Our performance indicator is:

$$J = \min_{u_k} \sum_{k=0}^{\infty} [\beta^k (y_k - r_k)^T Q(y_k - r_k) + u_k^T R u_k] \tag{4}$$

where $Q \geq 0$ and $R > 0$ are symmetric matrices, and β is a discount factor with $0 < \beta < 1$. The constraints are as follows:

$$\begin{cases} x_{k+1} = Ax_k + Bu_k \\ \quad y_k = Cx_k \\ \quad r_{k+1} = Fr_k \end{cases} \tag{5}$$

According to the performance index, we can define the optimal value function V^* as:

$$V^*(x_k, r_k) = \min_{u_k} \sum_{i=k}^{\infty} [\beta^k (y_k - r_k)^T Q(y_k - r_k) + u_k^T R u_k]$$
$$= \min_{u_k} \sum_{k=0}^{\infty} [\beta^k (Cx_k - r_k)^T Q(Cx_k - r_k) + u_k^T R u_k] \tag{6}$$

Then, the Q function can be expressed as:

$$Q(x_k, r_k, u_k) = (y_k - r_k)^T Q(y_k - r_k) + u_k^T R u_k + \sum_{i=k+1}^{\infty} [(y_i - r_i)^T Q(y_i - r_i) + u_i^T R u_i] \tag{7}$$

The optimal function Q^* can be expressed as:

$$Q^*(x_k, r_k, u_k) = (y_k - r_k)^T Q(y_k - r_k) + u_k^T R u_k + V^*(x_{k+1}, r_{k+1}) \tag{8}$$

For the convenience of calculation and understanding, (8) can be rewritten as:

$$Q^*(x_k, r_k, u_k) = \begin{bmatrix} y_k \\ r_k \\ u_k \end{bmatrix}^T \overline{Q} \begin{bmatrix} y_k \\ r_k \\ u_k \end{bmatrix} + V^*(x_{k+1}, r_{k+1}) \tag{9}$$

where $\overline{Q}$ can be written as:

$$\overline{Q} = \begin{bmatrix} Q & -Q & 0 \\ -Q & Q & 0 \\ 0 & 0 & R \end{bmatrix} \tag{10}$$

From the above formula, we can know that the relationship between the optimal value function V^* and the optimal function Q is:

$$V^*(x_k, r_k) = Q^*(x_k, r_k, u_k^*) \tag{11}$$

Since the input u_k of the control object is controllable, the optimal value function V^* can be expressed as [15]:

$$V^*(x_k, r_k) = \begin{bmatrix} x_k \\ r_k \end{bmatrix}^T P \begin{bmatrix} x_k \\ r_k \end{bmatrix} \tag{12}$$

The optimal Q function can be expressed as:

$$Q^*(x_k, r_k, u_k) = \begin{bmatrix} x_k \\ r_k \\ u_k \end{bmatrix}^T H \begin{bmatrix} x_k \\ r_k \\ u_k \end{bmatrix} \tag{13}$$

The matrix H can be expressed as follows:

$$H = \begin{bmatrix} H_{xx} & H_{xr} & H_{xu} \\ H_{rx} & H_{rr} & H_{ru} \\ H_{ux} & H_{ur} & H_{uu} \end{bmatrix} = \begin{bmatrix} A^T PA + C^T QC & A^T PF - C^T Q & A^T PB \\ F^T PA - QC & F^T PF + Q & F^T PB \\ B^T PA & B^T PF & B^T PB + R \end{bmatrix} \tag{14}$$

According to the necessary conditions to achieve optimal performance, implementing $\frac{\partial Q^*(x_k, r_k, u_k)}{\partial u_k} = 0$ yields the following forms.

$$\begin{cases} K_1 C = H_{uu}^{-1} H_{xu}^T \\ K_2 = H_{uu}^{-1} H_{ru}^T \end{cases} \tag{15}$$

From (15), we find that we cannot get K_1 if the matrix C is unknown. The output equation can be treated as follows.

$$(C^T C)^{-1} C^T y_k = x_k \tag{16}$$

By substituting (16) into (9) above, we can obtain a new optimal Q function equation:

$$Q^*(x_k, r_k, u_k) = (y_k - r_k)^T Q(y_k - r_k) + u_k^T R u_k + \begin{bmatrix} x_{k+1} \\ r_{k+1} \end{bmatrix}^T P \begin{bmatrix} x_{k+1} \\ r_{k+1} \end{bmatrix}$$

$$= \begin{bmatrix} y_k \\ r_k \\ u_k \end{bmatrix}^T \begin{bmatrix} Q & -Q & 0 \\ -Q & Q & 0 \\ 0 & 0 & R \end{bmatrix} \begin{bmatrix} y_k \\ r_k \\ u_k \end{bmatrix} + \begin{bmatrix} y_k \\ r_k \\ u_k \end{bmatrix}^T \begin{bmatrix} (C^T C)^{-1} C^T & 0 & 0 \\ 0 & I & 0 \\ 0 & 0 & I \end{bmatrix}^T \begin{bmatrix} A & 0 & B \\ 0 & F & 0 \end{bmatrix}^T P$$

$$\times \begin{bmatrix} A & 0 & B \\ 0 & F & 0 \end{bmatrix} \begin{bmatrix} (C^T C)^{-1} C^T & 0 & 0 \\ 0 & I & 0 \\ 0 & 0 & I \end{bmatrix} \begin{bmatrix} y_k \\ r_k \\ u_k \end{bmatrix} = \begin{bmatrix} y_k \\ r_k \\ u_k \end{bmatrix}^T \begin{bmatrix} \overline{H}_{yy} & \overline{H}_{yr} & \overline{H}_{yu} \\ \overline{H}_{ry} & \overline{H}_{rr} & \overline{H}_{ru} \\ \overline{H}_{uy} & \overline{H}_{ur} & \overline{H}_{uu} \end{bmatrix} \begin{bmatrix} y_k \\ r_k \\ u_k \end{bmatrix} \tag{17}$$

$$= Z_k^T \overline{H} Z_k$$

where

$$\overline{H} = \begin{bmatrix} \overline{H}_{yy} & \overline{H}_{yr} & \overline{H}_{yu} \\ \overline{H}_{ry} & \overline{H}_{rr} & \overline{H}_{ru} \\ \overline{H}_{uy} & \overline{H}_{ur} & \overline{H}_{uu} \end{bmatrix}$$

$$= \begin{bmatrix} [C^T(CC^T)^{-1}]^T(A^TPA)[C^T(CC^T)^{-1}] + Q & (C^TC)^{-1}C^TA^TPF - Q & (C^TC)^{-1}C^TA^TPB \\ (F^TPA)[C^T(CC^T)^{-1}] - Q & F^TPF + Q & F^TPB \\ B^TPA[C^T(CC^T)^{-1}] & B^TPF & B^TPB + R \end{bmatrix}$$

(18)

Implementing $\frac{\partial Q^*(x_k,r_k,u_k)}{\partial u_k} = 0$ yields the following forms

$$\begin{cases} K_1 = \overline{H}_{uu}^{-1}\overline{H}_{yu}^T \\ K_2 = \overline{H}_{uu}^{-1}\overline{H}_{ru}^T \end{cases}$$

(19)

According to the above relation, the Riccati equation for the optimal Q function is as follows:

$$Z_k^T\overline{H}Z_k = Z_k^T\overline{Q}Z_k + Z_{k+1}^T\overline{H}Z_{k+1}$$

(20)

3   Data Driven Q-Learning Algorithm

This section will give off-policy Q-learning algorithm for designing the output feedback optimal tracking control strategy, under which the system output can track the reference signal in an approximate optimal way.

First, the on-policy Q-learning algorithm is given, and then based on it the off-policy Q-learning algorithm is derived.

Algorithm 1: On-policy Q-learning algorithm.

1. Give a stablizing controller gain K_1^j and K_2^j, let the initial j value be 0, j represents the number of iterations. The control object input is defined as.

$$u_k^j = -K_1^j y_k - K_2^j r_k$$

(21)

2. Evaluate the control policy by solving the optimal Q function and matrix $\overline{H}$.

$$Z_k^T\overline{H}^{j+1}Z_k = (y_k - r_k)^TQ(y_k - r_k) + u_k^jRu_k^j + Z_{k+1}^T\overline{H}^{j+1}Z_{k+1}$$

(22)

3. Update the control policy.

$$
\begin{cases}
u_k^{j+1} = -K_1^{j+1} y_k - K_2^{j+1} r_k \\
K_1^{j+1} = (\bar{H}_{uu}^{j+1})^{-1} (\bar{H}_{yu}^{j+1})^T \\
K_2^{j+1} = (\bar{H}_{uu}^{j+1})^{-1} (H_{ru}^{j+1})^T
\end{cases}
\tag{23}
$$

4. If $\left\| \bar{H}^{j+1} - \bar{H}^j \right\| \le \varepsilon (\varepsilon$ is a small positive number) stops the iteration of the strategy. Otherwise, $j = j+1$, and return to Step 2.

In view of the advantages of off policy Q-learning algorithm, we will propose an off-policy algorithm based on Q-function, and adopt data-driven algorithm without model to solve the linear quadratic tracking problem of discrete-time system. We introduce the target control strategy into the system dynamics and get the following equation, where u_k is the behavior control strategy and u_k^j is the target control strategy.

$$
y_{k+1} = Cx_{k+1} = CAx_k + CBu_k = CAC^T(CC^T)^{-1} y_k + CBu_k
\tag{24}
$$

$$
\begin{bmatrix} y_{k+1} \\ r_{k+1} \end{bmatrix} = \begin{bmatrix} CAC^T(CC^T)^{-1} & 0 \\ 0 & F \end{bmatrix} \times \begin{bmatrix} y_k \\ r_k \end{bmatrix} + \begin{bmatrix} CB \\ 0 \end{bmatrix} \times u_k^j + \begin{bmatrix} CB \\ 0 \end{bmatrix} \times (u_k - u_k^j)
\tag{25}
$$

From Eq. (22), one has

$$
\begin{aligned}
& \begin{bmatrix} y_k \\ r_k \end{bmatrix}^T P^{j+1} \begin{bmatrix} y_k \\ r_k \end{bmatrix} - \begin{bmatrix} y_{k+1} \\ r_{k+1} \end{bmatrix}^T P^{j+1} \begin{bmatrix} y_{k+1} \\ r_{k+1} \end{bmatrix} + 2 \begin{bmatrix} y_{k+1} \\ r_{k+1} \end{bmatrix}^T P^{j+1} \begin{bmatrix} CB \\ 0 \end{bmatrix} \\[2mm]
& \times \begin{bmatrix} y_k \\ r_k \end{bmatrix}^T \times \begin{bmatrix} I & 0 \\ 0 & I \\ -K_1^j & -K_2^j \end{bmatrix}^T \bar{H}^{j+1} \begin{bmatrix} I & 0 \\ 0 & I \\ -K_1^j & -K_2^j \end{bmatrix} \times \begin{bmatrix} y_k \\ r_k \end{bmatrix} \\[2mm]
& - \left(\begin{bmatrix} y_{k+1} \\ r_{k+1} \end{bmatrix} - \begin{bmatrix} CB \\ 0 \end{bmatrix} \times (u_k - u_k^j) \right)^T \times \begin{bmatrix} I & 0 \\ 0 & I \\ -K_1^j & -K_2^j \end{bmatrix}^T \bar{H}^{j+1} \\[2mm]
& \times \begin{bmatrix} I & 0 \\ 0 & I \\ -K_1^j & -K_2^j \end{bmatrix} \left(\begin{bmatrix} y_{k+1} \\ r_{k+1} \end{bmatrix} - \begin{bmatrix} CB \\ 0 \end{bmatrix} \times (u_k - u_k^j) \right) \\[2mm]
& = (y_k - r_k)^T Q(y_k - r_k) + (u_k^j)^T R u_k^j
\end{aligned}
\tag{26}
$$

where

$$
P^{j+1} = \begin{bmatrix} I & 0 \\ 0 & I \\ -K_1^j & -K_2^j \end{bmatrix}^T \bar{H}^{j+1} \begin{bmatrix} I & 0 \\ 0 & I \\ -K_1^j & -K_2^j \end{bmatrix}
\tag{27}
$$

(26) can be rewritten as:

$$\begin{bmatrix} y_k \\ r_k \end{bmatrix}^T P^{j+1} \begin{bmatrix} y_k \\ r_k \end{bmatrix} - \begin{bmatrix} y_{k+1} \\ r_{k+1} \end{bmatrix}^T P^{j+1} \begin{bmatrix} y_{k+1} \\ r_{k+1} \end{bmatrix} + 2 \begin{bmatrix} y_{k+1} \\ r_{k+1} \end{bmatrix}^T P^{j+1} \begin{bmatrix} CB \\ 0 \end{bmatrix} \times (u_k - u_k^j)$$
$$- (u_k - u_k^j)^T (CB)^T P^{j+1} (CB)(u_k - u_k^j) = (y_k - r_k)^T Q(y_k - r_k) + (u_k^j)^T R u_k^j$$
$$\tag{28}$$

where

$$2 \begin{bmatrix} y_{k+1} \\ r_{k+1} \end{bmatrix}^T P^{j+1} \begin{bmatrix} CB \\ 0 \end{bmatrix} \times (u_k - u_k^j) = 2 \begin{bmatrix} CAC^T(CC^T)^{-1} y_k + CB u_k \\ F r_k \end{bmatrix}^T P^{j+1}$$
$$\times \begin{bmatrix} CB \\ 0 \end{bmatrix} (u_k - u_k^j) = 2 y_k^T [C^T(CC^T)^{-1}]^T (CA)^T P^{j+1} CB(u_k - u_k^j) \tag{29}$$
$$+ 2 u_k^T (CB)^T P^{j+1} CB(u_k - u_k^j) + 2 r_k^T F^T P^{j+1} CB(u_k - u_k^j)$$

Further, one has

$$\begin{bmatrix} y_k \\ r_k \end{bmatrix}^T P^{j+1} \begin{bmatrix} y_k \\ r_k \end{bmatrix} - \begin{bmatrix} y_{k+1} \\ r_{k+1} \end{bmatrix}^T P^{j+1} \begin{bmatrix} y_{k+1} \\ r_{k+1} \end{bmatrix}$$
$$+ 2 y_k^T [C^T(CC^T)^{-1}]^T (CA)^T P^{j+1} CB(u_k - u_k^j) + 2 u_k^T (CB)^T P^{j+1} CB(u_k - u_k^j) \tag{30}$$
$$+ 2 r_k^T F^T P^{j+1} CB(u_k - u_k^j) - (u_k - u_k^j)^T (CB)^T P^{j+1} (CB)(u_k - u_k^j)$$
$$= (y_k - r_k)^T Q(y_k - r_k) + (u_k^j)^T R u_k^j$$

Rewriting (30) yields below

$$\theta_k^j L^{j+1} = \rho_k^j \tag{31}$$

where

$$\rho_k^j = (y_k - r_k)^T Q(y_k - r_k) + (u_k)^T R u_k$$

$$L^{j+1} = \left[(vec(L_1^{j+1}))^T \ (vec(L_2^{j+1}))^T \ (vec(L_3^{j+1}))^T \right.$$
$$\left. (vec(L_4^{j+1}))^T \ (vec(L_5^{j+1}))^T \ (vec(L_6^{j+1}))^T \right]^T$$

$$L_1^{j+1} = \overline{H}_{yy}^{j+1}, \quad L_2^{j+1} = \overline{H}_{yr}^{j+1}, \quad L_3^{j+1} = \overline{H}_{yu}^{j+1}, \quad L_4^{j+1} = \overline{H}_{rr}^{j+1}, \quad L_5^{j+1} = \overline{H}_{ru}^{j+1},$$
$$L_6^{j+1} = \overline{H}_{uu}^{j+1},$$

$$\theta_k^j = \begin{bmatrix} \theta_1^j & \theta_2^j & \theta_3^j & \theta_4^j & \theta_5^j & \theta_6^j \end{bmatrix}$$

$$\theta_1^j = y_k^T \otimes y_k^T - y_{k+1}^T \otimes y_{k+1}^T$$

$$\theta_2^j = 2y_k^T \otimes y_k^T - 2y_{k+1}^T \otimes r_{k+1}^T$$

$$\theta_3^j = 2y_k^T \otimes u_k^T - 2y_{k+1}^T \otimes (u_{k+1}^j)^T$$

$$\theta_4^j = r_k^T \otimes r_k^T - r_{k+1}^T \otimes r_{k+1}^T$$

$$\theta_5^j = 2r_k^T \otimes u_k^T - 2r_{k+1}^T \otimes (u_{k+1}^j)^T$$

$$\theta_6^j = u_k^T \otimes u_k^T - (u_{k+1}^j)^T \otimes (u_{k+1}^j)^T$$

4 Simulation Experiment

In this section, we use the proposed algorithm to simulate the experiment, and use the experimental results to verify whether the algorithm is effective.

Example 1: Consider the following system:

$$x_{k+1} = \begin{bmatrix} -1 & 2 \\ 2.2 & 1.7 \end{bmatrix} x_k + \begin{bmatrix} 2 \\ 1.6 \end{bmatrix} u_k \tag{32}$$

$$y_k = \begin{bmatrix} 1 & -2 \\ -1 & 4 \end{bmatrix} x_k \tag{33}$$

The reference signal generator is:

$$r_{k+1} = \begin{bmatrix} -1 & 0 \\ 0 & -1 \end{bmatrix} r_k \tag{34}$$

Choose $Q = \begin{bmatrix} 1000 & 0 \\ 0 & 10 \end{bmatrix}$ and $R = 1$. The optimal matrix H and $\overline{H}$ can be obtained from (13) and (17), respectively, and the control gain K_1 and K_2 of the optimal tracking control can be obtained from (19).

$$H = \begin{bmatrix} 55014.1382 & -47201.9872 & -10241.4410 & 193.0046 & -48231.9264 \\ -47201.9872 & 124780.7167 & 9344.6976 & -162.08466 & 124282.7373 \\ -10241.4410 & 9344.6976 & 2067.8905 & 51.2434 & 9527.2908 \\ 193.0046 & -162.08466 & 51.2434 & 49.1044 & -166.5050 \\ -48231.9264 & 124282.7373 & 9527.2908 & -166.5050 & 123826.6900 \end{bmatrix}$$

$$\tag{35}$$

$$\overline{H} = \begin{bmatrix} 157847.7575 & 70420.4747 & -16810.5331 & 304.5860 & -34322.4842 \\ 70420.4747 & 39017.3301 & -5569.0922 & 101.5813 & 13909.4422 \\ -16810.5331 & -5569.0922 & 3067.8905 & 51.2434 & 9527.2908 \\ 304.5860 & 101.5813 & 51.2434 & 59.1044 & -166.5050 \\ -34322.4842 & 13909.4422 & 9527.2908 & -166.5050 & 123826.6900 \end{bmatrix}$$

(36)

$$\begin{cases} K_1 = [\,0.2772 & -0.1123\,] \\ K_2 = [\,0.0769 & 0.0013\,] \end{cases}$$

(37)

After 8 iterations, we find that the algorithm converges and the matrix $\overline{H}^8$ and the control gain K_1^8 and K_2^8 of the optimal tracking control are the following data.

$$\overline{H}^8 = \begin{bmatrix} 157847.7575 & 70420.4747 & -16810.5331 & 304.5860 & -34322.4842 \\ 70420.4747 & 39017.3301 & -5569.0922 & 101.5813 & 13909.4422 \\ -16810.5331 & -5569.0922 & 3067.8905 & 51.2434 & 9527.2908 \\ 304.5860 & 101.5813 & 51.2434 & 59.1044 & -166.5050 \\ -34322.4842 & 13909.4422 & 9527.2908 & -166.5050 & 123826.6900 \end{bmatrix}$$

(38)

$$\begin{cases} K_1^8 = [\,0.2772 & -0.1123\,] \\ K_2^8 = [\,0.0769 & 0.0013\,] \end{cases}$$

(39)

In the learning process, the optimal tracking controller gain is convergent, and the following figure shows its convergence process in the learning process (Fig. 1).

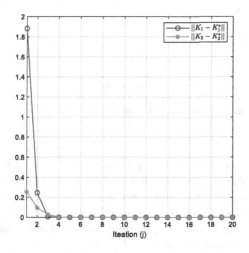

Fig. 1. Optimal control gain convergence process of tracking controller

In order to obtain the exact solution of Riccati Eq. (20) of the optimal Q-function under sufficient excitation conditions, it is necessary to add detection noise (Figs. 2 and 3).

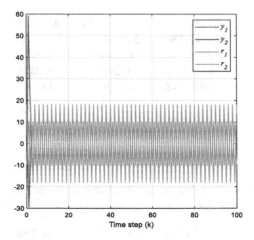

Fig. 2. System output and reference signal

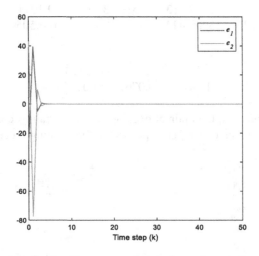

Fig. 3. Tracking errors using the learned controller

Example 2:

We select the water tank system made by ingenieurburo gurski Schramm company in Germany, which is a three tank water tank of TTS20 type, as our simulation object, and the water tank is shown in Fig. 4. This water tank system consists of a nonlinear multi input multi output system with two actuators and a digital controller, which meets our requirements for the system. The main structure and overall industrial process of TTS20 three tank water tank are shown in Fig. 5.

Fig. 4. Three tank of TTS20

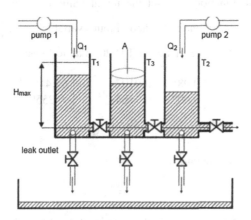

Fig. 5. Structure and industrial process of TTS20

TTS20 three tank device is composed of three plexiglass cylinders T1, T2 and T3 with Section A, which are connected in series with each other through cylindrical pipes with section Sn. There is a one-way valve in T2 glass pipe, and the outflow liquid will be collected in a reservoir to provide water for pump 1 and pump 2. H_{max} is the highest liquid level. If the level of T1 or T2 exceeds this value, the corresponding pumps 1 and 2 will automatically shut down. Q1 and Q2 represent the flow of pump 1 and pump 2. In addition, to simulate leakage, each tank has a circular opening with a manually adjustable ball valve on the cross section. The drain valve and leakage flow can describe the failure information of the water tank. The liquid extracted from the pool is injected into T1 and T2 by pump 1 (P1) and pump 2 (P2), respectively. Then their bottom valve and T3 drain valve discharge water into the reservoir for P1 and P2 recycling, forming a circuit. Among them, T1, T2 and T3 are measured by three pressure level sensors as the measuring elements of the system, and the flow of Q1 and Q2 is regulated by the digital controller.

For TTS20, we can design its model as follows:

$$
\begin{cases}
\begin{bmatrix} \dot{h}_1 \\ \dot{h}_3 \end{bmatrix} = \frac{1}{s} \begin{bmatrix} -Q_{13} \\ -Q_{13} - Q_{out} \end{bmatrix} Q_{in} \\
y = h_1
\end{cases}
\tag{40}
$$

In the model, h_1 and h_3 are the control variables, representing the water level height of water tanks T1 and T3, Q_n is selected as the control variable of the system as the flow of Q1, and the flow from T1 to T3 is $Q_{13} = az_1 S_n \mathrm{sgn}(h_1 - h_3)\sqrt{2g|h_1 - h_3|}$, the water flow from the bottom of T3 is $Q_{out} = az_2 S_1 \sqrt{2gh_2}$, $S_1 = S_n = 5 \times 10^{-5}\,\mathrm{m}^2$, $S = 0.154\,\mathrm{m}^2$, $H_{max} = 0.6\,\mathrm{m}$, Flow coefficient $az_1 = 0.48$, $az_2 = 0.58$, $\mathrm{sgn}(\bullet)$ is a symbolic function. We set the initial value of h_1 and h_3 is 0, and the relationship between state variable and input variable is $\begin{bmatrix} x_1(k) \\ x_2(k) \end{bmatrix} = \begin{bmatrix} h_1(k) \\ h_3(k) \end{bmatrix}$, $u(k) = Q_{in}(k)$. The state space model of TTS20 is as follows:

$$
x_{k+1} = \begin{bmatrix} 0.9850 & 0.0107 \\ 0.0078 & 0.9784 \end{bmatrix} x_k + \begin{bmatrix} 64.4453 \\ 0.2559 \end{bmatrix} u_k
\tag{41}
$$

$$
y_k = [1 \quad 0]x_k
\tag{42}
$$

The reference signal generator is:

$$
r_{k+1} = r_k
\tag{43}
$$

Select the value of reference signal water level as 0.5 m. Choose $Q = 10$ and $R = 5$, The optimal Q- function matrix H and $\overline{H}$ are obtained, The control gain K_1 and K_2 of the optimal tracking control are as follows:

$$
H = \begin{bmatrix} 17.7618 & -17.6205 & 507.6889 \\ -17.6205 & -131.2013 & -514.5708 \\ 507.6889 & -514.5708 & 33224.5750 \end{bmatrix}
\tag{44}
$$

$$\overline{H} = \begin{bmatrix} 17.7627 & -17.8809 & 507.8883 \\ -17.8809 & 18.0010 & -515.6235 \\ 507.8883 & -515.6235 & 33234.4549 \end{bmatrix} \tag{45}$$

$$\begin{cases} K_1 = [-0.0153] \\ K_2 = [0.0155] \end{cases} \tag{46}$$

After 10 iterations, we find that the algorithm converges and the optimal Q-function matrix $\overline{H}^{10}$ and the gain K_1^{10} and K_2^{10} of the optimal tracking control are the following data.

$$\overline{H}^{10} = \begin{bmatrix} 17.7627 & -17.8809 & 507.8883 \\ -17.8809 & 18.0010 & -515.6235 \\ 507.8883 & -515.6235 & 33234.4549 \end{bmatrix} \tag{47}$$

$$\begin{cases} K_1^{10} = [-0.0153] \\ K_2^{10} = [0.0155] \end{cases} \tag{48}$$

We find that in the learning process, the optimal tracking controller gain is convergent, and the following figure will show its convergence process in the learning process (Figs. 6, 7 8 and 9).

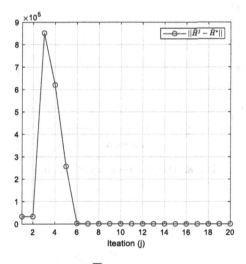

Fig. 6. Optimal $\overline{H}$ matrix convergence process

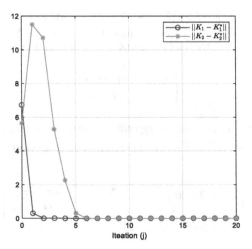

Fig. 7. Optimal control gain convergence process of tracking controller

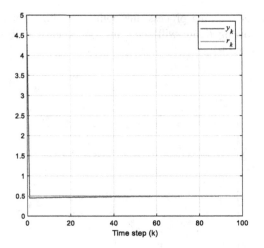

Fig. 8. Output trajectories of system

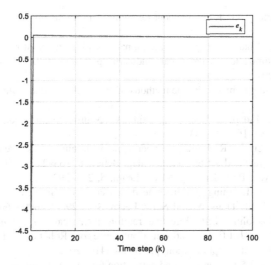

Fig. 9. Tracking errors using the learned controller

5 Conclusion

In this paper, a data-driven off policy Q-learning method is proposed to solve the linear quadratic tracking problem of discrete-time system based on the output feedback of the system. This paper introduces and compares the on policy Q-learning method and the off policy Q-learning method for the linear quadratic tracking problem of the discrete-time system, combines the dynamic programming with the Q-learning method, and uses the off policy Q-learning method to learn the optimal controller gain when the system environment is unknown. Finally, the simulation results show that the method is effective.

Acknowledgments. This work was supported by the National Natural Science Foundation of China under Grants 61673280, the Open Project of Key Field Alliance of Liaoning Province under Grant 2019-KF-03-06 and the Project of Liaoning Shihua. University under Grant 2018XJJ-005.

References

1. Lewis, F.L., Vrabie, D.L., Syrmos, V.L.: Optimal Control for Polynomial Systems. pp. 287–296, Wiley (2012)
2. Hengster-Movric, K., You, K., Lewis, F.L., Xie, L.: Synchronization of discrete-time multi-agent systems on graphs using riccati design. Automatica **49**(2), 414–423 (2013)
3. Stoorvogel, A.A., Weeren, A.J.T.M.: The discrete-time riccati equation related to the H∞ control problem. IEEE Trans. Autom. Control **39**(3), 686–691 (1994)
4. Kaelbling, L.P., Littman, M.L., Moore, A.W.: Reinforcement learning: a survey. J. Artif. Intell. Res. **4**(1), 237–285 (1996)

5. Sutton, R.S., Barto, A.G.: Reinforcement Learning: An Introduction. MIT Press, Cambridge (1998)
6. Li, J., Ding, J., Chai, T., Lewis, F.L.: Nonzero-sum game reinforcement learning for performance optimization in large-scale industrial processes. IEEE Trans. Cybern. **50**(9), 4132–4145 (2020)
7. Sutton, R.: Learning to predict by the methods of temporal difference. Mach. Learn. **3**(1), 9–44 (1988)
8. Bertsekas, D.P., Tsitsiklis, J.N., Volgenant, A.: Neuro-dynamic programming. Encycl. Optim. **27**(6), 1687–1692 (2011)
9. Santamaria, J.C., Sutton, R., Ram, A.: Experiments with reinforcement learning in problems with continuous state and action spaces. Adap. Behav. **6**, 163–217 (1997)
10. Watkins, C., Dayan, P.: Q-learning. Mach. Learn. **8**, 279–292 (1992)
11. Wang, D., Liu, D.: Learning and guaranteed cost control with event-based adaptive critic implementation. IEEE Trans. Neural Netw. Learn. Syst. **29**(12), 6004–6014 (2018)
12. Smart, W.D., Kaelbling, L.P.: Effective reinforcement learning for mobile robots. In: Proceedings of the IEEE International Conference on Robotics and Autoinforcement learning for performance optimization, pp. 3404–3410 (2002)
13. Beom, H.R., Cho, H.S.: A sensor-based navigation for a mobile robot using fuzzy logic and reinforcement learning. IEEE Trans. Syst. Man Cybern. **25**, 464–477 (1995)
14. Kondo, T., Ito, K.: A reinforcement learning with evolutionary state recruitment strategy for autonomous mobile robots control. Robot. Auton. Syst. **46**(2), 111–124 (2004)
15. Li, J.N., Ding, J.L., Chai, T.Y, Li, C., Lewis, F.L.: Nonzero-sum game reinforcement learning for performance optimization in large-scale industrial processes. IEEE Trans. Cybern. (2019). https://doi.org/10.1109/tcyb.2019.2950262
16. Kiumarsi, B., Lewis, F.L., Jiang, Z.P.: H∞ control of linear discrete-time systems: off-policy reinforcement learning. Automatic **37**(1), 144–152 (2017)
17. Kim, J.H., Lewis, F.L.: Model-free H∞ control design for unknown linear discrete-time systems via Q-learning with LMI. Automatica **46**(8), 1320–1326 (2010)
18. Al-Tamimi, A., Lewis, F.L., Abu-Khalaf, M.: Model-free Q-learning designs for linear discrete-time zero-sum games with application to H-infinity control. Automatica **43**(3), 473–481 (2007)
19. Li, J.N., Yin, Z.X.: Optimal tracking control of networked control systems based on off policy Q-learning. Control Decis. **34**(11), 2343–2349 (2019)
20. Xu, H., Sahoo, A., Jagannathan, S.: Stochastic adaptive event-triggered control and network scheduling protocol co-design for distributed networked systems. Control Theory Appl. IET. **8**(18), 2253–2265 (2014)
21. Li, J.N., Chai, T.Y., Lewis, F.L., Ding, Z.T., Jiang, Y.: Off-policy interleaved Q-learning: optimal control for affine nonlinear discrete-time systems. IEEE Trans. Neural Netw. Learn. Syst. **30**(5), 1308–1320 (2019)
22. Li, X.F., Xue, L., Sun, C.Y.: Linear quadratic tracking control of unknown discrete-time systems using value iteration algorithm. Neurocomputing **314**(7), 86–93 (2018)

Features of Interest Points Based Human Interaction Prediction

Xiaofei Ji[(⊠)], Xuan Xie, and Chenyu Li

School of Automation, Shenyang Aerospace University,
Shenyang 110136, Liaoning, China
jixiaofei7804@126.com

Abstract. The recognition and prediction of human interaction based on video has a broad application prospect in intelligent video monitoring and other fields, but the current integration algorithm of it is not mature, which greatly limits the application of the algorithm. A prediction method of human interaction based on the statistical characteristics of interest points is proposed, which established an integrated framework to the recognition and prediction of human interaction initially. First, the spatio-temporal interest points are extracted and performed as 3D-SIFT feature description, then the bag of words is used to represent the action video. In the training stage, Gaussian models are used to establish the action model for each action at different time scales. In the prediction stage, the bag of words representation is extracted and compared with the established prediction models of different time lengths to obtain similar prediction probabilities between the models for an action video of unknown length. Finally, the predictive probability of each action at different time scales is fused by weighted probabilities completing the recognition and prediction of interaction prediction. The experimental result on UT-interaction dataset demonstrated that the proposed approach is easy to implement and has a good predictive effect.

Keywords: Interaction prediction · Statistical features of interest points · Bag of words · Gaussian models · Probability fusion

1 Introduction

Human interactive behavior recognition and understanding based on videos is a frontier direction in the field of image processing and computer vision, which uses video analysis technology to detect, track, and recognize human and motion objects for understanding and describing from image sequences or videos containing people [1, 2]. The research results are mainly applied in the field of intelligent video surveillance. The human behavior analysis based on videos enables the monitoring system to intelligently analyze the captured image sequence, automatically identify the ongoing behavior and issue an early warning. So as to the system can quickly locate the crime scene, and respond to the crisis in a timely manner for reducing the crime rate when an abnormal situation occurs. At present, most of the research focuses on the post-event detection of the occurrence behavior. In many real-world scenarios where the safety factor is high, the system needs to be able to make early predictions of the ongoing and unfinished

© Springer Nature Singapore Pte Ltd. 2020
J. Qian et al. (Eds.): ICRRI 2020, CCIS 1336, pp. 305–316, 2020.
https://doi.org/10.1007/978-981-33-4932-2_22

behavior. The problem of human action prediction is defined as: a hypothesis that temporarily gives an unfinished action in a given and incomplete video, a probabilistic inference process that infers the type of action being performed based on the beginning of the behavioral action in the video [3] Unlike action classification, the motion prediction system needs to make a "what action behavior occurs" decision during the action execution, as shown in Fig. 1. The prediction of interaction between two people has great practical significance. For example, detecting the behavior of two people before they deteriorate will make the video surveillance system have the ability to prevent criminal behavior from happening, and make video resources play a greater role.

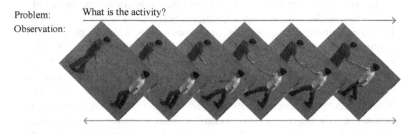

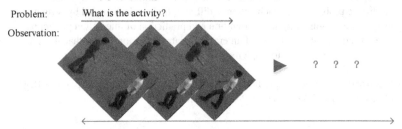

Fig. 1. Comparison of action recognition and action prediction

Ryoo [3] gives the definition of "action prediction", and first proposes the probabilistic method of using Bag of Word (BoW) to solve the prediction problem of double interaction behavior. The overall histogram form of space-time feature is used to perform the action representation. And then the distribution of features over time is modeled to achieve motion prediction effectively. This method is not only simple and easy, but also provides a basic framework for predicting the human interaction behavior, and successfully predicts the interaction behavior of two people. On this basis, Yu et al. [4] proposed a Spatial-Temporal Implicit Shape Model (STISM) to represent the spatio-temporal structure of local spatiotemporal features, and adopted a multi-class balanced random forest method to achieve human interaction prediction, but the algorithm implementation is more complicated. Raptis [5] proposed a new motion prediction model based on local key-frames, which uses the method of constructing key-frames features to represent and act as state nodes. This method successfully implements a unified framework for mutual interaction behavior recognition and prediction. However, this method uses a description of key frames, which relies on the accurate identification of sub-actions, and the accuracy of recognition and prediction is not high. Cao et al. [6] regard the human interaction prediction problem as the posterior probability maximization problem, in which the similarity at each moment is calculated by the sparse coding technique. This method can not only realize motion prediction, but also deal with the problem of interactive motion recognition in the case of missing video intermediate observation, but the computational complexity of this method is high. Li et al. [7], encode long-term activity into meaningful sequence of action units by monitoring the speed of motion, and then introduce Probabilistic Suffix Tree (PST) to represent Markov dependency relationship between action units. Finally, the Predictive Accumulative Function (PAF) is used to describe the predictability of various activities. All of the above methods are based on traditional probabilistic model statistics. Under the premise of ensuring the enforceability of the algorithm, the accuracy of recognition and prediction is not very high.

The human body's behavior is an orderly context-related model with certain structure, especially in the identification and prediction of two-person interaction behavior and multi-person interaction behavior, and a model is needed to represent the relationship between people. Although the human behavior analysis technology has made significant progress in recent years, due to the complexity and variability of human motion, there is no standard fixed classification method. Significant progress has been made in machine learning theory and algorithm research. The development of efficient and reliable machine learning algorithms has become a hot topic for many researchers. The Convolutional Neural Network (CNN)-based approach has begun to be applied to the field of two-person interaction behavior prediction. Ke et al. [8] first calculate optical flow images from successive video frames of partial sequences to capture the dependence of each RGB frame and each optical image on global and local context, respectively, and then use the Long Short Term Memory (LSTM) network learn the structural model including spatial and temporal information, and finally introduce the ranking score fusion method to automatically calculate the optimal weight of the score fusion of each model to finally predict the interaction category. But the choice of the best weight is random. Ke et al. [9] applied CNN to video stream encoded images to learn temporal information of human interaction prediction, and used several continuous optical flow image features to learn the law of variation with time. However, this method only utilizes the time feature, lacks the spatial feature information description of the human posture, and needs to learn a large number of samples, and the computational complexity is relatively high.

The data is based on the above analysis. The method proposed in [3] is simple and effective. The inadequacy is that the integration of prediction and recognition of interaction between two people is not ideal, and the accuracy of prediction and recognition is low. Based on the literature [3], this paper proposes a probabilistic fusion method between the bag of word model and the multi-time proportional action model, which realizes the integration of two-person interaction behavior prediction and recognition. Among them, the local interest points are invariant to the 3-dimensional scale of the neighborhood. The feature description (3D-Scale Invariant Feature

Transform, 3D-SIFT) [10] improves the characterization ability of local features. The specific process of the algorithm is shown in Fig. 2: First, the training set video is used for interest point extraction and 3D-SIFT feature description; then the training data is segmented into different time proportional data, and the bag of word method is used to obtain the video at different time ratios. Histogram representation; Finally, the Gaussian model is used to establish an action model for each time proportional data. When the test video of unknown length is given, the feature description is used to form a histogram representation of the bag of word, and the similarity probability of each Gaussian model under different time ratios is calculated. Finally, the action category of the test video is determined by probability fusion. Through a large number of experiments, the method has a good recognition effect while ensuring a certain prediction accuracy.

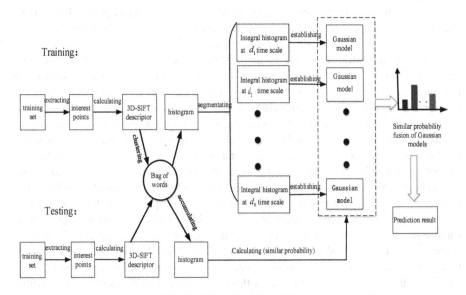

Fig. 2. Algorithm flow chart

2   Feature Point Feature Extraction and Description

Local features have the advantage of being able to describe information with significant changes in motion, so this article is used as a base feature. Local representation-based motion representations typically characterize human motion using points in the image sequence that meet certain criteria. These qualifying points are often referred to as local feature points (or points of interest [11]). By effectively describing the neighborhood of the local feature points, local information features representing the sequence of images can be obtained. The 3D-SIFT description operator is a three-dimensional space-time gradient direction histogram, which can accurately capture the essence of the spatiotemporal characteristics of video data. In order to make full use of the motion information of the context, this paper adopts the 3D-SIFT feature description of the interest point of the whole video. First, the interest points are extracted from the video frames with motion changes of the motion video, and then the 3D-SIFT feature description is performed on the points of interest, that is, 3D spherical volume blocks are established in the neighborhood of the interest points, and gradient accumulation is performed in each volume block. A specific example is shown in Fig. 3.

Fig. 3. Point of interest extraction (left) and 3D-SIFT feature description of the point of interest (right)

3 Action Prediction Model Establishment

3.1 Feature Representation of the Video

The combination of the interest point feature and the bag of word model can conveniently obtain the word bag histogram representation of the action video, and represent the human interaction by constructing the feature histogram of the video. The *k-means* method is usually used to cluster all local feature representations of the training data to form a dictionary. An example is shown in Fig. 4. Then all the local features of an action video are projected to the dictionary, and finally the frequency at which the visual words appear in the video in the dictionary is statistically formed to form a statistical histogram representation of the action video. This method has been widely used for human motion recognition, and the feature representation of the word bag histogram can well handle the influence of noise on the observed video at different time scales. In order to apply this framework to the prediction of two-person interaction behavior, this paper divides the training video into sub-training sets of different time lengths according to a certain proportion, and divides the training data described by the 3D-SIFT points of interest into segments of different time length according to a certain proportion. Training data, each sub-training data is represented by a word bag histogram. O_l. Represents an action video, d_i. Indicates the proportion of time, $h_{d_i}(O_l)$ Express d_i Time ratio O_l Child training data histogram. v_w The characteristic histogram of each sub-training data representing the w th visual word $h_{d_i}(O_l)$. The value of the w th word bag is expressed as:

$$h_{d_i}(O_l)[w] = \left| \{f | f \in v_w \wedge t_f < d_i\} \right| \tag{1}$$

Where f is a feature extracted from the video O_l, t_f indicates its time position. $h_{d_i}(O_l)$ describes the time ratio as d_i. The distribution of time-space feature histograms over time, an example is shown in Fig. 5.

Fig. 4. 3D spatiotemporal feature description (left) and visual word visualization (right) of the handshake action. Use the same shape description to group the same visual words.

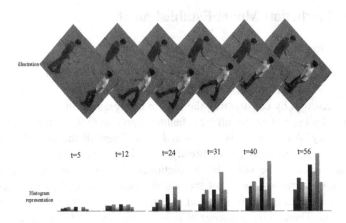

Fig. 5. Handshake action integral histogram distribution changes trend

3.2   A Video Model Building

The prediction of the interaction action category can be completed by modeling the bag of words histogram of the action video. Through a large number of experimental data, it is found that the bag of word histogram of the same class action video at the same time ratio satisfies the same distribution trend and roughly conforms to the Gaussian distribution. A brief example is shown in Fig. 6 and Fig. 7. Therefore, the Gaussian model can be used to establish the same-category action model under different time proportions. Experiments show that the model can better reflect the expression of the histogram of the word bag when a certain action is executed to a certain time node.

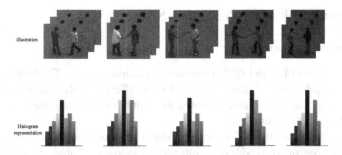

Fig. 6. Distribution of the same time ratio of the same category (handshake) action histogram

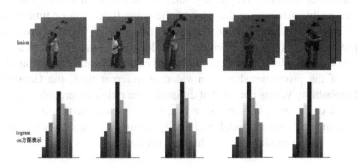

Fig. 7. Distribution of the same time ratio of the same category (hug) action histogram

This article will train the video in different time ratios d_1, d_2, d_3, d_4, d_5, d_6 Segmentation is performed, and a Gaussian model is established for the bag of word histogram of the segmented training video. Remember $h^{(1,d)}$, $h^{(2,d)}$, $h^{(3,d)}$, ..., $h^{(A,d)}$ Gaussian models of the A-type action with the current time ratio d, given an unknown action video O^{test} Calculation O^{test} The histogram represents the likelihood probability of each action Gaussian model with the current time ratio d, namely:

$$p(O^{test}|h^{(1,d)}), p(O^{test}|h^{(2,d)}), p(O^{test}|h^{(3,d)}), \cdots, p(O^{test}|h^{(A,d)});$$

$$p(O^{test}|h^{(a,d)}) = \frac{1}{\sqrt{2\pi\sigma^2}} e^{\frac{-(h(O^{test})-h_d(a))^2}{2\sigma^2}} \tag{2}$$

Where d is the current time ratio of the training video; A is the number of action model categories under the d time ratio; $h^{(a,d)}$. The Gaussian model corresponding to the histogram of action a, $h(O^{test})$. Is an unknown action video O^{test} histogram gaussian model, σ^2 Described is the same parameter of the Gaussian model of action a at time ratio d.

4 Probabilistic Fusion Prediction Strategy

4.1 A Subsection Sample

Given a test video of length t O^{test} (t unknown), calculation O^{test}. The histogram shows the similarity probability of Gaussian model with different kinds of actions at different time ratios. The rate value discriminates the similarity between the test video and various actions in the training set.

In this experiment, the UT-interaction database was used to test the method of combining the bag of word with the multi-time proportional Gaussian model. Take the handshake as the test video as an example. The example is shown in Fig. 8. It can be seen from the experimental results of Fig. 8 that as the proportion of the time of the test video increases, the similarity probability value of the Gaussian model of the same type of action gradually increases in the case of the same time ratio or time ratio, and is further verified. The validity of the Gaussian modeling of the word bag histogram feature representation of the action video. Considering the randomness of the length of the unknown test video time, in order to fully and effectively utilize the Gaussian model information of the proportional motion video at different time, the Gaussian model likelihood probability values obtained at different time ratios are probabilistically fused. The final result of the probabilistic fusion is used to characterize the similarity between the test video and each action model, and finally the unknown action is identified as the category of the action model with the highest similarity,

$$A^* = \arg\max_{1 \leq a \leq A}(\sum_{1 \leq d \leq D}^{D} \frac{p(O^{test}|h^{(a,d)})}{\sum_{1 \leq a \leq A}^{A} p(O^{test}|h^{(a,d)})}) \tag{3}$$

Where d is the time ratio of the current video, and a is the current action model category under the d time scale.

This paper designs the following three ways of probability fusion:

Take the Highest Probability: Regardless of the weight relationship of each time ratio, directly select the action category label corresponding to the highest similarity probability in each time scale as the action category of the unknown test video.

Maximum Probability Superposition: Without considering the weight relationship under each time scale, directly adding the similarity probability of the same class action Gaussian model under each time scale, and obtaining the similarity between the unknown motion test video and each action model under different time ratios, The action model category with the highest similarity is the unknown action category, namely:

$$A^* = \arg\max_{1 \leq a \leq A}(\sum_{1 \leq d \leq D}^{D} \frac{p(O^{test}|h^{(A,d)})}{\sum_{1 \leq a \leq A}^{a} p(O^{test}|h^{(A,d)})}) \tag{4}$$

Adjacent Highest Probability Superposition: Still do not consider the weight relationship under each time ratio, select the highest likelihood probability superposition of

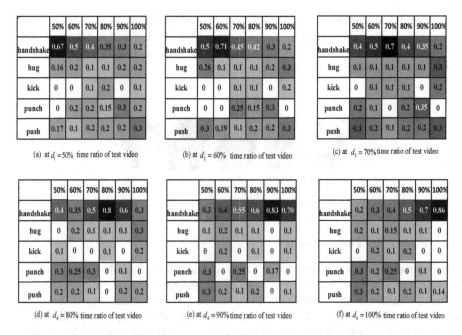

(a) at $d_1 = 50\%$ time ratio of test video

(b) at $d_2 = 60\%$ time ratio of test video

(c) at $d_3 = 70\%$ time ratio of test video

(d) at $d_4 = 80\%$ time ratio of test video

(e) at $d_4 = 90\%$ time ratio of test video

(f) at $d_4 = 100\%$ time ratio of test video

Fig. 8. Action recognition confusion matrix on the different time ratios

similar actions adjacent to each time ratio, and obtain the similarity between the unknown action and each action model under each time ratio, still take The category of the action model with the highest similarity belongs to the unknown action category,

$$A^* = \operatorname*{argmax}_{1 \le a \le A} \left(\sum_{1 \le d \le D}^{D} \frac{\max(P(O^{test}|h^{(a,d_i)}) + p(O^{test}|h^{(a,d_j)}))}{\sum_{1 \le a \le A}^{A} P(O^{test}|h^{(a,d)})} \right) \tag{5}$$

5 Experimental Results and Analysis

5.1 Database Introduction

This paper proposes the representation of the statistical features of the interest points of the action video, using the combination of the word bag model and the multi-time proportional Gaussian model. This method can basically predict the interaction behavior of unknown doubles. The database used in this experiment comes from the UT-interaction database, which is widely used in the research of two-person interactive behavior recognition and prediction algorithms. The experimental environment was completed on the Matlab 2014a software platform with a frequency of 2.40 GHz, a memory of 2 GB, and a 32-bit win7 operating system. In the experiment, the database was tested by leaving a cross-validation. A large number of experiments proved that the clustering visual word k = 800, $d_1, d_2, d_3, d_4, d_5, d_6$. The test results were best at 50%, 60%, 70%, 80%, 90%, and 100%, respectively. Since the interaction behavior of the two people requires rich behavior information, this experiment removes the "finger" action in the database. As shown in Fig. 9:

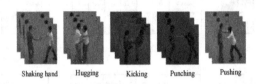

Shaking hand   Hugging   Kicking   Punching   Pushing

Fig. 9. UT-interaction database

5.2 Probability Fusion Result

By adopting three different probability fusion methods, a large number of test videos of different types of time and different types of actions are tested, and the experimental prediction results as shown in Table 1 are obtained. From the accuracy of the experimental prediction results in Table 1, it can be seen that the accuracy of prediction and recognition after probability fusion is higher than the maximum probability of direct selection: The method of large probability superposition is better than the method of superposition of the highest probability. That is, the experiment finally uses the method of maximum probability superposition to complete the representation of interactive action recognition and prediction results.

Table 1. Different time ratio test video final prediction results

	50%	60%	70%	80%	90%	100%
Maximu probability	0.50	0.63	0.67	0.75	0.80	0.80
Highest adjacent Probability superposition	0.60	0.67	0.70	0.78	0.83	0.83
Maximum probability Superimposed	0.67	0.70	0.73	0.78	0.83	0.86

5.3 Comparison of Different Forecasting Methods

Table 2 shows the prediction and recognition results of the interaction between two people in the public database in recent years. The method proposed in this paper is compared with other methods: the prediction and recognition rate of the method used in this paper are higher than those in the literature [3] and the literature [7], although the literature [4] and the literature [9] have higher prediction and recognition rates, but the algorithm complexity is very high, especially in the literature [9], the method of deep learning requires a large number of learning samples. The method of combining the word bag with the multi-time scale model proposed in this paper does not need to establish a complex prediction model, and the processing speed can reach 15 fps. And experiments show that the prediction and recognition accuracy is relatively high. The method adopted in this paper achieves good prediction results and good real-time performance, which can meet the basic requirements of double interaction behavior prediction.

Table 2. Prediction and recognition results of different algorithms on public databases

Source	Method	Accuracy w. half videos	Accuracy w. full videos
This article	3D-SIFT + IntegralBow + Gaussian model	67.0%	86.0%
Literature [3]	Integral Bow	65.0%	81.7%
Literature [7]	PST + PAF	55.0%	65.0%
Literature [4]	STISM	80.0%	91.7%
Literature [9]	CNN	88.3%	93.0%

6   Conclusion

The method based on the statistical features of interest points proposed in this paper realizes the prediction of different interaction behaviors. From the perspective of feature description and multi-time proportional model probability fusion, the combination of bag of word and Gaussian model is used to deal with the problem of predicting and identifying unknown actions with unknown time length. The method is simple to implement, meets real-time requirements, and has a good application background. However, there is a certain error in the motion prediction with similar action intervals. The next research focus will be on the prediction model optimization to improve the prediction accuracy of the interaction behavior of the two people.

References

1. Poppe, R.: A survey on vision-based human action recognition. Image Vis. Comput. **28**(6), 976–990 (2010)
2. Hassner, T.: A critical review of action recognition Benchmarks. In: Computer Vision and Pattern Recognition Workshops, pp. 245–250. IEEE (2013)
3. Ryoo, M.S.: Human activity prediction: early recognition of ongoing activities from streaming videos. In: International Conference on Computer Vision. IEEE Computer Society, pp. 1036–1043 (2011)
4. Yu, G., Yuan, J., Liu, Z.: Predicting human activities using spatio-temporal structure of interest points. In: ACM International Conference on Multimedia, pp. 1049–1052. ACM (2012)
5. Raptis, M., Sigal, L.: Methods and systems for action recognition using poselet keyframes, US20140294360 (2014)
6. Cao, Y., Wang, S., Barrett, D., et al.: Recognize human activities from partially observed videos. In: IEEE Conference on Computer Vision and Pattern Recognition. IEEE Computer Society, pp. 2658–2665 (2013)

7. Li, K., Hu, J., Fu, Y.: Modeling complex temporal composition of actionlets for activity prediction. In: Fitzgibbon, A., Lazebnik, S., Perona, P., Sato, Y., Schmid, C. (eds.) ECCV 2012. LNCS, vol. 7572, pp. 286–299. Springer, Heidelberg (2012). https://doi.org/10.1007/978-3-642-33718-5_21

8. Ke, Q., Bennamoun, M., An, S., et al.: Leveraging structural context models and ranking score fusion for human interaction prediction. IEEE Trans. Multimedia, **PP**(99), 1 (2016)

9. Ke, Q., Bennamoun, M., An, S., Boussaid, F., Sohel, F.: Human interaction prediction using deep temporal features. In: Hua, G., Jégou, H. (eds.) ECCV 2016. LNCS, vol. 9914, pp. 403–414. Springer, Cham (2016). https://doi.org/10.1007/978-3-319-48881-3_28

10. Weinland, D., Boyer, E., Ronfard, R.: Action recognition from arbitrary views using 3D exemplars. In: IEEE International Conference on Computer Vision, 1–7. IEEE (2007)

11. Schmid, C., Mohr, R.: Local Grayvalue invariants for image retrieval. IEEE Trans. Pattern Anal. Mach. Intell. **19**(5), 530–535 (1997)

Advanced Process Control
in Petrochemical Process

Robust Model Predictive Tracking Control for Chemical Process with Time-Varying Tracking Trajectory

Hui Li[1], Xueying Jiang[2], Huiyuan Shi[1(✉)], Bo Peng[1], and Chengli Su[1(✉)]

[1] School of Information and Control Engineering, Liaoning Shihua University, Fushun 113001, China
shy723915@126.com, sclwind@sina.com
[2] School of Information Science and Engineering, Northeastern University, Shenyang 110819, China

Abstract. A robust model predictive control method based on time-varying trajectory is proposed for discrete chemical process systems with uncertainties, bounded disturbances and time-varying target trajectories. Firstly, based on the traditional system state space model, the output tracking error is extended to the state space model, and the system controller is designed based on this extended model. Secondly, by means of Lyapunov stability theory and H-infinite performance index, the problems are solved for the process with model uncertainty, bounded disturbance and time-varying target trajectory, which can ensure the stability of the system. Then the linear matrix inequality (LMI) method is used to calculate the control law of the system. Finally, an example is given to verify the effectiveness and feasibility of the proposed method.

Keywords: Robust model predictive control · Uncertainty · Disturbances · Time-varying target trajectories · H-infinite

1 Introduction

With the increasing deepening of the industrialization process, the degree of automation is getting higher and higher, the control objects are getting more and more complex. Due to the complex uncertainty, the control of petrochemical process becomes more difficult. Traditional control methods have been unable to meet the increasingly complex needs of industrial production. In order to make the industrial production process more safe, reliable and effective, and to make up for the shortage of traditional control methods, more and more researchers focus on advanced control technologies. However, the robust model predictive control method has great advantages in dealing with uncertainties and disturbances, and it is considered as the most effective and developable advanced control method, which has attracted more and more attention from researchers [1–6].

In the actual industrial production process, the unknown external disturbance may lead to the poor control performance and the instability of the system. The robust

© Springer Nature Singapore Pte Ltd. 2020
J. Qian et al. (Eds.): ICRRI 2020, CCIS 1336, pp. 319–332, 2020.
https://doi.org/10.1007/978-981-33-4932-2_23

stabilization of discrete-time switched systems with state delay and parametric uncertainty via model predictive control approach proposed by Taghieh et al. [7] effectively solves the impact of model uncertainty on the system. By introducing the idea of dynamic output feedback into the rolling optimization problem, a minimax predictive controller design method based on linear matrix inequality (LMI) is proposed to ensure the stable operation of the system under the condition of small overshoot. However, its anti-disturbance ability is weak, and the system stability is poor in the presence of external disturbance.

In addition, the problem of the time-varying target trajectory also causes the stability of the system. The control method proposed by Liu et al. [8] can effectively solve the impact of model uncertainty and time delay on the system. By means of LMI method, the optimization problem of maximum and minimum values in infinite time domain is transformed into a linear programming problem, and the optimization tracking of the target is realized. However, the control effect for the time-varying tracking target is poor.

Aiming at the above problems, this paper focuses on the robust model predictive control method with uncertainties, unknown external bounded disturbance and time-varying tracking trajectories. Compared with other researches, the innovation points of this paper are as follows:

(1) The extended state space model is adopted in this paper. Based on the traditional state space model, the output tracking error is extended into the state space model. It not only improves the degree of freedom of the system but also reduces steady state error of the system.
(2) H-infinite performance index is used to deal with the unknown external bounded interference, which effectively improves the anti-disturbance ability of the system.
(3) The time-varying tracking trajectory of the system is considered and treated as a bounded disturbance. H-infinite constraint is constructed to reject the influence of the time-varying tracking target on the system.

The second chapter describes the problem of the robust model predictive control system with time-varying trajectory and establishes the extended model. The third chapter is the controller design of the control system, the fourth chapter is based on the simulation case of TTS20 water tank system, the fifth chapter is the conclusion of this paper.

2 Problem Description and Model Building

Consider the discrete system with uncertainty and unknown external disturbance described by the following equation of state space:

$$\begin{cases} x(k+1) = A(k)x(k) + B(k)u(k) + \omega(k) \\ y(k) = Cx(k) \end{cases} \tag{1}$$

where $x(k) \in R^{n_x}$, $u(k) \in R^{n_u}$, $y(k) \in R^{n_y}$ and $\omega(k)$ are system state, control input, system output and unknown external disturbance at discrete-time k, respectively $A(k) = A + \Delta_a(k)$ and $B(k) = B + \Delta_b(k)$, where A and B are constant matrices matched with the system dimension respectively. $\Delta_a(k)$ and $\Delta_b(k)$ are the uncertainty matrix matched with the system dimension. It is expressed as follows.

$$[\Delta_a(k) \quad \Delta_b(k)] = N\Delta(k)[H_a H_b] \tag{2}$$

$$\Delta(k)\Delta(k) \leq I \tag{3}$$

where N, H_a and H_b are constant matrices, $\Delta(k)$ is the unknown quantity in discrete time k. The form of the incremental state space can be expressed as follows.

$$\begin{cases} \Delta x(k+1) = A(k)\Delta x(k) + B(k)\Delta u(k) + \bar{\omega}(k) \\ \Delta y(k) = C\Delta x(k) \end{cases} \tag{4}$$

Defining the time-varying tracking trajectory of the system as $c(k)$, the system error is $e(k) = y(k) - c(k)$. It has:

$$e(k+1) = e(k) + \Delta e(k) = e(k) + C\Delta x(k+1) - \Delta c(k+1) \tag{5}$$

According to the above equations, the state space model with incremental state and system output tracking error is as follows.

$$\begin{cases} \bar{x}(k+1) = \bar{A}(k)\bar{x}(k) + \bar{B}\Delta u(k) + G\bar{\omega}(k) + L\Delta c(k+1) \\ \Delta y(k) = \bar{C}\bar{x}(k) \\ z(k) = e(k) = \bar{E}\bar{x}(k) \end{cases} \tag{6}$$

where
$$\bar{A}(k) = \bar{A} + \bar{\Delta}_a, \bar{A} = \begin{bmatrix} A & 0 \\ CA & I \end{bmatrix}, \bar{\Delta}_a = \begin{bmatrix} \Delta_a & 0 \\ C\Delta_a & I \end{bmatrix}, \bar{B}(k) = \bar{B} + \bar{\Delta}_b, \bar{B} = \begin{bmatrix} B \\ CB \end{bmatrix},$$

$$\bar{\Delta}_b = \begin{bmatrix} \Delta_b \\ C\Delta_b \end{bmatrix}, \bar{C} = [C \quad 0], G = \begin{bmatrix} I \\ C \end{bmatrix}, L = \begin{bmatrix} 0 \\ -I \end{bmatrix}, \bar{E} = [0 \quad I], \bar{H}_a = [H_a \quad 0],$$

$$\bar{H}_b = H_b, \bar{\Delta}_a = \bar{N}\Delta(k)\bar{H}_a, \bar{\Delta}_b = \bar{N}\Delta(k)\bar{H}_b.$$

Based on the above analysis, the control law of the system can be obtained as follows.

$$\Delta u(k) = \bar{K}\bar{x}(k) = \bar{K}\begin{bmatrix} \Delta x(k) \\ e(k) \end{bmatrix} \tag{7}$$

A robust model predictive controller is designed to meet the following performance indicators:

$$\min_{\Delta u(k+i), i \geq 0} \quad \max_{[A(k+i)\,B(k)]\in\Omega, i \geq 0} J_\infty$$

$$J_\infty(k) = \sum_{i=0}^{\infty} [(\bar{x}(k+i|k))^T \bar{Q}(\bar{x}(k+i|k)) + \Delta u(k+i|k)^T \bar{R}\Delta u(k+i|k)] \tag{8}$$

where $\bar{Q} > 0$ and $\bar{R} > 0$ are weighted matrices, $\Delta u(k+i|k)$ and $\bar{x}(k+i|k)$ represent the output prediction of the control input and state of the system at discrete-time k at time $k+1$, respectively.

3   Main Result

The main lemmas used in this paper are as follows:

Lemma 1: [9] (Schur complements lemma). Let W, L and V be matrices of appropriate dimensions in which W, V are real matrices, then for

$$L^T V L - W < 0 \tag{9}$$

if and only if

$$\begin{bmatrix} -W & L^T \\ L & -V^{-1} \end{bmatrix} < 0 \text{ or } \begin{bmatrix} -V^{-1} & L \\ L^T & -W \end{bmatrix} < 0 \tag{10}$$

Lemma 2: [10] Matrix D, F, E and M are given with appropriate dimensions, where M is symmetric, as follows:

$$M + DFE + E^T F^T D^T < 0 \tag{11}$$

where, for all matrices $F^T F \leq I$ satisfying F is true, and there is only one constant $\varepsilon > 0$, the following equation is true:

$$M + \varepsilon^{-1} DD^T + \varepsilon E^T E < 0 \tag{12}$$

Lemma 3: Consider a quadratic function $V(x(k|k)) = x^T P x$, $P > 0$ satisfies the following equation:

$$\begin{aligned} V(x(k+i+1|k)) - V(x(k+i|k)) \leq & - [(x(k+i|k)^T Q(x(k+i|k) \\ & - u(k+i|k)^T R u(k+i|k)] \end{aligned} \tag{13}$$

Under the condition of $V(\bar{x}(\infty)) = 0$ or $\bar{x}(\infty) = 0$, from $i = 0$ to ∞ on both sides of Eq. (13), then:

$$J_\infty(k) \leq V(x(k)) \leq \theta \tag{14}$$

where θ is the upper bound of $\bar{J}_\infty(k)$.

Lemma 4: Given a scalar $r > 0$, the asymptotic stability of the system with unknown disturbance $\omega(k)$ can be guaranteed, and the output $z(k)$ of the system can also satisfy $\|z\| \leq r\|\omega\|$. The discrete time robust H_∞ performance is considered.

Definition 1: $\Delta r^p(k + 1)$ is the change of time-varying set values. It is a known bounded variable that is not affected by system state variable, control input and other variables. It is treated here as bounded external interference.

Theorem 1. For the uncertain discrete system with external disturbance and time-varying tracking trajectory described in Eq. (6), when $\bar{\omega}(k) = 0$ and $c(k) = 0$ exist, there is a constant $\varepsilon > 0$, the matrix $W \in R^{m \times n}$ and the symmetric positive definite matrix $P_1 \in R^{m \times n}$ satisfying Eq. (15) and (16), so that the control parameters $\Delta u(k) = WP_1^{-1}\bar{x}(k)$ satisfies $V(\bar{x}(k|k))$ minimum.

$$
\begin{bmatrix}
-P_1 & P_1\bar{A}^T + W\bar{B}^T & P_1\bar{Q}^{1/2} & W\bar{R}^{1/2} & P_1\bar{H}_a^T & W\bar{H}_b^T \\
* & -P_1 + \varepsilon_1\bar{N}\bar{N}^T + \varepsilon_2\bar{N}\bar{N}^T & 0 & 0 & 0 & 0 \\
* & * & -\theta I & 0 & 0 & 0 \\
* & * & * & -\theta I & 0 & 0 \\
* & * & * & * & -\varepsilon_1 I & 0 \\
* & * & * & * & * & -\varepsilon_2 I
\end{bmatrix}
\tag{15}
$$

$$
s.t. \begin{bmatrix} 1 & \bar{x}^T(k) \\ \bar{x}(k) & P_1 \end{bmatrix} > 0
\tag{16}
$$

Proof 1: In order to ensure the stability of the robust model predictive control system, under the conditions of $\bar{\omega}(k) = 0$ and $c(k) = 0$. According to Lemma 3, there is a quadratic function $V(x(k|k)) = x^T Px$, $P > 0$ which satisfies the following equation:

$$
\begin{aligned}
V(\bar{x}(k + i + 1|k)) - V(\bar{x}(k + i|k)) \leq \\
-[(\bar{x}(k + i|k)^T\bar{Q}(\bar{x}(k + i|k) - \Delta u(k + i|k)^T\bar{R}\Delta u(k + i|k)
\end{aligned}
\tag{17}
$$

Under the condition of $V(\bar{x}(\infty)) = 0$ or $\bar{x}(\infty) = 0$, from $i = 0$ to ∞ on both sides of Eq. (15), then:

$$
\bar{J}_\infty(k) \leq V(\bar{x}(k)) \leq \theta
\tag{18}
$$

where θ is the upper bound of $\bar{J}_\infty(k)$. Let T and multiply both sides of Eq. (17) by θ^{-1} to get:

$$
\theta^{-1}\Delta V(\bar{x}(k + i|k)) + \theta^{-1}\bar{J}_\infty(k) \leq 0
\tag{19}
$$

Substituting $\Delta u(k|k) = \bar{K}\bar{x}(k|k)$ into Eq. (19) to get:

$$(\bar{A}(k) + \bar{B}(k)\bar{K})^T P_1^{-1}(\bar{A}(k) + \bar{B}(k)\bar{K}) - P_1^{-1} + \theta^{-1}\bar{Q} + \bar{K}^T\theta^{-1}\bar{R}\bar{K} < 0 \qquad (20)$$

According to Lemma 1, the following form can be obtained:

$$\begin{bmatrix} -P_1^{-1} & (\bar{A}(k)+\bar{B}(k)\bar{K})^T & \bar{Q}^{1/2} & \bar{R}^{1/2} \\ * & -P_1 & 0 & 0 \\ * & * & -\theta & 0 \\ * & * & * & -\theta \end{bmatrix} < 0 \qquad (21)$$

Based on $\bar{A}(k) = \bar{A} + \bar{\Delta}_a$ and $\bar{B}(k) = \bar{B} + \bar{\Delta}_b$, the following equation can be obtained:

$$\begin{bmatrix} -P_1^{-1} & (\bar{A}+\bar{B}\bar{K})^T & \bar{Q}^{1/2} & \bar{R}^{1/2} \\ * & -P_1 & 0 & 0 \\ * & * & -\theta & 0 \\ * & * & * & -\theta \end{bmatrix} + \begin{bmatrix} 0 & \bar{\Delta}_a^T & 0 & 0 \\ * & 0 & 0 & 0 \\ * & * & 0 & 0 \\ * & * & * & 0 \end{bmatrix} + \begin{bmatrix} 0 & \bar{\Delta}_b^T & 0 & 0 \\ * & 0 & 0 & 0 \\ * & * & 0 & 0 \\ * & * & * & 0 \end{bmatrix} < 0 \qquad (22)$$

Then, according to the Eq. (22) and Lemma 2. When $\bar{\Delta}_a$ and $\bar{\Delta}_b$ is a matrix that satisfies $\bar{\Delta}_a^T\bar{\Delta}_a \le I$ and $\bar{\Delta}_b^T\bar{\Delta}_b \le I$, it has a constant $\varepsilon_1 > 0$, $\varepsilon_2 > 0$ that make the following equation is true.

$$\begin{bmatrix} -P_1^{-1} & (\bar{A}+\bar{B}\bar{K})^T & \bar{Q}^{1/2} & \bar{R}^{1/2} \\ * & -P_1 & 0 & 0 \\ ** & -\theta & 0 \\ ** & * & -\theta \end{bmatrix} + \varepsilon_1 \begin{bmatrix} 0 \\ \bar{N} \\ 0 \\ 0 \end{bmatrix}\begin{bmatrix} 0 \\ \bar{N} \\ 0 \\ 0 \end{bmatrix}^T + \varepsilon_1^{-1}\begin{bmatrix} \bar{H}_a \\ 0 \\ 0 \\ 0 \end{bmatrix}\begin{bmatrix} \bar{H}_a \\ 0 \\ 0 \\ 0 \end{bmatrix}^T$$

$$+ \varepsilon_2 \begin{bmatrix} 0 \\ \bar{N} \\ 0 \\ 0 \end{bmatrix}\begin{bmatrix} 0 \\ \bar{N} \\ 0 \\ 0 \end{bmatrix}^T + \varepsilon_2^{-1}\begin{bmatrix} \bar{H}_b \\ 0 \\ 0 \\ 0 \end{bmatrix}\begin{bmatrix} \bar{H}_b \\ 0 \\ 0 \\ 0 \end{bmatrix}^T < 0 \qquad (23)$$

Based on Lemma 1 the following form can be obtained:

$$\begin{bmatrix} -P_1^{-1} & \bar{A}^T+\bar{B}^T\bar{K} & \bar{Q}^{1/2} & \bar{K}^T\bar{R}^{1/2} & \bar{H}_a^T & \bar{K}^T\bar{H}_b^T \\ * & -P_1+\varepsilon_1\bar{N}\bar{N}^T+\varepsilon_2\bar{N}\bar{N}^T & 0 & 0 & 0 & 0 \\ * & * & -\theta I & 0 & 0 & 0 \\ * & * & * & -\theta I & 0 & 0 \\ * & * & * & * & -\varepsilon_1 I & 0 \\ * & * & * & * & * & -\varepsilon_2 I \end{bmatrix} < 0 \qquad (24)$$

The left and right ends of Eq. (23) are multiplied by $\text{diag}[\,P_1\quad I\quad I\quad I\quad I\quad I\,]$, and carrying out inverse operation on Eq. (23), Eq. (15) can be obtained via $W = \bar{K}P_1$.

To get the system invariant set, take the maximum value of $\bar{x}(k)$, and set $P_1 = \theta P^{-1}$, Eq. (16) can be got based on $V(\bar{x}(k)) \le \bar{x}^T(k)P\bar{x}(k) \le \theta$ and Lemma 1.

Theorem 2. For the uncertain discrete system with external disturbance and time-varying tracking trajectory described in Eq. (6), when $\bar{\omega}(k)=0$ and $c(k)=0$ exist, there is a constant $\varepsilon > 0$, the matrix $W \in R^{m\times n}$ and the symmetric positive definite matrix $P_1 \in R^{m\times n}$ satisfying Eq. (25) and (26), so that the control parameters $\Delta u(k) = WP_1^{-1}\bar{x}(k)$ satisfies $V(\bar{x}(k|k))$ minimum.

$$\begin{bmatrix}
-P_1 & 0 & P_1\bar{A}^T + W^T\bar{B}^T & P_1\bar{E}^T & P_1\bar{Q}^{1/2} & W^T\bar{R}^{1/2} & P_1\bar{H}_a^T & W^T\bar{H}_b^T \\
* & -r^2I & G^T & 0 & 0 & 0 & 0 & 0 \\
* & * & -P_1 + \varepsilon_1\bar{N}\bar{N}^T + \varepsilon_2\bar{N}\bar{N}^T & 0 & 0 & 0 & 0 & 0 \\
* & * & * & -I & 0 & 0 & 0 & 0 \\
* & * & * & * & -\theta I & 0 & 0 & 0 \\
* & * & * & * & * & -\theta I & 0 & 0 \\
* & * & * & * & * & * & -\varepsilon_1 I & 0 \\
* & * & * & * & * & * & * & -\varepsilon_2 I
\end{bmatrix} < 0 \quad (25)$$

$$s.t.\begin{bmatrix} 1 & \bar{x}^T(k) \\ \bar{x}(k) & P_1 \end{bmatrix} > 0 \quad (26)$$

Proof 2: In the case of unknown external interference, in order to ensure the stability of the closed-loop system, on the basis of Theorem 1, the performance indexes are introduced as follows:

$$J = \sum_{k=0}^{\infty}[(z^T(k))z(k) - \gamma^2(\omega^T(k))\omega(k)] \quad (27)$$

Then for any $\bar{\omega}(k) \in l_2[0, \infty]$ with nonzero and according to Lemma 3, it has:

$$\bar{J}_{1\infty}(k) \le \sum_{k=0}^{\infty}[(z^T(k)z(k) - r^2(\omega^T(k)\omega(k) + \theta^{-1}\Delta V(\bar{x}(k+i|k)) + \theta^{-1}\bar{J}_{\infty}(k)] \quad (28)$$

Similar to Theorem 1, according to Lemma 1, the following form can be obtained:

$$\begin{bmatrix} \bar{x}(k) \\ \bar{\omega}(k) \end{bmatrix}^T \begin{bmatrix}
-P_1^{-1} & 0 & (\bar{A}(k)+\bar{B}(k)\bar{K})^T & \bar{E}^T & \bar{Q}^{1/2} & \bar{R}^{1/2} \\
* & -r^2 & G^T & 0 & 0 & 0 \\
* & * & -P_1 & 0 & 0 & 0 \\
* & * & * & -I & 0 & 0 \\
* & * & * & * & -\theta & 0 \\
* & * & * & * & * & -\theta
\end{bmatrix} \begin{bmatrix} \bar{x}(k) \\ \bar{\omega}(k) \end{bmatrix} < 0 \quad (29)$$

Based on $\bar{A}(k) = \bar{A} + \bar{\Delta}_a$ and $\bar{B}(k) = \bar{B} + \bar{\Delta}_b$, the following equation can be obtained:

$$
\begin{bmatrix}
-P_1 & 0 & \bar{A}^T + \bar{K}^T\bar{B}^T & P_1\bar{E}^T & \bar{Q}^{1/2} & \bar{K}^T\bar{R}^{1/2} \\
* & -r^2I & G^T & 0 & 0 & 0 \\
** & -P_1 + \varepsilon_1\bar{N}\bar{N}^T + \varepsilon_2\bar{N}\bar{N}^T & 0 & 0 & 0 \\
** & * & -I & 0 & 0 \\
** & ** & -\theta I & 0 \\
** & ** & * & -\theta I
\end{bmatrix}
+
\begin{bmatrix}
0 & 0 & \bar{\Delta}_a^T & 0 & 0 & 0 \\
* & 0 & 0 & 0 & 0 & 0 \\
** & 0 & 0 & 0 & 0 \\
** & * & 0 & 0 & 0 \\
** & ** & 0 & 0 \\
** & ** & * & 0
\end{bmatrix}
$$

$$
+
\begin{bmatrix}
0 & 0 & \bar{\Delta}_b^T & 0 & 0 & 0 \\
* & 0 & 0 & 0 & 0 & 0 \\
** & 0 & 0 & 0 & 0 \\
** & * & 0 & 0 & 0 \\
** & ** & 0 & 0 \\
** & ** & * & 0
\end{bmatrix}
< 0
\tag{30}
$$

Then according to the Eq. (30) and Lemma 2, when $\bar{\Delta}_a$ and $\bar{\Delta}_b$ is a matrix that satisfies $\bar{\Delta}_a^T\bar{\Delta}_a \le I$ and $\bar{\Delta}_b^T\bar{\Delta}_b \le I$, it has a constant $\varepsilon_1 > 0$, $\varepsilon_2 > 0$ that make the following equation is true.

$$
\begin{bmatrix}
-P_1 & 0 & \bar{A}^T + \bar{K}^T\bar{B}^T & P_1\bar{E}^T & \bar{Q}^{1/2} & \bar{K}^T\bar{R}^{1/2} \\
* & -r^2I & G^T & 0 & 0 & 0 \\
** & -P_1 + \varepsilon_1\bar{N}\bar{N}^T + \varepsilon_2\bar{N}\bar{N}^T & 0 & 0 & 0 \\
** & * & -I & 0 & 0 \\
** & ** & -\theta I & 0 \\
** & ** & * & -\theta I
\end{bmatrix}
+ \varepsilon_1
\begin{bmatrix} 0 \\ 0 \\ \bar{N} \\ 0 \\ 0 \\ 0 \end{bmatrix}
\begin{bmatrix} 0 \\ 0 \\ \bar{N} \\ 0 \\ 0 \\ 0 \end{bmatrix}^T
$$

$$
+ \varepsilon_1^{-1}
\begin{bmatrix} \bar{H}_a \\ 0 \\ 0 \\ 0 \\ 0 \\ 0 \end{bmatrix}
\begin{bmatrix} \bar{H}_a \\ 0 \\ 0 \\ 0 \\ 0 \\ 0 \end{bmatrix}^T
+ \varepsilon_2
\begin{bmatrix} 0 \\ 0 \\ \bar{N} \\ 0 \\ 0 \\ 0 \end{bmatrix}
\begin{bmatrix} 0 \\ 0 \\ \bar{N} \\ 0 \\ 0 \\ 0 \end{bmatrix}^T
+ \varepsilon_2^{-1}
\begin{bmatrix} \bar{H}_b \\ 0 \\ 0 \\ 0 \\ 0 \\ 0 \end{bmatrix}
\begin{bmatrix} \bar{H}b \\ 0 \\ 0 \\ 0 \\ 0 \\ 0 \end{bmatrix}^T
< 0
\tag{31}
$$

According to Lemma 1, the following form can be obtained:

$$
\begin{bmatrix}
-P_1 & 0 & \bar{A}^T + \bar{K}^T\bar{B}^T & \bar{E}^T & \bar{Q}^{1/2} & \bar{K}^T\bar{R}^{1/2} & \bar{H}_a^T & \bar{K}^T\bar{H}_b^T \\
* & -r^2 & G^T & 0 & 0 & 0 & 0 & 0 \\
* & * & -P_1 + \varepsilon_1\bar{N}\bar{N}^T + \varepsilon_2\bar{N}\bar{N}^T & 0 & 0 & 0 & 0 & 0 \\
* & * & * & -I & 0 & 0 & 0 & 0 \\
* & * & * & * & -\theta & 0 & 0 & 0 \\
* & * & * & * & * & -\theta & 0 & 0 \\
* & * & * & * & * & * & -\varepsilon_1 & 0 \\
* & * & * & * & * & * & * & -\varepsilon_2
\end{bmatrix}
\tag{32}
$$

The left and right ends of Eq. (32) are multiplied by $\text{diag}[\,P_1 \ \ I \ \ I \ \ I \ \ I \ \ I \ \ I \ \ I\,]$, and carrying out inverse operation on Eq. (32), Eq. (25) can be obtained via $W = \bar{K}P_1$. The H_∞ performance index $\|z\| \le r\|\bar{\omega}\|$ is guaranteed. So far, it completes proof 2.

Theorem 3. For the uncertain discrete system with external disturbance and time-varying tracking trajectory described in Eq. (6), when $\omega(k) \ne 0$ and $c(k) \ne 0$ exist, there is a constant $\varepsilon > 0$, the matrix $W \in R^{m \times n}$ and the symmetric positive definite matrix $P_1 \in R^{m \times n}$ satisfying Eq. (16), so that the control parameters $\Delta u(k) = WP_1^{-1}\bar{x}(k)$ satisfies $V(\bar{x}(k|k))$ minimum.

$$
\begin{bmatrix}
-P_1 & 0 & 0 & P_1\bar{A}^T + W^T\bar{B}^T & P_1\bar{E}^T & P_1\bar{E}^T & P_1\bar{Q}^{1/2} & W^T\bar{R}^{1/2} & P_1\bar{H}_a^T & W^T\bar{H}_b^T \\
* & -r^2 I & 0 & 0 & 0 & 0 & 0 & 0 & 0 & 0 \\
* & * & -o^2 I & 0 & 0 & 0 & 0 & 0 & 0 & 0 \\
* & * & * & -P_1 + \varepsilon_1\bar{N}\bar{N}^T + \varepsilon_2\bar{N}\bar{N}^T & 0 & 0 & 0 & 0 & 0 & 0 \\
* & * & * & * & -I & 0 & 0 & 0 & 0 & 0 \\
* & * & * & * & * & -I & 0 & 0 & 0 & 0 \\
* & * & * & * & * & * & -\theta I & 0 & 0 & 0 \\
* & * & * & * & * & * & * & -\theta I & 0 & 0 \\
* & * & * & * & * & * & * & * & -\varepsilon_1 I & 0 \\
* & * & * & * & * & * & * & * & * & -\varepsilon_2 I
\end{bmatrix}
$$

$$\tag{33}$$

$$
s.t. \begin{bmatrix} 1 & \bar{x}^T(k) \\ \bar{x}(k) & P_1 \end{bmatrix} > 0 \tag{34}
$$

Proof 3: Since the change in the time-varying target trajectories $c(k)$ can be considered as a known bounded disturbance, in terms of $\bar{\omega}(k) \ne 0$ and $c(k) \ne 0$, the following H_∞ performance index for time-varying tracking trajectory and bounded disturbance are introduced respectively.

$$
J = \sum_{k=0}^{\infty} [(z^T(k))z(k) - o^2(\Delta c(k+1)^T)\Delta c(k+1)] \tag{35}
$$

Then for any $\bar{\omega}(k) \in l_2[0, \infty]$ and $c(k) \in l_2[0, \infty]$, the following form can be obtained:

$$
\begin{aligned}
\bar{J}_{2\infty}(k) \le \sum_{k=0}^{\infty} & [(z^T(k)z(k) - r^2(\omega^T(k)\omega(k) + (z^T(k)z(k) \\
& - (r)^2(\Delta c(k+1)^T\Delta c(k+1)(k) + \theta^{-1}\Delta V(\bar{x}(k+i|k)) + \theta^{-1}\bar{J}_\infty(k)]
\end{aligned}
\tag{36}
$$

Through Lemma 3 and Lemma 1, the following form can be obtained:

$$
\begin{bmatrix} \bar{x}(k) \\ \bar{\omega}(k) \\ \Delta c(k+1) \end{bmatrix}^{\mathrm{T}}
\begin{bmatrix}
-P_1^{-1} & 0 & 0 & (\bar{A}(k)+\bar{B}(k)\bar{K})^{\mathrm{T}} & \bar{E}^{\mathrm{T}} & \bar{E}^{\mathrm{T}} & \bar{Q}^{1/2} & \bar{R}^{1/2} \\
* & -r^2 & 0 & G^{\mathrm{T}} & 0 & 0 & 0 & 0 \\
* & * & -o^2 & L^{\mathrm{T}} & 0 & 0 & 0 & 0 \\
* & * & * & -P_1 & 0 & 0 & 0 & 0 \\
* & * & * & * & -I & 0 & 0 & 0 \\
* & * & * & * & * & -I & 0 & 0 \\
* & * & * & * & * & * & -\theta & 0 \\
* & * & * & * & * & * & * & -\theta
\end{bmatrix}
\begin{bmatrix} \bar{x}(k) \\ \bar{\omega}(k) \\ \Delta c(k+1) \end{bmatrix} < 0
$$

(37)

Based on $\bar{A}(k) = \bar{A} + \bar{\Delta}_a$, $\bar{B}(k) = \bar{B} + \bar{\Delta}_b$ and Lemma 2, it has:

$$
\begin{bmatrix}
-P_1^{-1} & 0 & 0 & \bar{A}^{\mathrm{T}}+\bar{K}^{\mathrm{T}}\bar{B} & \bar{E}^{\mathrm{T}} & \bar{E}^{\mathrm{T}} & \bar{Q}^{1/2} & \bar{R}^{1/2} \\
* & -r^2 & 0 & G^{\mathrm{T}} & 0 & 0 & 0 & 0 \\
** & -o^2 & L^{\mathrm{T}} & 0 & 0 & 0 & 0 \\
** & * & -P_1+\varepsilon_1\bar{N}\bar{N}^{\mathrm{T}}+\varepsilon_2\bar{N}\bar{N}^{\mathrm{T}} & 0 & 0 & 0 & 0 \\
** & ** & -I & 0 & 0 & 0 \\
** & ** & * & -I & 0 & 0 \\
** & ** & ** & -\theta & 0 \\
** & ** & ** & * & -\theta
\end{bmatrix}
+ \varepsilon_1 \begin{bmatrix} 0 \\ 0 \\ 0 \\ \bar{N} \\ 0 \\ 0 \\ 0 \\ 0 \end{bmatrix}\begin{bmatrix} 0 \\ 0 \\ 0 \\ \bar{N} \\ 0 \\ 0 \\ 0 \\ 0 \end{bmatrix}^{\mathrm{T}}
+ \varepsilon_2 \begin{bmatrix} 0 \\ 0 \\ 0 \\ \bar{N} \\ 0 \\ 0 \\ 0 \\ 0 \end{bmatrix}\begin{bmatrix} 0 \\ 0 \\ 0 \\ \bar{N} \\ 0 \\ 0 \\ 0 \\ 0 \end{bmatrix}^{\mathrm{T}}
$$

$$
+ \varepsilon_1^{-1} \begin{bmatrix} \bar{H}_a \\ 0 \\ 0 \\ 0 \\ 0 \\ 0 \\ 0 \\ 0 \end{bmatrix}\begin{bmatrix} \bar{H}_a \\ 0 \\ 0 \\ 0 \\ 0 \\ 0 \\ 0 \\ 0 \end{bmatrix}^{\mathrm{T}}
+ \varepsilon_2^{-1} \begin{bmatrix} \bar{H}_b \\ 0 \\ 0 \\ 0 \\ 0 \\ 0 \\ 0 \\ 0 \end{bmatrix}\begin{bmatrix} \bar{H}_b \\ 0 \\ 0 \\ 0 \\ 0 \\ 0 \\ 0 \\ 0 \end{bmatrix}^{\mathrm{T}} < 0
$$

(38)

According to Lemma 1, it has:

$$
\begin{bmatrix}
-P_1 & 0 & 0 & \bar{A}^{\mathrm{T}}+\bar{K}^{\mathrm{T}}\bar{B}^{\mathrm{T}} & \bar{E}^{\mathrm{T}} & \bar{E}^{\mathrm{T}} & \bar{Q}^{1/2} & \bar{K}^{\mathrm{T}}\bar{R}^{1/2} & \bar{H}_a^{\mathrm{T}} & \bar{K}^{\mathrm{T}}\bar{H}_b^{\mathrm{T}} \\
* & -r^2 & 0 & G^{\mathrm{T}} & 0 & 0 & 0 & 0 & 0 & 0 \\
* & * & -o^2 & L^{\mathrm{T}} & 0 & 0 & 0 & 0 & 0 & 0 \\
* & * & * & -P_1+\varepsilon_1\bar{N}\bar{N}^{\mathrm{T}}+\varepsilon_2\bar{N}\bar{N}^{\mathrm{T}} & 0 & 0 & 0 & 0 & 0 & 0 \\
* & * & * & * & -I & 0 & 0 & 0 & 0 & 0 \\
* & * & * & * & * & -I & 0 & 0 & 0 & 0 \\
* & * & * & * & * & * & -\theta & 0 & 0 & 0 \\
* & * & * & * & * & * & * & -\theta & 0 & 0 \\
* & * & * & * & * & * & * & * & -\varepsilon_1 & 0 \\
* & * & * & * & * & * & * & * & * & -\varepsilon_2
\end{bmatrix} < 0
$$

(39)

The left and right ends of Eq. (39) are multiplied by $\mathrm{diag}[\,P_1\ I\ I\ I\ I\ II\ I\ I\ I\,]$, and carrying out inverse operation on Eq. (39), Eq. (33) can be obtained via $W = \bar{K}P_1$. The H_∞ performance index $\|z\| \le o\|\bar{\omega}\|$ is guaranteed. So far, it complets proof 3.

4 Simulation Case

4.1 A Subsection Sample

This study uses the model of Ingenieurburo GurskiSchramm's TTS20 water tank system and uses MATLAB to simulate some or whole of the controlled objects in actual industrial production. Figure 1 is the schematic of TTS20 the water tank.

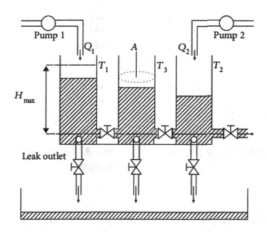

Fig. 1. Structure and process flow of TTS20

4.2 Process Model

Liquid level is one of the most important control parameters in petrochemical production. Taking TTS20 water tank level as an example, the proposed control method is used to control the liquid level. The state space model of liquid level for TTS20 water tank is as follows [11].

$$A = \begin{bmatrix} 0.9850 & 0.0107 \\ 0.0078 & 0.9784 \end{bmatrix}, B = \begin{bmatrix} 64.4453 \\ 0.2559 \end{bmatrix}, C = \begin{bmatrix} 1 & 0 \end{bmatrix},$$
$$N = \begin{bmatrix} 0.1 & 0 \\ 0 & 0.1 \end{bmatrix}, H_a = \begin{bmatrix} 0.1 & 0 \\ 0 & 0.2 \end{bmatrix}, H_b = \begin{bmatrix} 0.2 \\ 0.1 \end{bmatrix}.$$

The initial state of the system is $x(0) = \begin{bmatrix} 0 & 0 \end{bmatrix}^T$, The weighted matrix of performance indicators are $Q = \text{diag}[10\,5\,1]$, $R = 0.1$. The boundary of the unknown disturbance is 0.2.

In order to describe the system output error conveniently, the following formula is defined:

$$D(k) = \sqrt{e^T(k)e(k)} \tag{40}$$

The controller input of the time-varying trajectory robust predictive control system is shown in Fig. 2. The traditional control input is shown in Fig. 3. From the comparison between Fig. 2 and Fig. 3, the disturbance of the traditional control method is obviously greater than that of the control method proposed in this paper.

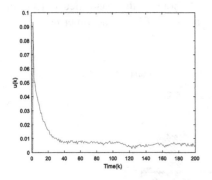

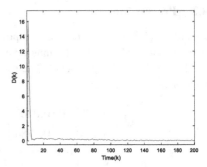

Fig. 2. Controller input with the proposed method

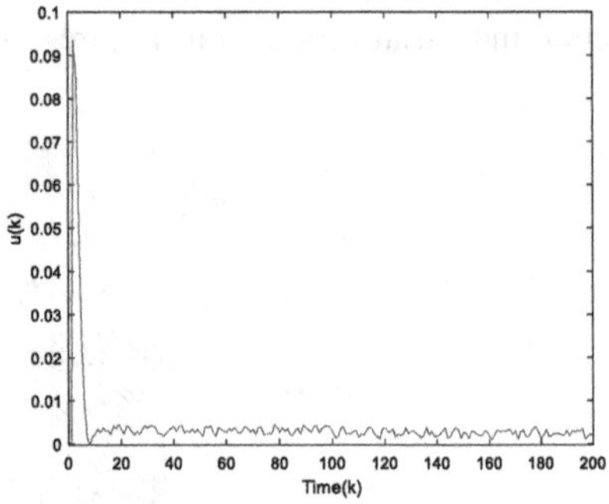

Fig. 3. Control input with the traditional method

Figure 4 and Fig. 5 respectively show the tracking performance of the proposed method and the tracking performance of the traditional method, as shown below.

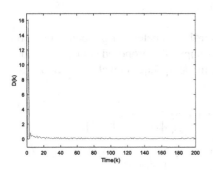

Fig. 4. System tracking performance with the proposed method

Fig. 5. System tracking performance with the traditional control method

Compared with Fig. 6 and Fig. 7, the control method proposed in this paper is more effective than the traditional control method. The comparison of the output response of the system is as follows.

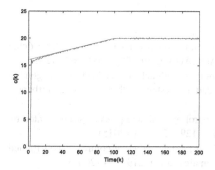

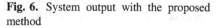

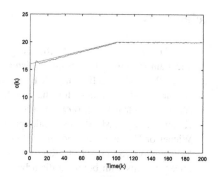

Fig. 6. System output with the proposed method

Fig. 7. System output with the traditional method

Through the above simulation examples, it can be seen that, for uncertainty, unknown external disturbance and time-varying tracking track on system, the control method proposed can not only guarantee the stability of the system but also effectively avoid the influence of uncertainties on the control effect of the system. Moreover, the system has faster response speed, smaller overshoot and better control performance, which is more suitable for the actual industrial production process.

5 Conclusions

In this paper, a robust predictive control method is designed for chemical processes with uncertainties, unknown external disturbances and time-varying trajectories. Based on the traditional state space model, the output tracking error is extended to reduce the steady-state error and improve the degree of freedom of the system. By means of Lyapunov Krasovskii stability theory, linear matrix inequality method and H-Infinite performance index, the model uncertainty, bounded disturbance and target trajectory time-varying disturbance are effectively suppressed while ensuring the stable operation of the system. The simulation results show that the proposed method can guarantee the stability of chemical system in the presence of the time-varying tracking trajectory.

In future studies, the time-varying tracking trajectory of multi-phase batch processes will be considered. Because the tracking trajectory of the target usually changes according to the production requirements in the actual production process, there are also related requirements in the multi-phase batch process.

Acknowledgments. This work was supported by the National Natural Science Foundation of China under Grant 61673199, 61703191, 61803191, Natural Science Foundation of Liaoning Province 20180550905, Natural Science Fund Project of Liaoning province 2019-KF-03-05, the Program of Innovative Talents in Universities LR2019037 and the Open Research Project of the State Key Laboratory of Industrial Control Technology, Zhejiang University, China (ICT20059).

References

1. Su, C.L., Zhao, J.C., Li, P.: Robust predictive control for a class of multiple time delay uncertain systems with nonlinear disturbance. Acta Autom. Sin. **5**(39), 644–649 (2013)
2. Ding, B.C., Gao, C.B., Ping, X.B.: Dynamic output feedback robust MPC using general polyhedral state bounds for the polytopic uncertain system with bounded disturbance. Asian J. Control **18**(2), 699–708 (2016)
3. Khani, F., Haeri, M.: Robust model predictive control of nonlinear processes represented by Wiener or Hammerstein models. Chem. Eng. Sci. **129**, 223–231 (2015)
4. Mayne, D.Q., Seron, M.M., Raković, S.V.: Robust model predictive control of constrained linear systems with bounded disturbances. Automatica **2**(41), 219–224 (2005)
5. Wu, S., Jin, Q.B., Zhang, R.D., Zhang, J.F., Gao, F.R.: Improved design of constrained model predictive tracking control for batch processes against unknown uncertainties. ISA Trans. **69**, 273–280 (2017)
6. Bououden, S., Chadli, M., Karimi, H.R.: A robust predictive control design for nonlinear active suspension systems. Asian J. Control **1**(18), 122–132 (2016)
7. Taghieh, A., Hashemzadeh, F., Shafiei, M.H.: Robust stabilization of discrete-time switched systems with state delay and parametric uncertainty via model predictive control approach. In: 6th International Conference on Control, Instrumentation and Automation, pp. 1–6. IEEE, Sanandaj (2019)
8. Liu, Z.L., Zhang, J., Pei, R.: Robust model predictive control of time-delay systems. In: Proceedings of 2003 IEEE Conference on Control Applications, pp. 470–473. IEEE, Istanbul (2003)
9. Boyd, S., Ghaoui, L.E., Feron, E., Balakrishnan, V.: Linear matrix inequalities in system and control theory. Society for industrial and applied mathematics, Philadelphia (1994)
10. Yu, K.W., Lien, C.H.: Stability criteria for uncertain neutral systems with interval time-varying delays. Chaos Soliton Fractals **3**(38), 650–657 (2008)
11. Shi, H.Y., Li, P., Wang, L.M., Su, C.L., et al.: Delay-range-dependent robust constrained model predictive control for industrial processes with uncertainties and unknown disturbances. Complexity **2019**, 1–15 (2019)

Hybrid Integrated Model of Water Quality in Wastewater Treatment Process via RBF Neural Network

Qiumei Cong[✉], Xiaoxun Xi, Chengli Su, Shuxian Deng,
and Yu Zhao

School of Information and Control Engineering, Liaoning Shihua University,
Fushun, China
cong_0828@163.com

Abstract. Mechanism model of activated sludge process (ASM) is hard to predict the water qualities because of the frequent operating conditions of wastewater treatment process (WWTP). This paper presented an integrated soft sensor of effluent COD, which is composed of simplified mechanism model and RBF neural network (RBFNN) as sub-models. The output of the integrated model is the sum of the outputs of sub-models. Simplified mechanism model can describe the dynamic mechanism characteristics of WWTP and RBFNN is used to compensate the modeling error of the mechanism model. The integrated model has the simplified structure and satisfying real-time capability. Simulations showed the hybrid integrated model has high predicting accuracy.

Keywords: Modeling · WWTP · Integrated model · RBF neural network

1 Introductions

WWTP is a typical nonlinear process with varying operating conditions. WWTP based on activated slusge process to remove the organic contaminants to purify the wastewater. Some key water qualities of activated sludge process are difficult to be measured online, which is not conducive for the control and optimization of WWTP. The prediction accuracy of ASM series models is very poor when they are only adopted to predict water qualities because of high dimensions, complex structure, parameter invariability with varying operating conditions. Moreover, the high-dimensionality of detailed phenomenological models results in an enormous computational requirements and ill-conditioned problems due to the interaction between fast and slow dynamics [1]. From the perspective of modeling, ASM series models are difficult to identify in real time due to the large number of unknown parameters, which is not conducive to the practical application [2]. Existing soft sensors of water quality based on the mechanism model adopted simplified or reduced-order model to decrease the calculation cost, including linearization of ASM models, the simplification of ASM models based on mechanism analysis and the simplification of ASM models based on the separation of time scales.

J. Qian et al. (Eds.): ICRRI 2020, CCIS 1336, pp. 333–341, 2020.
https://doi.org/10.1007/978-981-33-4932-2_24

Soft sensor of water quality based on data-driven model is investigated widely, where neural network, SVM, feedforward neural networks, adaptive and recursive extensions of PCR and PLS [3], fuzzy neural network are used to predict water qualities of WWTP. Data-driven model cannot indicate the biochemical reactions of WWTP, and require the high quality and adequate quantity of the modeling data. Bates and Granger presented the multi-models idea to improve the prediction accuracy and robustness by means of integrating several models [4].

Hybrid modeling method gains wide attention recently through integrating the mechanism model including biochemical reaction expressions and intelligent models. Literature [5] presented the hybrid soft sensor for COD, MLSS and cyanide concentration, which ASM1 and error compensation model are arranged in parallel. But the order of ASM1 model is very high resulting to the long computation time. The parameters of error compensation model are numerous and difficult to be identified. Literature [6] presented a hybrid soft sensor for COD based on mechanism model and linear polynomial models, and linear models are used to compensate the modeling error of mechanism model. The number of linear models can be achieved by synchronous clustering algorithm that the time interval between input and output data and the relevance of adjacent data is considered.

A hybrid integrated model for effluent COD of the activated sludge A/O process is presented in this paper. The hybrid model is composed of two sub-models that are simplified ASM1 model and RBFNN. The output of the hybrid model is the sum of those of sub-models, simplified ASM1 model describes the biochemical reactions of A/O process and the RBFNN is used to compensate the error of the simplified ASM1 model. The presented model has compact structure and satisfying real-time capability. Simulations based on real data showed the hybrid integrated model has high predicting accuracy.

2   Descriptions of A/O Process

The flow diagrams of A/O process is shown as Fig. 1.

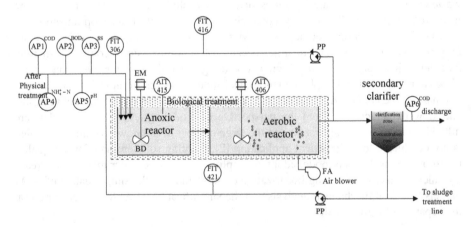

Fig. 1. Framework of A/O process

In A/O process, pretreated wastewater flows into anoxic tank, and mixes with recycled sludge and mixed liquor from aerobic tank. In anoxic tank, denitrification removes nitrogenous pollutant and portion of carbonic pollutant. Then, wastewater flows into aerobic tank where nitration and carbon degradation accur and most of carbonic pollutant is removed. The DO (Dissolved Oxygen) of aerobic tank is controlled by aeration blowers. Portion of wastewater from aerobic tank is recycled back to anoxic tank to participate in denitrification, the remainder flows into secondary clarifier where the upper clarified water is recycled as industrial consumption or discharged to receiving waterbody.

The variables of influencing effluent COD of A/O process are influent SS (Suspended Solid concentration), influent COD, ammonia nitrogen $NH_4^+ - N$, flowrate Q_{in}, DO of aerobic tank. The relationship of above variables can be expressed as follows:

$$\hat{y}_{COD} = f(SS, NH_4^+ - N, Q_{in}, COD, DO) \tag{1}$$

Here, $f(\cdot)$ denotes unknown dynamic nonlinear function. Hybrid model with the relationship $f(\cdot)$ of effluent COD and related variables can be used to calculate the estimated COD.

3 The Structure of Hybrid Integrated Model

The hybrid integrated model presented in this paper is composed of data acquisition and preprocessing, simplified mechanism model of effluent COD and RBFNN, shown as Fig. 2.

The raw data from real WWTP should be processed by filtering, recognition and correction of outliers. A/O process model based on simplified ASM1 and RBFNN are the submodels of hybrid integrated model whose output is the sum of the outputs of submodels. The error of mechanism model is compensated by RBFNN.

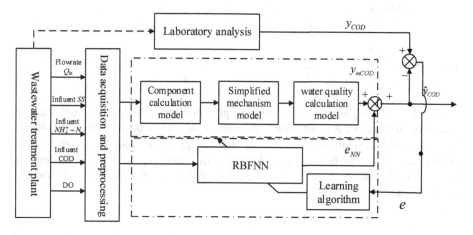

Fig. 2. The structure of hybrid integrated model

The output of hybrid integrated model can be written by:

$$\hat{y}_{COD} = y_{mCOD} + e_{NN} \tag{2}$$

Here, y_{mCOD} is the calculated COD of mechanism model, e_{NN} the output of RBFNN.

3.1 Mechanism Model of A/O Process Based on Simplified ASM1 Model

Simplification assumptions of ASM1 refer to [6] and SASM1 (Simplified ASM1 model) in matrix format is built, see Table 1. SASM1 model is used to describe the mechanism of A/O process, and the parameters of SASM1 use the default values of ASM1 model at 20 °C [7]. Component calculation model proposed in [6] is adopted to calculate components concentration corresponding to SASM1 model from influent COD, SS and $NH_4^+ - N$. COD is calculated by water quality calculation model using the components concentration obtained by A/O process model, expressed as y_{mCOD}.

Ss is readily biodegradable substance; X_{IP} is inert particulate organic matter; X_S is slow biodegradable substance; X_{BH} is heterotrophic bacteria; X_{BA} is autotrophic bacteria; S_{NO} is nitrogen in the form of nitrate and nitrite nitrogen; S_{NH} is ammonia nitrogen; S_O is dissolved oxygen, using actual measured DO concentration.

Table 1. SASM1 model

Components i		1 S_S	2 X_{IP}	3 X_S	4 X_{BH}	5 S_{NO}	6 S_{NH}	Process rate ρ_j
j	Process							
1	Aerobic growth of heterotrophic bacteria	$-\frac{1}{Y_H}$			1		$-i_{XB}$	$\mu_{mH} = \frac{S_S}{K_S + S_S} \frac{S_O}{K_{OH} + S_O} X_{BH}$
2	Anoxic growth of heterotrophic bacteria	$-\frac{1}{Y_H}$			1	$\frac{Y_H - 1}{2.86 Y_H}$	$-i_{XB}$	$\mu_{mH} = \frac{S_S}{K_S + S_S} \frac{K_{OH}}{K_{OH} + S_O} \frac{S_{NO}}{K_{NO} + S_{NO}} \eta_g X_{BH}$
3	Decay of heterotrophic bacteria		f_P	$1 - f_P$	-1			$b_H X_{BH}$
Conversion Rate (M/L^3 · T)	$r_j = \sum_j v_{ij}\rho_j$							

μ_{mH} is heterotrophic maximum specific growth rate, b_H attenuation coefficient of heterotrophic bacteria, K_S half-saturation constant of readily biodegradable substance, K_{OH} oxygen half-saturation constant of heterotrophic bacteria, K_{NO} nitrates half-saturation constant of heterotrophic bacteria, Y_H heterotrophic yield coefficient, f_P proportion of particulate microorganisms, i_{XB} proportion of microbial cells for the nitrogen content.

3.2   RBF Neural Network

In order to improve the modeling accuracy and trace the real COD, the modeling error of mechanism model should be compensated. RBFNN is used to compensate the modeling error of A/O process model in order that the accuracy of hybrid integrated model can be improved and the dynamic trend of WWTP be traced accurately. As a submodel of the integrated model, the inputs of SORBFNN are $X = [Q_{in}, SS, NH_4^+ - N, COD, DO]$, and the output is e_{NN}.

3.2.1   RBF Neural Network

RBFNN is a typical feedforward neural network and has the uniform approximation capability. RBFNN is widely used in pattern classification and function approximation and fit for soft sensor of industrial process. The following single-output RBFNN can describe nonlinear system:

$$\hat{e}_{NN}(k + 1) = \mathbf{W}(k) \, \Phi \, [\mathbf{x}(k)] \tag{3}$$

Here, $\mathbf{x}(k) \in \mathbf{R}^I$ denote the inputs, $\mathbf{W} \in \mathbf{R}^{1 \times H}$ weights of output layer, I, H the node numbers of input layer and hidden layer, $\Phi(x)$ the vector of radial basis function.

$$\Phi_i(x) = exp\left(-\frac{(\mathbf{x} - c_i)^2}{\sigma_i^2}\right), \, i = 1, 2, \ldots, H \tag{4}$$

Here, $\mathbf{c} = [c_i]$ denotes the center of Gaussian function, $\sigma = [\sigma_i]$ the width of Gaussian function.

The number of hidden nodes can be fixed or adjusted, and the centers and widths be learned during the learning process of RBFNN. In this paper, the parameters of RBFNN is learned by stable learning algorithm to avoid the drift of parameters and the unstablility of modeling error.

3.3   The Learning Algorithm

The following performance index is defined:

$$J = \frac{1}{2N} \sum_{k=1}^{N} [e(k)]^2 \tag{5}$$

Here, $e(k) = \hat{y}_{COD}(k) - y_{COD}(k)$ denotes modeling error, $y_{COD}(k)$ real COD, $\hat{y}_{COD}(k)$ the estimated COD of integrated model, N the number of training samples.

The weights, centers and widths of RBFNN are learned by the following equations:

$$W(k + 1) = W(k) - \eta(k)e(k + 1)\Phi^{\mathrm{T}}(k) \tag{6}$$

$$c_i(k+1) = c_i(k) - \eta(k)\frac{2w_i}{\sigma_i^2}e(k+1)\Phi_i(k)\sum_{j=1}^{I}(x_j - c_i) \tag{7}$$

$$\sigma_i(k+1) = \sigma_i(k) - \eta(k)\frac{w_i}{\sigma_i^3}e(k+1)\Phi_i(k)\sum_{j=1}^{I}(x_j - c_i)^2 \tag{8}$$

Here, $\eta(k)$ is stable learning rate [8],

$$\eta(k) = \frac{\eta_0}{1 + \|\Phi(k)\|^2 + \|W_C(k)\|^2 + \|W_S(k)\|^2}.$$

Here, $0 < \eta_0 \leq 1$,

$$W_C(k) = \left[\frac{2w_1}{\sigma_1^2}\Phi_1 E^T(x - Ec_1), \cdots, \frac{2w_H}{\sigma_H^2}\Phi_H E^T(x - Ec_H)\right],$$

$$W_S(k) = \left[\frac{w_1}{\sigma_1^3}\Phi_1 E^T(x - Ec_1)^2, \cdots, \frac{w_H}{\sigma_H^3}\Phi_H E^T(x - Ec_H)^2\right],$$

$$E = [1, \cdots, 1]^T.$$

Above stable learning algorithm can limit the modeling error of integrated model within a bounded scope.

4 Simulations

In order to verify the hybrid integrated model, real data from A/O process of Shenyang north wastewater treatment plant are used in the simulations, and 120 samples are taken as training data, 130 samples to test the hybrid model. Missing data caused by instrument faults are estimated based on the similarity with other data and historical data.

the kinetic parameters and stoichiometric parameters of SASM1 model are selected as those of ASM1 at 20 °C. that is, $\mu_{mH} = 6.0$, $b_H = 0.62$, $K_S = 20$, $K_{OH} = 0.20$, $K_{NO} = 0.5$, $Y_H = 0.67$, $f_P = 0.08$, $i_{XB} = 0.086$.

There is no systematic method to determine the number of hidden-layer nodes of RBFNN except empirical approach or trial and error method. Here the number H of hidden-layer nodes of RBFNN satisfies $H = \sqrt{I+1} + \varsigma$, here ς is an integer between 0–10. H is selected as $H = 10$, then the structure of RBFNN is selected as 5-10-1 by experience; learning rate $\eta_0 = 0.9$, the initial values of W, C and σ are random numbers between [0,1].

Figure 3 shows the comparison results of real COD and estimated COD of mechanism model based on SASM1. The estimated COD and real COD have the same trend, which indicates mechanism model based on SASM1 can express the basic dynamic characteristics of real WWTP. The differences between them are obvious, and the modeling error cannot be ignored yet which should be compensated further. The errors of some estimated values are higher than others because of the sensor fault or deviation of data preprocessing.

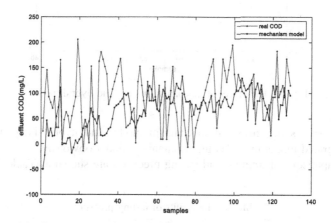

Fig. 3. The comparison results of real COD and estimated COD of mechanism model

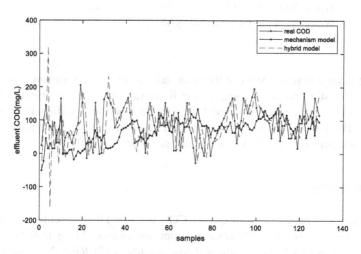

Fig. 4. The comparison results of real COD and estimated COD of hybrid integrated model

Figure 4 shows the comparison results of real COD and estimated COD of hybrid integrated model. The predicted accuracy of hybrid model is improved greatly after error compensation.

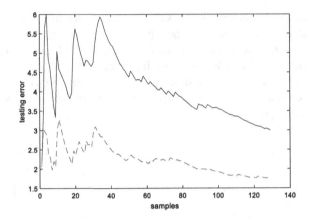

Fig. 5. The errors of mechanism model and hybrid model

Figure 5 shows the errors of mechanism model and hybrid integrated model. Hybrid integrated model precedes the mechanism model obviously.

The comparisons of training and testing precision are shown as Table 2.

Table 2. The results of testing precision

Method	Training *RMSE*	Testing *RMSE*
A/O model based on SASM1	8.35	8.54
Hybrid integrated model	7.20	7.51
Hybrid model in Literature [6]	8.02	8.22

There is no difference between the error indices of A/O mechanism model during training and testing phases because of the fixed values of the parameters. After the modeling error of mechanism model is compensated, the prediction accuracy is improved. Hybrid model in Literature [6] is a little inferior to integrated model because the error compensation model is built by linear model.

5   Conclusions

Hybrid integrated model of water quality COD for WWTP is presented in this paper, the following conclusions can be obtained from simulations: (1) mechanism model has the defects of complex calculation and low accuracy; (2) RBFNN is integrated to compensate the error, which the parameters are learned in real time. (3) the learning rate and the structure of RBFNN are selected by the experiences, and the method to seek the optimal parameters should be researched further to enhance the precision.

Acknowledgement. This work was supported by the National Natural Science Fund of China (61803191, 61673199); Natural Science Fund Project of Liaoning province (2019-KF-03-05).

References

1. Weijers, S.R.: Modelling, identification and control of activated sludge processes for nitrogen removal. Technical university of eindhoven, Netherlands (2000)
2. Du, S.X.: Modeling and control of biological wastewater treatment processes. Control Theory Appl. **19**(5), 660–666 (2002)
3. Kadlec, P., Grbie, R., Gabrys, B.: Review of adaptation mechanisms for data-driven soft sensor. Comput. Chem. Eng. **35**(1), 1–24 (2011)
4. Bates, J.M., Granger, C.W.: The combination of forecasts. Oper. Res. Q. **20**(4), 451–468 (1969)
5. Lee, D.S., Vanrolleghem, P.A., Park, J.M.: Parallel hybrid modeling methods for a full-scale cokes wastewater treatment plant. J. Biotechnol. **115**(3), 317–328 (2005)
6. Cong, Q.M., Zhang, B.W., Yuan, M.Z.: On-line soft sensor for water quality of wastewater based on synchronous clustering. Comput. Eng. Appl. Comput. Eng. Appl. **51**(24), 27–33 (2015)
7. IWA task group of mathematical modeling for design and operation of biological wastewater treatment. Activated Sludge Models ASM1, ASM2, ASM2d and ASM3. Tongji University Press, Shanghai (2002)
8. Cong, Q.M., Deng, S.X., Zhao, Y., Wang, Y.: Stable soft sensor based on RBF neural network and its applications. Control Eng. China **25**(5), 823–825 (2018)

Optimal Control for Cracking Outlet Temperature (COT) of SC-1 Ethylene Cracking Furnace by Off-Policy Q-Learning Approach

Yihan Zhang, Zhenfei Xiao, and Jinna Li[✉]

School of Information and Control Engineering, Liaoning Shihua University,
Fushun 113001, Liaoning, China
lijinna_721@126.com

Abstract. In this paper, a novel off-policy Q-learning is developed for solving optimal tracking problem of cracking outlet temperature (COT) of SC-1 ethylene cracking furnace, using only the measured data along the system trajectories. This paper takes the outlet temperature of ethylene cracking furnace as the background, taking the state space model as the basis, and combines the data-driven off-policy Q-learning algorithm. A novel off-policy Q-learning algorithm is presented by introducing behavior control policy and combining dynamic programming with Q-learning, such that the optimal tracking controller gain is learned with no need of knowledge of system dynamics enabling the tracking of the target. Simulation results are given to verify the effectiveness of the proposed method.

Keywords: Cracking outlet temperature · Off-policy Q-learning · Data-driven · Optimal tracking control

1 Introduction

Ethylene unit is one of the most energy-consuming units in the petrochemical industry. The main objective of the ethylene cracking furnace is to crack petroleum hydrocarbons at high temperatures to obtain many products, such as propylene, ethylene, butene, etc. [1]. The control of COT is very important, because it not only affects the stable operation of subsequent units, but also affects the ethylene absorption rate. The high temperature of COT will result in fast speed of coking on the inner tube surface and shorten tube life. The low temperature of COT will influence the cracking effect. In practical applications, nonlinear performance with the deposition of coke on the inner tube surface, the traditional PID control has been proven insufficient and incapable for such complex industrial process. Therefore, some advanced control methods [1–8] have been developed.

However, the application of off-policy Q-learning based on extended state space model to ethylene cracking furnace has not yet been reported. Therefore, this paper reports an application using the state space model of the COT system of two 270 k tone/year SC-1 ethylene cracking furnace in Petro China Daqing Petrochemical

© Springer Nature Singapore Pte Ltd. 2020
J. Qian et al. (Eds.): ICRRI 2020, CCIS 1336, pp. 342–355, 2020.
https://doi.org/10.1007/978-981-33-4932-2_25

Company by off-policy Q-learning. Q-learning, also known as action-dependent heuristic dynamic programming (ADHDP), has always been adopted by approximate dynamic programming (ADP) schemes, which combine adaptive critics, a reinforcement learning technique with dynamic programming to solve optimal control problems [9].

In this paper, the off-policy Q-learning is employed into an extended state space formulation of COT systems. The main merit of the proposed approach lies in that the optimal tracking controller can be found using the measured augmented variables including input, output and their past values together tracking errors rather than utilizing the system model parameters or state of systems, since they are hard to be available in practice.

The rest of paper is given as follows. Section 2 discusses the production process of ethylene cracking furnace and total feed flow lifting/lowering control. Section 3 proposes the off-policy Q-learning design algorithm for the cracking outlet temperature of ethylene cracking furnace. Section 4 verifies the effectiveness and no bias of solution for the proposed method. Conclusions are stated in Sect. 5.

2   The Ethylene Cracking Furnace

2.1   Process Description

The overall process of the 270k tone/year SC-1 type ethylene cracking furnace is shown in Fig. 1 [1]. It has a single chamber. The chamber is made of eight sets of pipes and 24 furnace tubes. This cracking furnace can crack naphtha (NAP) or hydrocracking tail oil (HTO). There are two cracking furnaces that are labeled M and N furnace in practice, respectively. Taking M furnace (EH111M) as an example, the raw material is NAP. The measurement of the chamber temperature is TI1015M. For the fuel gas system, the fuel pressure controller is PIC1022M, and the COT controller is TC1022M, where PV is process variable, SV is set-point, MV is manipulated variable, AUTO or RCAS is operation mode of controller. The final objective is to control COT within a stable range based on process requirement. The process flow is as follows.

The flow of NAP is separated into eight branches and sent into the convection room of the furnace (EH111M) to be heated. In the first time, the heated temperature (HC PREHEAT1) is about 350.6 °C. In the second time, the heated temperature (HC PREHEAT2) is about 531.9 °C. Each branch joins together with heated dilution steam (DS) by the way of ratio control and then flows into the radiation room of the furnace. DS is injected, two times, into NAP. The first time is HC+DS1. The second time is HC +DS2 that is mixed by a hybrid plant (M-1122A/C). In the radiation section, the hydrocarbon is cracked into a combination of target products and other heavier hydrocarbons [1].

After leaving the radiation section of the furnace, the cracked gas cools rapidly to stop the unwanted reaction. For the fuel gas system, the control mode which is between the designed controller and the fuel pressure controller (PID) is remote cascade (RCAS). The basic pressure controller is output by the advanced controller to a set value and the PID controller achieves the control requirement of COT by adjusting the fuel pressure (PIC1022M). The control objective of the COT (TC1022M) is 860 °C.

Here for the designed controller, MV is the set-point of pressure controller, CV is the set-point of COT, and PV is a calculated value of feed flow and COT of eight branches, respectively. For the pass flow system, DS and pass flow are mixed using a double closed-loop ratio controller where CV is the setpoint of feed flow [1]. They adopt PID control method.

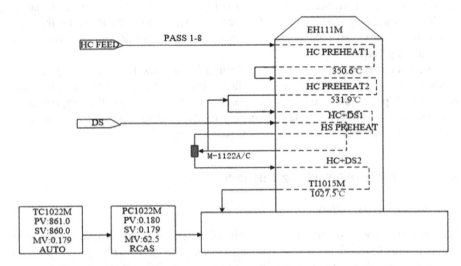

Fig. 1 Process of the SC-1 type ethylene cracking furnace.

2.2   Total Feed Flow Lifting/Lowering Control

The design principle of the feed flow lifting/reducing control system of the cracking furnace is that when the raw material processing quantity needs to be changed, the control system can reasonably distribute the flow of each group of the cracking furnace according to the current operation condition, so as to minimize the impact of the flow change on the outlet temperature of the cracking furnace. Therefore, the lifting/lowering control system realizes the specified lifting/lowering within the specified time, and works in coordination with the cot temperature balance control system of each group, so as to minimize the adverse effects caused by the change of the feeding quantity. In this way, the lifting/lowering control can be realized smoothly, which not only ensures the stable operation of cracking furnace, but also changes the production load. The above control scheme is used for single feed operation. In the single feed operation, the average furnace tube outlet temperature is controlled by adjusting the combustion of cracking furnace. By adjusting the hydrocarbon flow to the furnace tube, the temperature balance at the outlet of the furnace tube and the flow rate of the cracking furnace are controlled in the zone feeding state. The same control scheme is used for the main raw materials as the single feeding state. For the secondary raw materials, the outlet temperature of the furnace tube is controlled by regulating the hydrocarbon flow of the furnace tube. For the secondary feed furnace tube, the flow rate cannot be controlled because there is no redundant adjustable variable. The maximum mass flow of the feed is defined as the main raw material. This feed determines the combustion and total flow rate of the cracking furnace [10].

3 The Off-Policy Q-Learning Design for the COT of Ethylene Cracking Furnace

In the past, we used the advanced control strategy based on the state space model for the temperature control of ethylene cracking furnace, but we still need to know the state space of the system. In this section, we use the Q-learning algorithm and do not need to know the system model parameters or states. We only use the measured data and do not need to design the state observer to get the approximate optimal tracker so as to realize the tracking purpose of 860 °C.

3.1 Problem Description

The COT problem of DT linear systems is formulated by using an extended state space expression, such that the differenced input becomes the decision variable that drives the systems to follow the target reference signal [1].

$$
\begin{aligned}
y(k+1) &+ F_1 y(k) + F_2 y(k-1) + \cdots + F_n y(k-n+1) \\
&= H_1 u(k) + H_2 u(k-1) + \cdots + H_n u(k-n+1)
\end{aligned}
\tag{1}
$$

where $y(k)$ and $u(k)$ are the output and input of system at sampling time instant. $y(k)$ represents the COT of ethylene cracking furnace, and $u(k)$ represents the COT controller TC1022M. $H_i (i = 1, 2, \cdots, n)$ are the matrices with appropriate dimensions.

Inspired by [12], now we define two back shift operators Δ to (1), i.e. $\Delta y(k+1) = y(k+1) - y(k)$ and $\Delta u(k) = u(k) - u(k-1)$, then (1) can be rewritten as

$$
\begin{aligned}
\Delta y(k+1) &+ F_1 \Delta y(k) + F_2 \Delta y(k-1) + \cdots + F_n \Delta(k-n+1) \\
&= H_1 \Delta u(k) + H_2 \Delta u(k-1) + \cdots + H_n \Delta u(k-n+1)
\end{aligned}
\tag{2}
$$

Define an augmented vector

$$
\Delta x_m(k) = \left[\Delta y(k)^T \Delta y(k-1)^T \cdots \Delta y(k-n+1)^T \Delta u(k-1)^T \Delta u(k-2)^T \cdots \Delta u(k-n+1)^T \right]
\tag{3}
$$

Then, can be transformed as the following form

$$
\begin{aligned}
\Delta x_m(k+1) &= A_m \Delta x_m(k) + B_m \Delta u(k) \\
\Delta y(k+1) &= C_m \Delta x_m(k+1)
\end{aligned}
\tag{4}
$$

where

$$A_m = \begin{bmatrix} -F_1 & -F_2 & \cdots & -F_{n-1} & -F_n & H_2 & \cdots & H_{n-1} & H_n \\ I_q & 0 & \cdots & 0 & 0 & 0 & \cdots & 0 & 0 \\ 0 & I_q & \cdots & 0 & 0 & 0 & \cdots & 0 & 0 \\ \vdots & \vdots & \cdots & \vdots & \vdots & \vdots & \cdots & \vdots & \vdots \\ 0 & 0 & \cdots & I_q & 0 & 0 & \cdots & 0 & 0 \\ 0 & 0 & \cdots & 0 & 0 & 0 & \cdots & 0 & 0 \\ 0 & 0 & \cdots & 0 & 0 & I_p & \cdots & 0 & 0 \\ \vdots & \vdots & \cdots & \vdots & \vdots & \vdots & \cdots & \vdots & \vdots \\ 0 & 0 & \cdots & 0 & 0 & 0 & 0 & I_p & 0 \end{bmatrix}$$

$$B_m = \begin{bmatrix} H_1^T & 0 & 0 & \cdots & 0 & I_P & 0 & 0 \end{bmatrix}$$
$$C_m = \begin{bmatrix} I_q & 0 & 0 & \cdots & 0 & 0 & 0 & 0 \end{bmatrix}$$

Suppose the target output is r to be a constant reference signal, then the output tracking error can be defined as $e(k) = y(k) - r$ By (3), one has

$$e(k+1) = e(k) + C_m A_m \Delta x_m(k) + C_m B_m \Delta u(k) \tag{5}$$

Let $z(k) = \begin{bmatrix} \Delta x_m(k) \\ e(k) \end{bmatrix}$, compacting (4) and (5) yields an extended state space model below

$$z(k+1) = \bar{A} z(k) + \bar{B} \Delta u(k) \tag{6}$$

where

$$\bar{A} = \begin{bmatrix} A_m & 0 \\ C_m A_m & I \end{bmatrix}, \bar{B} = \begin{bmatrix} B_m \\ C_m B_m \end{bmatrix}$$

Remark 1: By compacting the differenced variable $\Delta x_m(k)$ with the tracking error $e(k)$, the extended state space model (6) will be employed for handling the optimal tracking control problem in the dynamic programming combined with RL framework, rather than using the general state and output equations of systems like [13–17].

The target of this paper is to find the optimal tracking control policy by designing the differenced input, such that the output $y(k)$ of the system (1) follows the target reference signal and meanwhile minimizing the following performance index:

$$J = \sum_{k=0}^{\infty} \left((z(k)^T Q z(k) + \Delta u(k)^T R \Delta u(k) \right) \tag{7}$$

where $Q \geq 0$ and $R > 0$ are symmetric matrices. Thus, the optimal tracking control problem can be formulated below:

$$J^* = \min_{\Delta u(k)} \sum_{k=0}^{\infty} \left((z(k)^T Q z(k) + \Delta u(k)^T R \Delta u(k) \right) \tag{8}$$

s. t. (6)

The sequence is going to find the optimal tracking control policy by solving the optimization problem shown in (8) using only data without requiring the knowledge of system.

3.2   Off-Policy Q-Learning

An off-policy Q-learning algorithm is developed for finding the optimal tracking controller without the of system dynamics for the COT of ethylene cracking furnace.

Suppose $\Delta u(k) = -Kz(k)$, one can respectively define the optimal value function and the optimal Q-function based on the specific cost function (7) as

$$V^*(z(k)) = \min_{\Delta u(k)} \sum_{k=0}^{\infty} \left(z(k)^T Q z(k) + \Delta u(k)^T R \Delta u(k) \right) \tag{9}$$

and

$$Q^*(z(k), \Delta u(k)) = \left(z(k)^T Q z(k) + \Delta u(k)^T R \Delta u(k) \right) + V^*(z(k+1)) \tag{10}$$

where K is an appropriate matrix. Thus, the following relation holds

$$V^*(z(k)) = \min_{\Delta u(k)} Q^*(z(k), \Delta u(k)) = Q^*(z(k), \Delta u^*(k)) \tag{11}$$

Lemma 1. [18, 19] For a stabilizing control policy, the optimal value function and Q-function have the quadratic forms of

$$V^*(z(k)) = z(k)^T P z(k) \tag{12}$$

and

$$Q^*(z(k), \Delta u(k)) = \begin{bmatrix} z(k) \\ \Delta u(k) \end{bmatrix}^T H \begin{bmatrix} z(k) \\ \Delta u(k) \end{bmatrix} \tag{13}$$

with

$$H = \begin{bmatrix} H_{zz} & H_{zu} \\ (H_{zu})^T & H_{uu} \end{bmatrix}$$
$$= \begin{bmatrix} \bar{A}^T P \bar{A} + Q & \bar{A}^T P \bar{B} \\ (\bar{A}^T P \bar{B})^T & \bar{B}^T P \bar{B} + R \end{bmatrix} \tag{14}$$

$$P = \begin{bmatrix} I \\ -K \end{bmatrix}^T H \begin{bmatrix} I \\ -K \end{bmatrix} \tag{15}$$

where $\xi(k) = \begin{bmatrix} z(k)^T & \Delta u(k)^T \end{bmatrix}^T$. The Q-function based Bellman equation can be derived below

$$\xi(k)^T H \xi(k) = z(k) Q z(k) + \Delta u(k)^T R \Delta u(k) + \xi(k+1)^T H \xi(k+1) \tag{16}$$

Implementing $\frac{\partial Q^*(\xi(k), \Delta u(k))}{\partial \Delta u(k)} = 0$ yields

$$\Delta u^*(k) = (H_{nu})^{-1} (H_{zu})^T z(k) \tag{17}$$

From (16), one has

$$\begin{bmatrix} I \\ -K^j \end{bmatrix} H^{j+1} \begin{bmatrix} I \\ -K^j \end{bmatrix}$$
$$= Q + (K^j)^T R^j + (\bar{A} - \bar{B}^j)^T \begin{bmatrix} I \\ -K^j \end{bmatrix} H^{j+1} \begin{bmatrix} I \\ -K^j \end{bmatrix} (\bar{A} - \bar{B}^j) \tag{18}$$

Introducing an auxiliary variable $\Delta u^j(k) = -K^j z(k)$ into system (6) yields

$$z(k+1) = A_c z(k) + \bar{B}\big(\Delta u(k) - \Delta u^j(k)\big) \tag{19}$$

where $A_c = \bar{A} - \bar{B} K^j$, $\Delta u(k)$ is called as the behavior policy to generate data and $\Delta u^j(k)$ is viewed as the target policy needed to be updated. Along the trajectory of (19) and combined with (18), one has

$$Q^{*j+1}(z(k), u^j(k)) - z(k)^T A_c^T \begin{bmatrix} I \\ -K^j \end{bmatrix}^T H^{j+1} \begin{bmatrix} I \\ -K^j \end{bmatrix} A_c z(k)$$
$$= z(k)^T \begin{bmatrix} I \\ -K^j \end{bmatrix}^T H^{j+1} \begin{bmatrix} I \\ -K^j \end{bmatrix} z(k) - (z(k+1) - \bar{B}(\Delta u(k) - \Delta u^j(k)))^T$$
$$\begin{bmatrix} I \\ -K^j \end{bmatrix}^T H^{j+1} \begin{bmatrix} I \\ -K^j \end{bmatrix} (z(k+1) - \bar{B}(\Delta u(k) - \Delta u^j(k)))$$
$$= z(k)^T \big(Q + (K^j)^T R K^j\big) z(k) \tag{20}$$

Since P^{j+1} and H^{j+1} have the relationship shown in (14) and (15), one has a Q-function based iterative Bellman equation as

$$
z(k)^T \begin{bmatrix} I \\ -K^j \end{bmatrix}^T H^{j+1} \begin{bmatrix} I \\ -K^j \end{bmatrix} z(k)
$$
$$
- z(k+1)^T \begin{bmatrix} I \\ -K^j \end{bmatrix}^T H^{j+1} \begin{bmatrix} I \\ -K^j \end{bmatrix}
$$
$$
z(k+1) + 2(\bar{A}z(k) + \bar{B}\Delta u(k))P^{j+1}\bar{B}(\Delta u(k) - \Delta u^j(k)) - (\Delta u(k) - \Delta u^j(k))^T
$$
$$
\bar{B}^T P^{j+1}\bar{B}(\Delta u(k) - \Delta u^j(k)) = z(k)^T \Big(Q + (K^j)^T R K^j\Big) z(k)
$$

(21)

Further, one has,

$$
z(k)^T \begin{bmatrix} I \\ -K^j \end{bmatrix}^T H^{j+1} \begin{bmatrix} I \\ -K^j \end{bmatrix} z(k)
$$
$$
- z(k+1)^T \begin{bmatrix} I \\ -K^j \end{bmatrix}^T H^{j+1} \begin{bmatrix} I \\ -K^j \end{bmatrix} z(k+1) + 2z(k)H_{zu}^{j+1}(\Delta u(k) + K^j z(k))
$$
$$
+ \Delta u(k)(H_{uu} - R)(\Delta u(k) + K^j z(k)) - (K^j z(k))^T (H_{uu} - R)
$$
$$
(\Delta u(k) + K^j z(k)) = z(k)^T \Big(Q + (K^j)^T R K^j\Big) z(k)
$$

(22)

Properly manipulating (22) yields the following form

$$
\theta^j(k)L^{j+1} = \rho_k^j
$$

(23)

$$
\rho_k^j = z(k)^T Q z(k) + \Delta u(k)^T R \Delta u(k)
$$

$$
L^{j+1} = \Big[\big(vec\big(L_1^{j+1}\big)\big)^T \quad \big(vec\big(L_2^{j+1}\big) \quad \big(vec\big(L_3^{j+1}\big)\big)^T \Big]
$$

$$
L_1^{j+1} = H_{zz}^{j+1}, \ L_2^{j+1} = H_{zu}^{j+1}, L_3^{j+1} = H_{uu}^{j+1}
$$

$$
\theta^j(k) = \begin{bmatrix} \theta_1^j & \theta_2^j & \theta_3^j \end{bmatrix}
$$

$$
\theta_1^j = z(K)^T \otimes z(K) - z(K+1)^T \otimes z(K+1)
$$

$$
\theta_2^j = 2z(K+1)^T \otimes (K^j z(K+1)) + 2z(K)^T \otimes \Delta u(K)^T
$$

$$
\theta_3^j = \Delta u(K)^T \otimes \Delta u(K)^T - (K^j z(K+1))^T \otimes (K^j z(K+1))^T
$$

K^{j+1} can be calculated if finding L_1^{j+1} and L_3^{j+1}

$$
K^{j+1} = -\Big(L_1^{j+1}\Big)^{-1}\Big(L_3^{j+1}\Big)^T
$$

(24)

Then, the approximately optimal tracking control policy can be presented below:

$$\Delta u^{j+1}(k) = \Delta u(k-1) - K^{j+1}z(k) \tag{25}$$

Now, a data-driven off-policy reinforcement learning algorithm is proposed to solve the tracking control of linear systems without the information of the system dynamics. Instead of directly solving K from (16), Algorithm 1 presents a policy iteration algorithm where the value of K is numerically approximated by solving recursive least squares (RLS).

Algorithm 1: Off-policy Q-learning algorithm

1: Data collection: Collect system data $x_m(k)$ and store them in the sample sets $\theta^j(k)$ and $\rho_k^{\ j}(k)$ by using an arbitrary stabilizing behavior control policy $\Delta u(k)$;

2: Initiation: Choose the initial gain k^0, such that system (4) can steadily track the r. Let $j = 0$;

3: Implementing Q-learning: By using recursive least squares (RLS) or batch least squares (BLS) methods, $L_i^{j+1}(i = 1, 2, 3)$ can be estimated using the collected data in Step 1, and then K^{j+1} can be updated in terms of (24);

4: If $\left\| H^j - H^{j+1} \right\| \leq l$ (l is some small positive numbers), then stop the iteration and the Optimal control policy has been obtained. Otherwise, let $j = j+1$ and go back to step 3.

Remark 2: If the real value L^{j+1} can be solved correctly, then the $\Delta u^j(k) = -K^j z(k)$ learned by Algorithm 1 will definitely converge to the optimal $\Delta u^*(k)$ as $j \to \infty$, i.e. $\lim_{j \to \infty} \Delta u^j(k) = \Delta u^*(k)$ indicating $\lim_{j \to \infty} u^{j+1}(k) = u^*(k)$ which has been proven in [18, 20].

Remark 3: Algorithm 1 uses a simple extended state space equation to simplify the solving COT of SC-1 ethylene cracking furnace off-policy Q-learning method. The main merits include: (a) The optimal tracking problems of COT are handled by solving linear quadratic regulation problem for the extended state space model of those systems using only measured data, which is a different way from [21] where states were estimated by output and input; (b) Rather than designing state observer [22] and employing output regulation methods [13, 23] where partial system parameters still need to be known, the measured difference output, difference input and tracking error can be directly utilized to learn the optimal tracking control strategy.

4   Simulation Results

In the last section, we describe the ethylene cracking furnace in the state space, and propose the off-policy Q-learning algorithm. In this section, the simulation results show that the algorithm 1 can guarantee to track the COT of 860 °C.

The model of the generalized process of the SC-1 ethylene cracking furnace system is as follow

$$G(s) = \frac{700e^{-180s}}{720s + 1} \tag{26}$$

Discretizing (26) by adding a zero-order holder using sampling time $T_s = 60\,\mathrm{s}$ yields the following form

$$G(z) = z^{-4}\frac{55.97}{1 - 0.92z^{-1}} \tag{27}$$

Let
$$\Delta x_m(k) = [\Delta y(k)^T \Delta y(k-1)^T \cdots \Delta y(k-n+1)^T$$
$$\Delta u(k-1)^T \Delta u(k-2)^T \cdots \Delta u(k-n+1)^T]^T$$
Then, the extended state space expression is derived below

$$\Delta x_m(k+1) = \begin{bmatrix} 0.92 & 0 & 0 & 55.97 \\ 1 & 0 & 0 & 0 \\ 0 & 1 & 0 & 0 \\ 0 & 0 & 1 & 0 \end{bmatrix} \Delta x_m(k)$$

$$+ \begin{bmatrix} 0 \\ 0 \\ 1 \\ 0 \end{bmatrix} \Delta u(k)$$

$$\Delta y(k+1) = [1 \ 0 \ 0 \ 0]\Delta x_m(k+1)$$

Choose $Q = \mathrm{diag}(6,6,6,6,5)$ and $R = 1$. First, suppose the parameters of the system are known, then the optimal Q-function matrix H^* and the optimal tracking controller gain K^* can be respectively obtained in terms of (14) and (17) using command "dare" in Matlab.

$$H^* = \begin{bmatrix} 60 & 290 & 1160 & 2640 & 20 & 290 \\ 170 & 6770 & 6590 & 9920 & 50 & 6760 \\ 170 & 13240 & 50150 & 66930 & 380 & 13240 \\ 2640 & 1674 & 6952 & 59280 & 960 & 16740 \\ 20 & 50 & 350 & 920 & 20 & 50 \\ 170 & 6760 & 6590 & 9920 & 50 & 6760 \end{bmatrix}$$

$$K^* = [-0.0283 \ -1.0000 \ -1.3676 \ -1.7233 \ -0.0082]$$

Implementing Algorithm 1 yields the optimal Q-function matrix and the optimal tracking controller gain after 10 iterations.

$$H^{10} = \begin{bmatrix} 60 & 290 & 1160 & 2640 & 20 & 290 \\ 170 & 6770 & 6590 & 9920 & 50 & 6760 \\ 170 & 13240 & 50150 & 66930 & 380 & 13240 \\ 2640 & 1674 & 6952 & 59280 & 960 & 16740 \\ 20 & 50 & 350 & 920 & 20 & 50 \\ 170 & 6760 & 6590 & 9920 & 50 & 6760 \end{bmatrix}$$

$$K^{10} = [\, -0.0283 \quad -1.0000 \quad -1.3676 \quad -1.7233 \quad -0.0082 \,]$$

Figure 1 shows the difference between the controller gain K^j and the optimal tracking controller gain K^* when the probing noise is $e_k = 0.6\left(1.5 \sin^2(2.0k)\right)$ $\cos(10.1k) + 0.9 \sin^2(1. \quad 102k) \cos(4.001k)$. during learning process. Figure 2 shows the tracking result under the learned control policy using the proposed off-policy learning algorithm. Figure 3 shows the control input trajectory by using off-policy Q-learning algorithm. From Fig. 2 and Fig. 3, the target temperature can be tracked well under the learned control policy using the proposed algorithm (Fig. 4).

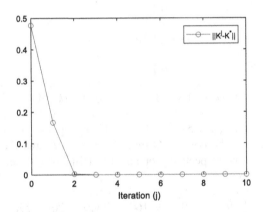

Fig. 2. Convergence result of K^j

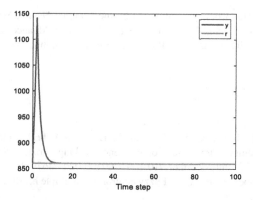

Fig. 3. The tracking results by using off-policy Q-learning algorithm

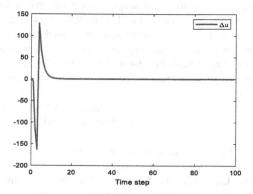

Fig. 4. The control input trajectory by using off-policy Q-learning algorithm

5   Conclusion

To solve the problem of temperature control at the exit of ethylene cracking furnace, a data-driven off-policy Q-learning algorithm is proposed based on the extended state space model. The advantage of this algorithm is that it combines the RL framework, uses the extended state space model to deal with the optimal tracking control problem in dynamic programming, uses the measured data and does not need to design the state observer, and can directly use the measured differential output, differential input and tracking error to learn the optimal tracking control target. Combined with the RL framework, the extended state space model is used to deal with the optimal tracking control problem in dynamic programming. Accurate and effective control of COT is of great significance to improve the automatic control level of cracking furnace, effectively achieving the goal of increasing production, saving energy and reducing consumption.

Acknowledgments. This work was supported by the National Natural Science Foundation of China under Grants 61673280, the Open Project of Key Field Alliance of Liaoning Province under Grant 2019-KF-03-06 and the Project of Liaoning Shihua. University under Grant 2018XJJ-005.

References

1. Shi, H., Peng, B., Jiang, X., Su, C., Cao, J., Li, P.: A hybrid control approach for the cracking outlet temperature system of ethylene cracking furnace. Soft Comput. **24**(16), 12375–12390 (2020). https://doi.org/10.1007/s00500-020-04679-0
2. Ramyirez, M., Haber, R., Pena, V.: Fuzzy control of a multiple hearth furnace. Comput. Ind. **54**(1), 105–113 (2014)
3. Chai, T.Y., Ding, J.L., Wu, F.H.: Hybrid intelligent control for optimal operation of shaft furnace roasting process. Control Eng. Pract. **19**(3), 264–275 (2011)
4. Jana, A.K.: Differential geometry-based adaptive nonlinear control law: application to an industrial refinery process. IEEE Trans. Ind. Inform. **9**(4), 2014–2022 (2013)
5. Shi, H.Y., Su, C.L., Cao, J.T., Li, P.: Nonlinear adaptive predictive functional control based on the Takagi-Sugeno model for average cracking outlet temperature of the ethylene cracking furnace. Ind. Eng. Chem. Res. **54**(6), 1849–1860 (2015)
6. Li, P., Li, T., Cao, J.: Advanced process control of an ethylene cracking furnace. Meas. Control 2(48), 50–53 (2015)
7. Tawai, A., Panjapornpon, C.: Input-output linearizing control strategy for an ethylene dichloride cracking furnace using a coupled PDE-ODE model. Ind. Eng. Chem. Res. **55**, 683–691 (2016)
8. Luan, X.L., Min, Y., Liu, F.: Distributed pass balancing and tracking control of feed heater. ACTA Autom. Sinica **43**(6), 1056–1064 (2017)
9. Li, J.N., Ding, J.L., Chai, T.Y., Li, C., Lewis, F.L.: Nonzero-sum game reinforcement learning for performance optimization in large-scale industrial processes. IEEE Trans. Cybern. **99**, 1–14 (2019)
10. Li, P., Li, Q.A., Lei, R.X., Chen, A.J., Ren, L.L., Cao, W.: Development and application of advanced process control system for ethylene cracking heaters. CIESC J. **62**(8), 0438–1157 (2011)
11. Zhang, R.D., Tao, J.L.: Data-driven modeling using improved multi-objective optimization based neural network for coke furnace system. IEEE Trans. Ind. Electron. **64**, 3147–3155 (2017)
12. Wei, Q., Liu, D.: Data-driven neuro-optimal temperature control of water-gas shift reaction using stable iterative adaptive dynamic programming. IEEE Trans. Ind. Electron. **61**(11), 6399–6408 (2014)
13. Kamalapurkar, R., Dinh, H., Bhasin, S., Dixon, W.E.: Approximate optimal trajectory tracking for continuous-time nonlinear systems. Automatica **51**, 40–48 (2015)
14. Li, J.N., Xiao, Z.F., Li, P., Ding, Z.T.: Networked controller and observer design of discrete-time systems with inaccurate model parameter. ISA Trans. **98**, 75–86 (2020)
15. Farnaz, A.Y., Svante, G., Lewis, F.L.: Output regulation of unknown linear systems using average cost reinforcement learning. Automatica **110**, 108549 (2019)
16. Jiang, Y., Kiumarsi, B., Fan, J., Chai, T., Li, J., Lewis, F.L.: Optimal output regulation of linear discrete-time systems with unknown dynamics using reinforcement learning. IEEE Trans. Cybern. **50**(7), 1–10 (2019)

17. Lewis, F.L., Vrabie, D.L., Syroms, V.L.: Optimal Control for Polynomial Systems, pp. 287–296. John Wiley (2012)
18. Lewis, F.L., Modares, H., Kiumarsi, B.: Reinforcement Q-learning for optimal tracking control of linear discrete-time systems with unknown dynamics. Automatica 50(4), 167–1175 (2014)
19. Li, J.N., Chai, T.Y., Lewis, F.L., Ding, Z.T., Jiang, Y.: Off-policy interleaved Q-learning: optimal control for affine nonlinear discrete-time systems. IEEE Trans. Neural Netw. Learn. Syst. 30(5), 1308–1320 (2019)
20. Li, J., Kiumarsi, B., Chai, T., Lewis, F.L., Fan, J.: Off-policy reinforcement learning: optimal operational control for two-timescale industrial processes. IEEE Trans. Cybern. 47(12), 547–4558 (2017)
21. Kiumarsi, B., Lewis, F.L., Naghib-Sistani, M.B.: Optimal tracking control of unknown discrete-time linear systems using input-output measured data. IEEE Trans. Cybern. 45(12), 2770–2779 (2015)
22. Modares, H.: Optimal tracking control of uncertain systems: on-policy and off-policy reinforcement learning approaches. Control Complex Syst. 165–186 (2016)
23. Modares, H., Lewis, F.L., Jing, Z.P.: Optimal output-feedback control of unknown continuous-time linear systems using off-policy reinforcement learning. IEEE Trans. Cybern. 46(11), 1–10 (2016)

Verification, Validation and Accreditation and Credibility Evaluation of Aerospace Product Performance Prototype

Shan Gao[✉], Huifeng Xue, Pu Zhang, Xuan Zuo, and Hao Zheng

Northwestern Polytechnical University, Xi'an, Shaanxi, China
gaoshan@mail.nwpu.edu.cn

Abstract. In this paper, we give a new definition to the aerospace product performance prototype on the foundation of lately studies and emphasize on the importance of the credibility of the performance prototype according to the definition and characteristics of it. Then we design a VV&A scheme for the aerospace product performance prototype according to the characteristics and development process of it and emphasize that the VV&A activities should run through the whole lifecycle of it. The VV&A scheme is designed based on the general VV&A process of M&S and corresponding to the characteristics and development process of the performance prototype. Then we introduce the VV&A scheme in details step by step and analyze appropriate V&V techniques for each V&V activity. This research gives a macroscopic description for the VV&A of the aerospace product performance prototype and can be the foundation for the further researches on the implementation of the VV&A of aerospace product performance prototype and choosing and optimizing the appropriate V&V techniques for different V&V activities.

Keywords: Performance prototype · Verification · Validation · Accreditation · Credibility evaluation

1 Introduction

The performance evaluation of the aerospace product is a complex system engineering project and usually needs to be done by implementing flight tests. Comparing to traditional flight tests, digital prototype technology can significantly shorten the development cycle and reduce the development costs, so it is necessary to build up digital performance prototype for the aerospace product in the early phase of its design and development cycle.

In recent years, some domestic scholars have done a lot of researches on the digital performance prototype for the aerospace product but mostly focused on the modeling theories and modeling methods. Professor Zhang and his team studied several appropriate and effective modeling methods for the aerospace product performance prototype such as integrated multi-objective optimization modeling method [1], collaborative modeling method [2] and reliability modeling method based on characteristic model [3]. The modeling theories and modeling methods of the digital performance prototype are important, but the credibility of the digital performance prototype is also very

J. Qian et al. (Eds.): ICRRI 2020, CCIS 1336, pp. 356–366, 2020.
https://doi.org/10.1007/978-981-33-4932-2_26

important. An M&S (modeling and simulation) without credibility evaluation is unreliable and meaningless. The core of the performance prototype is the simulation model of the aerospace product that built in computer environment. Therefore the credibility of the M&S represents the credibility of the performance prototype. A performance prototype without credibility evaluation may has defects and may possibly lead to the failure of the development, or even further risks and financial loss in the manufacturing of the real product.

This paper does a preliminary study on the VV&A of the performance prototype to ensure the high credibility and validity of the performance prototype. Section 2 gives a new definition of the aerospace product performance prototype. Section 3 introduces the concepts of VV&A and credibility evaluation. Section 4 first introduces a VV&A scheme for the aerospace product performance prototype and then explains each step in detail along with the V&V techniques appropriate for them. Section 5 is the summary of the article.

2 Definition of the Aerospace Product Performance Prototype

The national definition of the digital prototype in GB/T26100-2010 i.e. "General principles of digital mock-up for mechanical products" is: the digital description for functional independent subsystems or overall machine of the product which reflects not only the geometric attributes of the product but also the functions and performance in at least one certain domain [4]. The digital prototype is a product of the development of computer design and simulation technology. Compared to the traditional physical prototype, the digital prototype is a digital simulation model in virtual environment implemented by computer techniques and can present the outlook, structure, function, performance and behavior of the product as well as the physical prototype does.

The digital prototype can be categorized into geometric prototype, function prototype and performance prototype based on their application. The geometric prototype focuses on the geometric description of the product, the function prototype focuses on the function description of the product and the performance prototype focuses on the performance description of the product (see Fig. 1) [5].

There is no uniform definition for the aerospace product performance prototype domestically so far. Based on the present studies, this paper defines the aerospace product performance prototype as follow. The aerospace product performance prototype is an aggregation of simulation models, virtual environment and simulation data formed by analysis, validation and optimizing the performance indicators of each development phase of the aerospace product based on the geometric prototype and the function prototype using CAE/CFD software as core techniques.

The major purpose of developing the digital prototype is to replace the physical prototype to analyze and evaluate the performance of the aircraft under different flight conditions by simulate and calculate the form, structure, and aerodynamic parameters in computer environment.

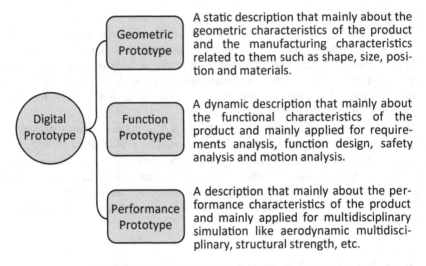

Fig. 1. The classification of the digital prototype based on application

3 The Concepts of VV&A and Credibility Evaluation

VV&A is the acronym of verification, validation, and accreditation and is a generic term of a series of activities that run through the whole lifecycle of the M&S to inspect the validity and credibility of the simulation model. Model verification solves the problem of "did we build the model rightly" i.e. ensures that the transformation process of the model from one form to another is with sufficient accuracy as intended, for example the process of transformation from the requirements documentation to the conceptual model or the transformation from the conceptual model to the computer model. Model validation solves the problem of "did we build the right model" i.e. ensures the model behaves with sufficient accuracy consistent with the application objectives within its domain of applicability. Accreditation is the official certification that a model or simulation is acceptable for use for a specific purpose [6].

The credibility is the extent of the consistency between the simulation model and the simulation purpose decided by the similarity between the simulation system and the original system [7]. The credibility is an important indicator to measure the validity of the simulation model and to decide whether the model is acceptable. The credibility can be implemented by VV&A and in some sense the credibility is VV&A. Substantially, VV&A is not only an activity but a process runs through the whole lifecycle of the simulation model. The purpose of VV&A is to improve the credibility of the simulation model, and implementing the VV&A activities through the lifecycle of the simulation model effectively ensures the high credibility of the simulation model.

4 A VV&A Scheme for the Aerospace Product Performance Prototype

The development cycle of an aerospace product performance prototype is a complex task which relates to several disciplines and contains designs and developments of multiple subsystems along with their integration and application. On the perspective of the overall system, the development process of the performance prototype can be divided into six phases which are requirements analysis, building conceptual model, prototype design, prototype implementation, simulation results output and prototype acceptance (see Fig. 2). Corresponding to each phase of the development cycle of the performance prototype, the VV&A activities can be divided into seven steps which are requirement V&V and VV&A planning corresponding to the requirements analysis phase, conceptual model V&V corresponding to the conceptual model building phase, design verification corresponding to the prototype design phase, implementation verification corresponding to the prototype implementation phase, simulation results validation corresponding to the simulation results output phase, credibility evaluation corresponding to the prototype acceptance phase and at last accreditation.

In Fig. 2 we can see, during the development cycle of the performance prototype, every development phase has a V&V corresponding activity. If flaws or defects are found in one phase during the V&V activity, the developer should go back to the phase before to modify them and redo the V&V activity until the V&V result is satisfactory. So the development of the performance prototype is an iteration process which is modified and optimized repeatedly until the expected requirements are satisfied. All the data, operations, modifications, inputs, outputs etc. have to be recorded and organized to form reports during all the V&V activities and would be the additional information and important reference for the final accreditation. Only if the system passes the final accreditation can it be assumed acceptable and conforming to the application requirements.

The contents and methods of the seven VV&A steps will be introduced and discussed in the following sections form Sects. 4.1, 4.2, 4.3, 4.4, 4.5, 4.6 and 4.7.

4.1 Requirements V&V for Performance Prototype and VV&A Planning

Requirements definition is the first step of the development cycle of the performance prototype which is the criteria and foundation of the whole development process. If the description of the requirements is unclear, ambiguous or hard to understand, it will raise the difficulty for the developers to do the following work. If the description of the requirements is not consistent with the real system, the model that built afterward based on them would be invalid and unacceptable. So it is necessary to verify and validate the requirement specification documentation during the requirement analysis phase. Requirements V&V is also an important part among all the VV&A activities.

The requirements V&V is primarily about reviewing and checking the requirement specification documentation generated during the requirement analysis phase. The contents of requirements V&V include the integrity of the requirement description, the clarity and unambiguity of the requirement description language, and the traceability

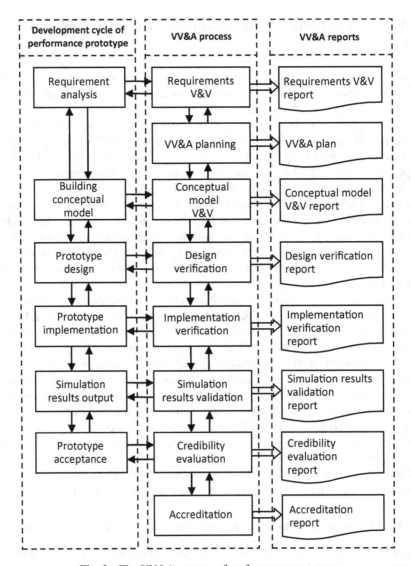

Fig. 2. The VV&A process of performance prototype

and testability of the requirement model. The review should also focus on the intended use, acceptability criteria for model fidelity, risk assessment and configuration management to ensure that all the requirements are clearly defined, testable, consistent, and complete. All the related work and results should be recorded and documented and an overall requirement V&V report should be composed after the requirement V&V phase is completed [8].

The techniques of requirements V&V include static techniques and dynamic techniques. Static techniques are methods of reviewing and inspecting the requirement documentations according to certain criteria without using computer tools. Dynamic techniques are methods of using simulation execution tools to execute the requirement model in order to analyze the accuracy and feasibility of it directly. Static techniques do not require machine execution of the model, while mental execution can be used [9].

Another important activity that should be carried out during the requirement analysis phase along with the requirement V&V is the VV&A planning. First, an accreditation plan which contains system overview, system development scheme, main models and key data, acceptability criteria and accreditation agency, etc. should be made according to the expected application purpose. Then, a specific VV&A plan which contains schedule for VV&A process, main tasks of each VV&A phase, main VV&A objects and techniques, organizational structure and work distribution of the VV&A team and the resources and preparations, etc. should be formulated. The VV&A plan can ensure that the whole VV&A process is well supervised and managed which means the VV&A tasks are reasonably distributed, the resources are adequately used, the schedule is primely controlled and all the members of the VV&A team know their jobs and responsibilities.

4.2   Conceptual Model V&V for Performance Prototype

The conceptual model plays a very important role in the lifecycle of performance prototype as the first step of the transforming process from the real system to the performance prototype. The conceptual model is the first abstraction of the real system and is an expression of model between the real system and the computer simulation model. The conceptual model is a series of data models that can be recognized and saved by the corresponding data base by analyzing and organizing the feature information of the real system. These data models are not simulation models that can be executed directly by computer but are just models at the conceptual level. The conceptual model is not usually expressed in a single pattern because of the complexity of the model but in different patterns including informal language, formal language, mathematical formulas, sheets or figures according to the contents to be abstracted.

The conceptual model of the performance prototype should contain all the data information including definitions, designs, figures, formulas, relationships, work flows, criteria, algorithms that related to the organizational structure, appearance, structure, subsystem structure, subsystem performance, system performance of the performance prototype. Whether the conceptual model can satisfy all the requirements that generated in the requirement analysis phase and can completely express the appearance, structure, function, behavior, performance of the real system are the key issues that should be concerned during building the conceptual model, so the conceptual model needs verification and validation. Only if the conceptual model passes the conceptual model V&V can it be the foundation of the development of the computer model.

The conceptual model V&V is to examine and inspect the conceptual model using specific techniques and tools and then modify and correct the flaws and errors that have been found to finally get a satisfactory conceptual model. The effect of the conceptual model V&V is to make sure the conceptual model that generated during the conceptual model building phase can precisely express the requirements of the real system.

Different techniques can be used according to different V&V objects including informal techniques, static techniques, and formal techniques. Informal techniques such as desk checking, Turing test, documentation checking, reviews and walkthroughs are used to deal with conceptual models described in informal language. Formal techniques such as inductions, inductive assertions, lambda calculus, logical deduction and predicate transformation are used to deal with conceptual models described in strict mathematical formulas and mathematical expressions. Static techniques such as data analysis, control flow analysis, control analysis, structural analysis and symbolic evaluation are used to deal with conceptual models described in formal language or charts such as flow charts, class diagrams and interaction diagrams [10].

4.3   Design Verification for Performance Prototype

The design of the performance prototype is a process that turning the requirements and the conceptual model into a specific implementable design scheme including appearance design, parameter design, structure design, subsystem design, function design, interface design, etc. The design scheme has to be extremely accordance with the requirements and the conceptual model and must completely transform and express the contents and goals of them. To ensure the consistency and traceability between the design scheme and the conceptual model, the design scheme needs verification. The design verification is a process of checking and inspecting the validity, consistency and traceability of the design documentations in order to get an accurate and traceable design scheme.

Informal techniques and static techniques can be used in the design verification according to the specific contents of the design scheme. These two classes of techniques have been introduces in Sect. 3.2 and the V&V staff can choose the appropriate techniques according to the contents of the design scheme that they are going to verify.

4.4   Implementation Verification for Performance Prototype

The implementation process of the performance prototype is the process that the developers building up the computer model using computer aided tools according to the design scheme. The commonly used computer aided tools for modeling are CAE (Computer Aided Engineering) software and CFD (Computer Fluid Dynamics) software. The CAE software is mainly used to analyze and calculate the structural parameters of the performance prototype and the CFD software is mainly used to analyze and calculate the aerodynamic parameters of the performance prototype.

The implementation verification for the performance prototype is to verify the process of the implementation of the performance prototype i.e. to test and examine whether the computer model that developed by the developers is correct and accurate. Since the implementation of the performance prototype is multidisciplinary and involves multiple subsystems, the developers should follow unified criteria and build the subsystems first then integrate the subsystems into a complete system via the unified interfaces and the multidisciplinary integrated modeling and simulation platform. So the implementation verification for the performance prototype contains three parts which are subsystem verification, interface verification and integration verification (see Fig. 3). The V&V staff can choose the corresponding techniques for different verification contents. The process of the implementation of the performance prototype is the process of coding and executing the code in the computer environment by the developers, so the implementation verification is mainly about checking and testing the computer codes. There are two types of techniques that can be used which are static techniques and dynamic techniques.

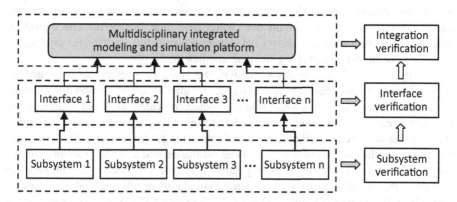

Fig. 3. The implementation verification for the performance prototype

The static techniques are to review and analyze the code without running it and find the potential flaws and errors in a static condition. The static techniques are usually implemented by human beings, or sometime with the help of software tools. The static techniques can be done by the programmer themselves or by organizing a review group to hold review meetings. The details of the verification are mainly about the consistency between the code and the design scheme, the accuracy and readability of the code, the accuracy of the code logic and the rationality of the code structure.

The dynamic techniques are implemented by running the code to check out whether the computational results of the code matching the expected purposes. The dynamic techniques usually consist of three steps witch are building test cases, execute testing programs and analyzing testing results. Some of the commonly used dynamic techniques are execution testing, module testing, interface testing, unit testing and integration testing, etc.

4.5 Simulation Results Validation for Performance Prototype

The simulation results are a series of calculation results that obtained by running the simulation model and calculate the performance parameters under certain simulation environment and flight conditions. The simulation results validation is to compare the data of calculation results with the reference data which are results of real flight tests or sets of ideal data to judge if the performance prototype simulation model matches the real system and can replace the real system to be tested and analyzed to satisfy the expected application purposes by using scientific techniques.

The simulation results validation is the most important part of the whole VV&A process, because the matching degree between the prototype and the real system is the most intuitive expression of the validity of the performance prototype. The comparison results and matching degree between the simulation results data and the reference data can directly tell if the performance prototype satisfy the criteria of replacing the real system and the expected application purposes.

The simulation results can be divided into two types of data which are static performance parameters and dynamic performance parameters according to the performance assessment criteria. Static parameters are outputs that won't change over time and dynamic parameters are outputs that change over time. For the aerospace product, most of the performance parameters are dynamic parameters such as lift coefficient, resistance coefficient and pitching moment coefficient.

During the simulation results validation, the static parameters are evaluated by static techniques and relatively the dynamic parameters are evaluated by dynamic techniques. Static parameters validation techniques are calculating and analyzing the simulation results data and the reference data to see if they have the same distribution characters in statistical significance i.e. evaluating the population distribution consistency between the two sets of data. Parameter estimation and hypothesis testing are two commonly used static techniques. Dynamic parameters validation techniques are evaluating the overall consistency between the simulation results data and the reference data within the same time series. Dynamic parameters validation techniques can be divided into two types of techniques which are time domain techniques and frequency domain techniques. The commonly used time domain techniques are subjective comparison, Theil inequality coefficient, grey relational analysis, error analysis, sensitivity analysis, etc. The frequency domain techniques are deducing the consistency of the data series from the frequency spectrum characteristic according to the feature that the probability distribution and the frequency spectrum characteristic of the same data series both show the same characteristics [10].

4.6 Credibility Evaluation for Performance Prototype

The credibility evaluation is a process of summarizing and refining the previous V&V steps to get the credibility of the performance prototype. If all the V&V steps that introduced before have been strictly and effectively executed and all the errors and flaws have been corrected and modified, the performance prototype can be considered highly credible. But the degree of the credibility needs to be presented in a direct and clear manner which can be obtained by implementing the credibility evaluation. During the credibility evaluation process the evaluators formulate a set of evaluating and scoring criteria, then evaluate the credibility of the performance prototype using scientific techniques and finally get a specific evaluation result.

The first step of the credibility evaluation is to decide the evaluation pattern and clarify the evaluation details. There are two patterns of credibility evaluation, one is evaluating the credibility of each phase, and another is evaluating the credibility of each subsystem. The second step is to build up a set of credibility evaluation criteria which must be correlative, scientific, complete, limited and measurable. The last step is to choose the appropriate techniques according to the features of the performance prototype. The quantitative techniques such as confidence assessment, similarity evaluation and artificial neural network can be used to process and analyze the criteria which have specific output data and reference data. The combination techniques of qualitative and quantitative such as fuzzy comprehensive evaluation, analytic hierarchy and grey comprehensive evaluation can be used to process and analyze the criteria which do not have specific output data and reference data [10].

4.7 Accreditation

Accreditation is the last step of the VV&A activities and the official conclusion of whether the performance prototype can be put into application. The accomplishment of the accreditation represents the accomplishment of the development cycle of the performance prototype and the finish of the VV&A activities. The accreditation staff (authorities, domain experts) review the credibility evaluation report, give the conclusion of whether the performance prototype satisfies the expected application purposes and submit the accreditation report with the reference of all the V&V reports generated in each VV&A step, all the documents and records in the development process and the operating environment and configuration documents [11].

There is one thing that needs to be paid attention to is that the accreditation plan and the accreditation criteria are not made until the V&V activities are finished but should be made at the very beginning of the VV&A process during the requirements V&V phase. The purpose of the accreditation is to decide whether the performance prototype satisfies the expected application purposes, so the accreditation criteria should consistent with the requirements and should not be changed after the requirement V&V is done [8].

5   Conclusions

This paper proposes a VV&A process for the aerospace product performance prototype and a preliminary scheme for the credibility evaluation. The VV&A activities should run through the whole lifecycle of the performance prototype to ensure the high credibility of it. But this paper only discussed the VV&A scheme in a macroscopic perspective, and the further researches should focus on the specific tasks of each step of the VV&A process and the techniques that can be used to perform V&V activities. Also, to improve the efficiency of the VV&A activities and reduce the workloads of the VV&A crews, a specialized computer-aided tool should be developed to support the VV&A and the credibility evaluation.

References

1. Zhang, F., Xu, Y., Xue, H.: New integrated multi-objective optimization model for aerodynamic spacecraft performance prototype. Control Eng. China **24**(8), 1670–1678 (2016)
2. Zhang, F., Xue, H., Xu, Y.: Collaborative modeling method of performance prototype for aerospace products based on ontology. Comput. Integr. Manuf. Syst. **22**(8), 1887–1899 (2016)

3. Zhang, F., Xu, Y., Xue, H.: A method of reliability modelling based on characteristic model for performance digital mock-up of hypersonic vehicle. Comput. Modell. New Technol. **18** (4), 238–242 (2014)
4. National technical committee on standardization of technical product documents: GB/T26100—2010 General principles of digital mock-up for mechanical products. National technical committee on standardization of technical product documents, Beijing (2010)
5. Zheng, D., Liu, K., Liu, J.: Research on the technology and system of aircraft behavior digital mockup. Aeronaut. Sci. Technol. **26**(3), 05–09 (2015)
6. Balci, O.: Verification, validation and accreditation of simulation models. In: Proceedings of the 1997 Winter IEEE, Simulation Conference (1998)
7. Zhang, W., Wang, X.: Simulation credibility. J. Syst. Simul. **13**(3), 312–314 (2001)
8. Chew, O., Sullivan, C.: Verification, validation, and accreditation in the life cycle of models and simulations. In: Proceedings of the 2000 Winter Simulation Conference, USA, pp. 813–818. IEEE Press, Orlando (2010)
9. Tang, J., Zhang, M., Ye, L.: Research on verification and validation of requirement analysis model in process of M&S. J. Syst. Simul. **20**(18), 43–45 (2008)
10. Tang, J.: Research on the Credibility of Warfare Simulation System. National University of Defense Technology (2009)
11. Zhao, S., Xu, H., Zhang, Q.: Missile simulation model validation techniques and platform design. Ship Electron. Eng. **32**(9), 107–109+135 (2012)

Distributed Model Predictive Control Algorithm Based on PSO-IGA Processing System Structure Decomposition

Zhenbo Liu and Xin Jin[✉]

School of Information and Control Engineering, Liaoning Shihua University,
Fushun 113001, China
liuzhenbo1993@163.com, jinxin@lnpu.edu.cn

Abstract. A novel distributed model predictive control (DMPC) approach based on genetic immune algorithm through particle swarm optimization (PSO-IGA) to find out the optimal system decomposition structure is proposed. The PSO-IGA is used to solve decomposition problems for input clustering decomposition (ICD) and input-output pairing decomposition (IOPD), which can minimize the impact of input-output coupling between systems, and then DMPC algorithm is used to control the decomposed system. This approach effectively reduces the coupling between subsystems, and reduces the communication load of the system. Finally, a case study of a heavy oil fractionation chemical process is presented to demonstrate the effectiveness of the proposed approach.

Keywords: Distributed model predictive control · Genetic immune algorithm · System decomposition

1 Introduction

Model predictive control (MPC), as a type of computer control algorithm [1], has better control performance and strong robustness, and is widely used in process industry [2]. However, with the increasing complexity of industrial processes and the improvement of economic performance of industrial technology, traditional centralized predictive control algorithms are difficult to meet the requirements of control performance [3], and then decentralized predictive control algorithm is proposed [4, 5], the problem of solving a large system is transformed into a solution of several independent subsystems. The advantage of this algorithm is that the structure is simple and there is no communication burden between controllers. When the coupling between the subsystems is relatively strong, decentralized control systems may not achieve satisfactory control performance [6, 7]. In recent years, with the development of computer network technology, DMPC algorithm has emerged as the times require [8], and has received the attention and research of a wide range of experts and scholars.

Stewart et al. introduced the cooperative DMPC algorithm [9] based on the Jacobian method [10]. This algorithm can effectively solve the problem of coupling input constraints under output feedback, but it also greatly aggravates the communication

J. Qian et al. (Eds.): ICRRI 2020, CCIS 1336, pp. 367–385, 2020.
https://doi.org/10.1007/978-981-33-4932-2_27

between subsystems and cannot guarantee the algorithm convergence speed. For the communication problem between systems, Liu et al. proposed a series DMPC algorithm [11] for a system with a series structure. Any subsystem only needs to communicate with the upstream and downstream systems, and does not need to consider communicating with other subsystems.

When the above scholars solve the problems of insufficient convergence speed, complicated communication, and complicated stability conditions in the DMPC algorithm, they rarely consider the impact of system decomposition [12] on such problems, and designing a good system decomposition method can effectively alleviate The communication burden between systems can also greatly improve the efficiency of DMPC online implementation. Aiming at the problem of dynamic prediction coupling in the system decomposition of DMPC, Xing et al. proposed a method for structural decomposition of distributed predictive control system based on genetic algorithm (GA) [13], which is divided into input grouping (ICD) Paired with input and output (IOPD) two stages, and use GA to solve this combined optimization problem, thereby effectively reducing the coupling between subsystems and improving the efficiency of the DMPC algorithm. However, GA, as a general optimization algorithm, has defects such as poor local search ability and "premature maturity", which cannot guarantee the convergence of the algorithm. At the same time, this method has not been well solved for some indelible coupling between subsystems.

This paper proposes a distributed model prediction algorithm based on PSO-IGA to solve the structural decomposition of the system. The method is divided into two parts: In the first part, the method of literature [13] is followed, and the system structure decomposition problem is divided into two stages: ICD and IOPD to solve, but this paper uses PSO-IGA to optimize these two stages objective function. The PSO-IGA algorithm is a combined optimization algorithm with global search capabilities. The evolutionary equation of particles in PSO is introduced into the immune operation of IGA, making it adaptive, random, and diverse in population, which can overcome the problem of "premature" is inevitable in the general optimization process. In the second part, after the PSO-IGA decomposes the system structure, the DMPC algorithm is used to perform distributed control of the system under constraints. Finally, through the experimental study on the heavy oil fractionation chemical process, the effectiveness and speed of the algorithm are verified.

2 Structure Decomposition of DMPC System Based on PSO-IGA

There are two issues to consider when decomposing a large system: ICD and IOPD. The ICD problem is to find the input corresponding to each subsystem, and the IOPD problem is to solve the ICD problem, and then consider the corresponding output according to the corresponding input of each system.

2.1   ICD for DMPC

According to literature [13], a input decomposition matrix H and an objective function $J_{coupling}$ representing the degree of coupling between subsystems are defined here. The matrix H is defined as follows:

$$H = [\, h_1 \quad \cdots \quad h_M \,] \in \mathbb{R}^{m \times M}, \tag{1}$$

where m is the number of system inputs, M is number of the subsystems, and $M \leq m$. The entry of matrix H satisfies:

$$h_{ij} = \begin{cases} 1 & \text{if input } i \text{ belongs to group } j \\ 0 & \text{otherwise} \end{cases} \tag{2}$$

The objective function representing the degree of system coupling between subsystems is defined as follows:

$$J_{coupling} = \frac{\|V - V^*\|}{\|V\|}, \tag{3}$$

where $V = \begin{bmatrix} H_{11} & \cdots & H_{1M} \\ \vdots & \ddots & \vdots \\ H_{M1} & \cdots & H_{MM} \end{bmatrix}$; V^* is the off-diagonal parts of matrix V.

By minimizing $J_{coupling}$, the input of the given large system is grouped.

2.2   IOPD for DMPC

In this section, based on the input clustering result earned in the previous section, the IOPD solution to DMPC considered, that is, to solve an objective function representing the coupling relationship between the input and output.

Similar to the input decomposition matrix of Eq. (1), define an output decomposition matrix H' as follows:

$$H' = \begin{bmatrix} h'_1, & \cdots & h'_M \end{bmatrix} \in \mathbb{R}^{P \times M}, \tag{4}$$

where p is the number of the output, the entry of matrix H' satisfies:

$$h'_{ij} = \begin{cases} 1 & \text{if output } i \text{ belongs to group } j \\ 0 & \text{otherwise} \end{cases} \tag{5}$$

The objective function representing the degree of input-output coupling between subsystems can be defined as follows:

$$J'_{coupling} = \frac{\left\| V' - (V')^* \right\|}{\left\| V' \right\|},$$ (6)

where $V' = \begin{bmatrix} H'_{11} & \cdots & H'_{1M} \\ \vdots & \ddots & \vdots \\ H'_{M1} & \cdots & H'_{MM} \end{bmatrix}$, $(V')^*$ is the off-diagonal parts of matrix V'.

By minimizing $J'_{coupling}$, pair the input and output of the system.

2.3   PSO-IGA Based on Decomposition of DMPC

For solving ICD and IOPD problems, it is essentially the process of solving minimization Eq. (3) and Eq. (6). For the elements of the matrix H and matrix H' to be optimized are both 0 and 1. The optimization method used here is PSO-IGA, because PSO-IGA is easy to encode the variables 0 and 1.

In order to use PSO-IGA to solve the objective function, it is necessary to redefine its operations such as encoding, immune selection, cloning and mutation.

2.3.1   Implement PSO-IGA to Solve the ICD

① *Encoding*

A 0–1 binary of row vector with mM elements is defined to represent each chromosomes of antibody, that is input decomposition matrix: $H = \begin{bmatrix} h_1^T & \cdots & h_M^T \end{bmatrix}$. For example, 8 inputs are partitioned into two subsystems, which can be represented by 0–1 binary:

$$\begin{bmatrix} \underbrace{1\ 1\ 1\ 1\ 0\ 0\ 0\ 0}_{\text{subsystem1}} & \underbrace{0\ 0\ 0\ 0\ 1\ 1\ 1\ 1}_{\text{subsystem2}} \end{bmatrix},$$ (7)

where, the 1st, 2nd, 3rd and 4th inputs, the 5th, 6th, 7th and 8th inputs are located into subsystem 1 and 2. In general, the chromosomes of each antibody can be divided into subsystems, and the binary 1 at the corresponding position indicates that the input is selected into the current subsystem.

② *Initial population*

The initial population is the initial antibody population produced after antigen recognition. Each population must follow the constraint, each input can only belong to one subsystem.

Here, the evolutionary equation of particles in PSO is used to define the antibody, so that the antibody has two properties of "speed" and "position". The evolution equation of the particle is:

$$v_{ij}(t+1) = w \cdot v_{ij}(t) + c_1 r_1(t)[p_{ij}(t) - x_{ij}(t)] + c_2 r_2(t)[p_{gi}(t) - x_{ij}(t)] \qquad (8)$$

where w is the inertia weight, c_1, c_2 is the learning factor.

③ *Affinity function*

Affinity characterizes the binding strength of immune cells to antigens. After the population is generated, each antibody is used as an input to the affinity function, and the output is the affinity evaluation result. For the minimization problem Eq. (3), The affinity function of each antibody can be written as the reciprocal of the objective function:

$$affinity_{ICD} = \frac{1}{\delta + \frac{\|V - V^*\|}{\|V\|}}, \qquad (9)$$

where, δ is a small number. It ensures that the denominator of the Eq. (9) is not 0. According to Eq. (9), the smaller value of the objective function means the larger affinity function value.

④ *Antibody concentration function*

The antibody concentration indicates the diversity of the antibody population. The high concentration of antibody indicates that there are a large number of similar antibodies in the population, which will limit the search for optimization. Therefore, in the PSO-IGA, antibodies with excessive concentrations will be inhibited to ensure the diversity of antibodies.

Antibody concentration is usually defined as:

$$den(a_i) = \frac{1}{N} \sum_{j=1}^{N} S(a_i, a_j), \qquad (10)$$

where N is antibody population size, $S(a_i, a_j)$ is the similarity between antibodies, which can be expressed as:

$$S(a_i, a_j) = \begin{cases} 1, & \alpha(a_i, a_j) > \xi_s \\ 0, & \alpha(a_i, a_j) \leq \xi_s \end{cases}, \qquad (11)$$

where a_i is the i th antibody of the population, ξ_s is the similarity threshold value, $\alpha(a_i, a_j)$ is the affinity between antibody i and antibody j. For the 0–1 encoding algorithm, the calculation method for the affinity between antibodies is based on the Hamming distance calculation method.

Based on Hamming distance, antibody-antibody affinity calculation method is following as:

$$\alpha(a_i, a_j) = \sum_{k=1}^{L} \partial_k, \qquad (12)$$

where $\partial_k = \begin{cases} 1, & a_{i,k} = a_{j,k} \\ 0, & a_{i,k} \neq a_{j,k} \end{cases}$, $a_{i,k}$ and $a_{j,k}$ are the k th position of antibody i and the k th position of antibody j.

⑤ *Excitation Function*

Antibody excitation function is the final evaluation result of antibody quality. Its purpose is to retain the antibody with high affinity and low concentration.

The antibody excitation function can be expressed as:

$$sim(a_i) = m \cdot \alpha(a_i) - n \cdot \alpha(a_i), \tag{13}$$

where $sim(a_i)$ is the excitation degree of antibody i, n, m is the calculation parameter.

⑥ *Cloning*

Determine which antibodies enter clone selection based on antibody excitation, the antibody with high excitation degree has better quality and is more likely to be selected for clone selection.

The clone function can be expressed as:

$$T_c(a_i) = clone(a_i), \tag{14}$$

where $clone(a_i)$ is the set of w_i clone antibodies identical to a_i. w_i is the number of antibody clones.

⑦ *Mutation*

Mutation operation is to expand the search and optimization space, so as to produce a new antibody. For the 0–1 coding PSO-IGA, the mutation strategy is to randomly select one or more positions from the mutation source antibody. The mutation defined here is to replace 1 that represents the corresponding input position with 0 in the antibody, and then replace 0 with 1 at the other corresponding positions. For example:

$$[1\ 1\ 1\ (1)\ 0\ 0\ 0\ 0\ |\ 0\ 0\ 0\ 0\ 1\ 1\ 1\ 1] \Rightarrow [1\ 1\ 1\ 0\ 0\ 0\ 0\ |\ 0\ 0\ 0\ (1)\ 1\ 1\ 1\ 1] \tag{15}$$

In subsystem 1, entry 1 in brackets is replaced by 0, that means 4th input is not included in subsystem 1. Then in subsystem 2, entry 0 in the fourth position is replaced by 1 in brackets, that means 4th input is included in subsystem 2.

In the optimization process of solving the minimization Eq. (9) problem, immune selection, cloning, and mutation can all generate new antibodies, and these operations need to be iterated continuously to finally meet a convergence condition.

Thus, an optimal solution H^* is obtained to construct the optimal input grouping matrix for ICD problem.

2.3.2   Implement PSO-IGA to Solve the IOPD

Like IGA operators of the ICD problem, all the operators of the IOPD problem are same, except for encoding and affinity function.

① *Encoding*

A 0–1 binary of row vector with pM elements is defined to represent each chromosomes of antibody, that is input-output decomposition matrix: $H' = \left[(h'_1)^T \cdots (h'_M)^T \right]$. For example, an 6 outputs are partitioned into three subsystems, which can be represented by 0–1 binary:

$$
\left[\underbrace{1\,1\,0\,0\,0\,0}_{\text{subsystem1}} \Big| \underbrace{0\,0\,1\,1\,0\,0}_{\text{subsystem2}} \Big| \underbrace{0\,0\,0\,0\,1\,1}_{\text{subsystem3}} \right],
\tag{16}
$$

where, the 1st and 2nd outputs, the 3rd and 4th outputs, the 5th and 6th outputs are located into subsystem 1, 2 and 3. In general, the chromosomes of each antibody can be divided into M subsystems, and the binary 1 at the corresponding position indicates that the output is selected into the current subsystem.

② *Affinity function*

For the minimization problem Eq. (6), The affinity function of each antibody can be written as the reciprocal of the objective function:

$$
affinity_{IOPD} = \frac{1}{\delta + \frac{\| v' - (v')^* \|}{\| v' \|}}
\tag{17}
$$

According to Eq. (17), the smaller value of the objective function means the larger affinity function value.

Therefore, the ICD problem and the IOPD problem are solved accordingly, and two grouping matrices are obtained: $H^* = \begin{bmatrix} h_1^* & \cdots & h_M^* \end{bmatrix}$ and $(H')^* = \begin{bmatrix} (h'_1)^* \cdots (h'_M)^* \end{bmatrix}$.

3   DMPC Algorithm Based on PSO-IGA Processing System Structure Decomposition

The previous section introduced how a large system with multiple inputs and multiple outputs can be decomposed into multiple subsystems. When there is an unsolvable coupling problem between multiple subsystems, a DMPC algorithm is needed to solve this problem. In this section, on the premise that the system structure has been decomposed, the state space prediction model of each subsystem in the DMPC and the implementation process of the DMPC algorithm based on the PSO-IGA processing system structure decomposition will be introduced.

3.1   DMPC Controller Design

For a distributed system M containing S subsystems, the state space model of its subsystem $S_i (i = 1, \cdots, M)$ can be written as:

$$
\begin{cases}
x_i(k+1) = A_i x_i(k) + B_{ii} u_i(k) + \sum_{j=1, j \neq i}^{M} B_{ij} u_j(k) \\
y_{i(k)} = C_i x_i(k)
\end{cases}
\tag{18}
$$

where $B_{ij}(i \neq j)$ represents the coupled input matrix of subsystem S_j to subsystem S_i, u_j is the input sequence of subsystem S_j at time instant k.

Suppose that starting from the time instant k, the state vector and output vector of the subsystem are predicted in n-steps.

The model represented by Eq. (18) can derive the following matrix-vector form:

$$
\begin{bmatrix} x_i(k+1|k) \\ \vdots \\ x_i(k+N|k) \end{bmatrix} = \begin{bmatrix} A_i \\ \vdots \\ A_i^{N-1} \end{bmatrix} x_i(k|k) + \begin{bmatrix} B_{ii} & 0 & \cdots & 0 \\ A_i B_{ii} & B_{ii} & \cdots & 0 \\ \vdots & & & \vdots \\ A_i^{N-1} B_{ii} & A_i^{N-2} B_{ii} & \cdots & B_{ii} \end{bmatrix} \begin{bmatrix} u_i(k|k) \\ u_i(k+1|k) \\ \vdots \\ u_i(k+N-1|k) \end{bmatrix}
$$

$$
+ \begin{bmatrix} B_{ij} & 0 & \cdots & 0 \\ A_i B_{ij} & B_{ij} & \cdots & 0 \\ \vdots & & & \vdots \\ A_i^{N-1} B_{ij} & A_i^{N-2} B_{ij} & \cdots & B_{ij} \end{bmatrix} \begin{bmatrix} u_j(k|k) \\ u_j(k+1|k) \\ \vdots \\ u_j(k+N-1|k) \end{bmatrix}
\tag{19}
$$

recorded as:

$$
X_i(k) = \begin{bmatrix} x_i(k+1|k) \\ x_i(k+2|k) \\ \vdots \\ x_i(k+N|k) \end{bmatrix}, \quad \eta_i = \begin{bmatrix} A_i \\ A_i^2 \\ \vdots \\ A_i^{N-1} \end{bmatrix}
\tag{20}
$$

$$
U_i(k) = \begin{bmatrix} u_i(k|k) \\ u_i(k+1|k) \\ \vdots \\ u_i(k+N-1|k) \end{bmatrix}, \quad U_j(k) = \begin{bmatrix} u_j(k|k) \\ u_i(k+1|k) \\ \vdots \\ u_j(k+N-1|k) \end{bmatrix}
\tag{21}
$$

$$
\Gamma_{ii} = \begin{bmatrix} B_{ii} & 0 & \cdots & 0 \\ A_i B_{ii} & B_{ii} & \cdots & 0 \\ \vdots & & & \vdots \\ A_i^{N-1} B_{ii} & A_i^{N-2} B_{ii} & \cdots & B_{ii} \end{bmatrix}, \quad \Gamma_{ij} = \begin{bmatrix} B_{ij} & 0 & \cdots & 0 \\ A_i B_{ij} & B_{ij} & \cdots & 0 \\ \vdots & & & \vdots \\ A_i^{N-1} B_{ij} & A_i^{N-2} B_{ij} & \cdots & B_{ij} \end{bmatrix}
\tag{22}
$$

The Eq. (19) can be rewritten as:

$$X_i(k) = \eta_i x_i(k) + \Gamma_{ii} U_i(k) + \sum_{j=1, j\neq i}^{M} \Gamma_{ij} U_j(k) \tag{23}$$

The objective function of subsystem S_i is defined as:

$$\min \Phi_i(k) = \gamma_i \left(\|x_i(k) - r_i(k)\|_{Q_i}^2 + \|u_i(k)\|_{R_i}^2 \right) + \sum_{j=1, j\neq i}^{M} \gamma_j \left(\|x_j(k) - r_j(k)\|_{Q_j}^2 + \|u_j(k)\|_{R_j}^2 \right)$$
$$s.t. \quad u_i(k+l|k) \in \tau_i, \quad 0 < l \leq N - 1 \tag{24}$$

Combining Eq. (23), the optimal solution of the objective function Eq. (24) can be written as:

$$\min \Phi_i(k) = U_i(k)^T \psi_i U_i(k) + 2U_i(k)^T \left(\Theta_i(k) - \Pi_{ii}(k) + \sum_{j=1, j\neq i}^{M} \Omega_{ij} U_j(k) \right) \tag{25}$$
$$s.t. \quad U_i \in \tau_i$$

where,

$$\psi_i = \gamma_i \Gamma_{ii}^T Q_i \Gamma_{ii} + \gamma_i R_i + \sum_{j=1, j\neq i}^{M} \gamma_j \Gamma_{ji}^T Q_j \Gamma_{ji} \tag{26}$$

$$\Theta_i(k) = \gamma_i \Gamma_{ii}^T Q_i \eta_i x_i(k) + \sum_{j=1, j\neq i}^{M} \gamma_j \Gamma_{ji}^T Q_j \eta_j x_j(k) \tag{27}$$

$$\Pi_i(k) = \gamma_i \Gamma_{ii}^T Q_i r_i(k) - \sum_{j=1, j\neq i}^{M} \gamma_j \Gamma_{ji}^T Q_j r_j(k) \tag{28}$$

$$\Omega_{ij} = \sum_{g=1}^{M} \gamma_g \Gamma_{gi}^T Q_g \Gamma_{gi} \tag{29}$$

The optimal control sequence U_i of subsystem S_i at time instant k is obtained by the quadratic programming algorithm.

3.2 Implementation Process of Distributed Predictive Control Algorithm Based on PSO-IGA System Structure Decomposition

The structural decomposition method for large systems was given in the previous section. In combination with the DMPC algorithm in this section, a DMPC algorithm based on PSO-IGA system structure decomposition is proposed. The specific implementation process is:

Step 1. Given a large system with m inputs and p outputs, and the system needs to be divided into M subsystems:

a. In ICD, use IGA to find the maximum value of objective function Eq. (9) and get the input decomposition matrix H^* for partitioning the input M subsystems.
b. In IOPD, use IGA to find the maximum value of objective function Eq. (17) and get the input-output decomposition matrix for partitioning the output M subsystems.

Step 2. At time instant k, obtain the state space model and output Eq. (18) of each subsystem S_i. And given the prediction horizon N, state variable weight Q_i, control variable weight R_i, Objective function weight γ_i, Initial value of control variable $u_{i,0}(k)$.

Step 3. Through network communication, subsystem S_i transfers the control variables to other subsystems, and obtains the control variables of other subsystems.

Step 4. Parallel solution: the iterative solution method is adopted. At the iterative time $q(q \geq 1)$. According to Eq. (20)–(23), (25)–(29), Solving the optimal control sequence $U_i^q(k)$ of each subsystem in parallel under the condition of satisfying the constraints. Given the convergence accuracy ξ and Maximum number of iterations q_{max}. If the control input of each subsystem satisfies the inequality $\left\| U_i^q(k) - U_i^{q-1} \right\| < \xi$ or the number of iterations satisfies $q > q_{max}$, then go to step 5. otherwise, let $q = q + 1$ and go to step 3.

Step 5. Take the first term of the optimal control sequence $U_i^q(k)$ of the subsystem as the control input:

$$u_i(k) = [I \ 0 \ \cdots \ 0] U_i^q(k), \tag{30}$$

and applied to each subsystem.

Step 6. Let $k = k + 1$, the process of solving the control variables is optimized to the next moment, and returns to step 2.

4 Case Study and Discussion

In this paper, the model of heavy oil fractionator [14] is used to verify the effectiveness of the proposed PSO-IGA-DMPC algorithm. The model of heavy oil fractionator is a multi-input and multi-output control problem with constraints (Table 1, Table 2, Fig. 1). In this case:

Table 1. Input and output of heavy oil fractionator

Manipulated variables	Variable number	Controlled variables	Variable number
Top draw	u_1	Top end point	y_1
Side draw	u_2	Side end point	y_2
Inter reflux duty	u_3	Upper reflux temp	y_3
Upper reflux duty	u_4	Side draw temp	y_4

Table 2. Transfer function models for heavy oil fractionator

Outputs	Inputs			
	u_1	u_2	u_3	u_4
y_1	$\dfrac{4.05e^{-27s}}{50s+1}$	$\dfrac{1.77e^{-28s}}{60s+1}$	$\dfrac{1.20e^{-27s}}{15s+1}$	$\dfrac{1.44e^{-27s}}{40s+1}$
y_2	$\dfrac{5.39e^{-18s}}{50s+1}$	$\dfrac{5.72e^{-14s}}{60s+1}$	$\dfrac{1.52e^{-15s}}{25s+1}$	$\dfrac{1.83e^{-15s}}{20s+1}$
y_3	$\dfrac{5.92e^{-11s}}{12s+1}$	$\dfrac{2.54e^{-12s}}{17s+1}$	$\dfrac{1.73}{5s+1}$	$\dfrac{1.79}{19s+1}$
y_4	$\dfrac{4.13e^{-5s}}{8s+1}$	$\dfrac{2.38e^{-7s}}{19s+1}$	$\dfrac{1.19}{19s+1}$	$\dfrac{1.17}{24s+1}$

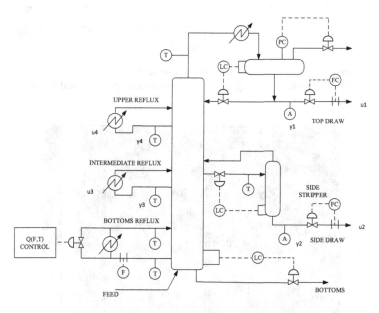

Fig. 1. Heavy oil fractionator process flowsheet.

The First Part: PSO-IGA to Solve DMPC System Decomposition Problem

- *PSO-IGA for ICD of DMPC*

Given that the system has four inputs, and system is divided into two subsystems, the prediction horizon is $K = 20$ and the sampling time is $T_s = 1$ s. The weight of the inputs and outputs are $Q = I$ and $R = I$. Each antibody population in IGA contains 30 members.

The maximum number of antibody population is 100. The number of immune selection antibodies is 15. The probability for the mutation is 0.5. The number of cloning is 15. The evolutions of the objective function are shown in Fig. 2 (Fig. 3).

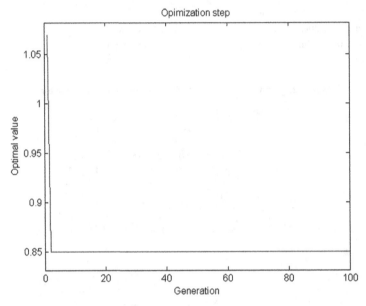

Fig. 2. The performance of the ICD structure across generations in case. The initial affinity value of the first generation of decomposition objective function is 1.06, and the optimal affinity value is 0.85 after 4 generations.

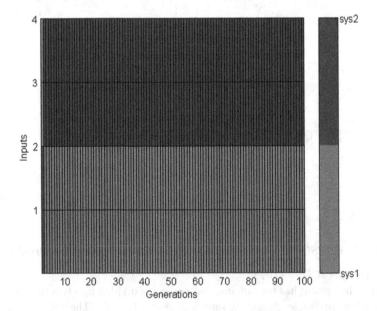

Fig. 3. The decomposition pattern for ICD across generations in case. It shows the best input clustering antibody in each generation and the two colors represent two subsystems.

The resulting grouping can be represented by the following 0–1 binary:

$$v = [1\ 1\ 0\ 0\ 0\ 0\ 1\ 1] \tag{31}$$

Get the 1st and 2nd inputs divided into subsystem1, the 3rd and 4th inputs divided into subsystem2.

- *PSO-IGA for IOPD of DMPC*

After input grouping, the IOPD problem is solved on this basis. According to the result of input grouping in Eq. (31), the input grouping matrix can be obtained:

$$H = \begin{bmatrix} 1\ 1\ 0\ 0 \\ 0\ 0\ 1\ 1 \end{bmatrix} \tag{32}$$

Based on input decomposition matrix H, The IOPD problem is solved to obtain the input-output pairing matrix H'. The evolutions of the objective function are shown in Fig. 4 (Fig. 5).

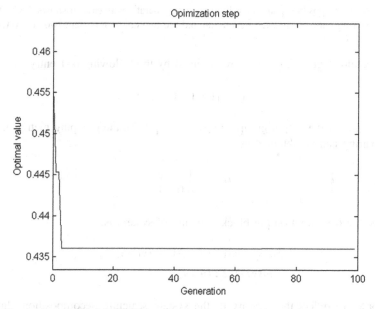

Fig. 4. The performance of the IOPD structure across generations in case. The initial affinity value of the first generation of decomposition objective function is 0.455, and the optimal affinity value is 0.436 after 4 generations.

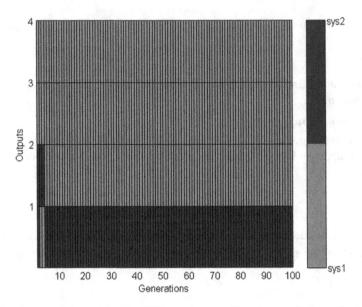

Fig. 5. The decomposition pattern for IOPD across generations in case. It shows the best input-output clustering antibody in each generation and the two colors represent two subsystems.

The resulting grouping can be represented by the following 0–1 entry:

$$v' = [1\ 1\ 1\ 0\ 0\ 0\ 0\ 1] \qquad (33)$$

According to the output grouping result of Eq. (33), the grouping matrix of input-output pairing can be obtained as:

$$H' = \begin{bmatrix} 1\ 1\ 1\ 0 \\ 0\ 0\ 0\ 1 \end{bmatrix} \qquad (34)$$

Thus, the best input-output block structure of system is:

$$\begin{aligned} Subsystem1 &: (u_1, u_2) \rightarrow (y_1, y_2, y_3) \\ Subsystem2 &: (u_3, u_4) \rightarrow (y_4) \end{aligned} \qquad (35)$$

In order to reflect the rapidity of the system structure decomposition algorithm using PSO-IGA, this paper compares with the GA used in the literature [13] (Fig. 6):

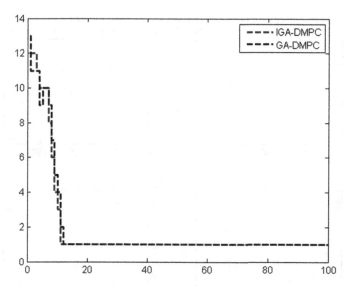

Fig. 6. It shows the number of iterations at each time point of the two algorithms. It can be seen that the number of iterations at each time point under the PSO-IGA is significantly less than the number of iterations at each time point under the GA.

The Second Part: Distributed Control of Decomposed System by DMPC Algorithm

It is known that the system structure of heavy oil fractionation process is divided into subsystem 1 with 1st, 2nd input and 1st, 2nd,3rd output. subsystem 2 with 3rd, 4th input and 4th output. The main control parameters in distributed control are:

a. The sampling time is $T_s = 1$ s;
b. The simulated temporal is $N = 200$, the control horizon is $M = 5$, the prediction horizon is $P = 20$;
c. The weight matrix $R_i = I$, The value of weight matrix Q_i is:

$$Q_i = diag\left(q_i \cdots q_i, q_i^*\right) \tag{36}$$

d. d. In DMPC simulation, the weights of two subsystems are: $\gamma_1 = 0.8$, $\gamma_2 = 0.2$;
e. e. For each subsystem, the optimization proposition is:

$$\min \Phi_i(k) = \gamma_i\left(\|x_i(k) - r_i(k)\|_{Q_i}^2 + \|u_i(k)\|_{R_i}^2\right) + \sum_{j=1,j\neq i}^{M} \gamma_j\left(\|x_j(k) - r_j(k)\|_{Q_j}^2 + \|u_j(k)\|_{R_j}^2\right)$$

$$s.t. \quad |u_i(k)| \leq 0.5$$

$$\tag{37}$$

In order to verify the effectiveness of the algorithm, this paper uses literature [13] (GA-MPC) for comparative experiments, and obtains the control trajectories respectively. Figures 7, 8, 9 and 10 shows the trajectories of inputs and outputs for these two algorithms:

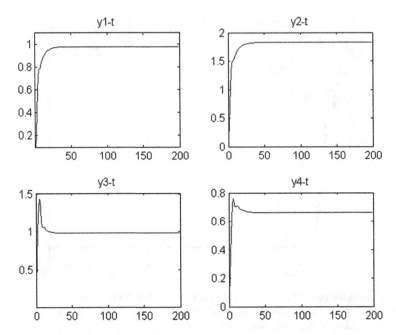

Fig. 7. Outputs trajectories based on PSO-IGA-DMPC algorithm.

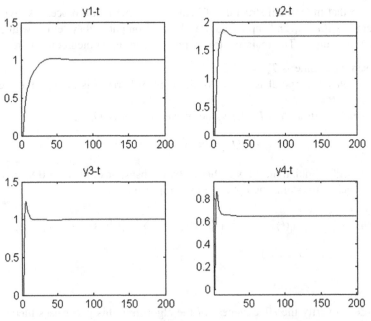

Fig. 8. Outputs trajectories based on GA-MPC algorithm.

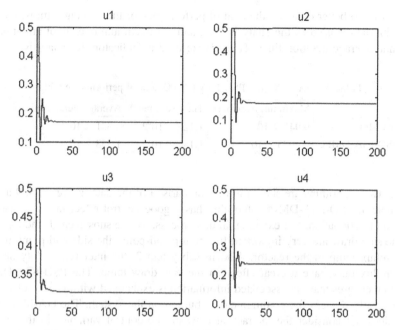

Fig. 9. Inputs trajectories based on PSO-IGA-DMPC algorithm.

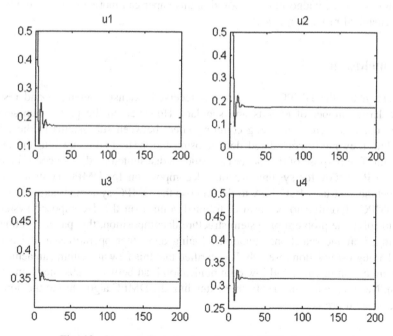

Fig. 10. Inputs trajectories based on GA-MPC algorithm.

In order to better compare the control performance of the two algorithms, the Top end point data is used as the analysis data, and the maximum overshoot amount, rise time, and average iteration time (Table 3) are used as indicators for analysis:

Table 3. GA-MPC and PSO-IGA-DMPC control performance table

	Maximum overshoot	The rise time(s)	Average iteration time(s)
GA-MPC	9.943×10^{-1}	1.123×10^{-1}	3.321×10^{-2}
PSO-IGA-DMPC	9.9975×10^{-1}	1.184×10^{-1}	1.538×10^{-2}

The two algorithms are shown in the comparison of the data in Fig. 7, 8, 9 and 10 and Table 3, PSO-IGA-DMPC algorithm has a good control effect in the process of heavy oil fractionation. For example, in this process, for the subsystem 1, the top draw and the side draw are very important for the top end point, the side end point and the upper reflux temp in the reactor. Also in subsystem 2, the inter reflux duty and the upper reflux duty have a great effect on the side draw temp. The PSO-IGA-DMPC algorithm ensures that the associated information is exchanged within each subsystem, which greatly reduces the communication burden on the system. The GA-MPC algorithm does not consider the interaction between the control variables in the system. Under the same control effect, PSO-IGA-DMPC takes significantly less time than GA-MPC in iteration time, which greatly improves the working efficiency of the algorithm. In summary, the new algorithm proposed in this paper can more effectively control the actual chemical process in case.

5 Conclusion

In recent years, the DMPC algorithm has received extensive attention and research from a large number of experts and scholars. However, in the past, when scholars pointed out that there is a strong coupling effect between subsystems in the DMPC algorithm, they rarely considered decomposing the system structure into the DMPC algorithm. The purpose of this paper is to study a distributed predictive control method based on PSO-IGA for system structure decomposition for DMPC control systems. This method first uses PSO-IGA to decompose the DMPC system structure, and then uses DMPC algorithm to perform distributed control on the decomposed system. In order to solve the problem of system structure decomposition, this paper redefines the encoding, immune selection, mutation, cloning and other operations in the adopted IGA. Through simulation research, It is verified that this new algorithm can achieve the minimum coupling effect and low communication load between subsystems in DMPC system. Next, we will consider further extending the DMPC algorithm to the nonlinear DMPC system problem.

References

1. Richalet, J., Rault, A., Testud, J.L., Papon, J.: Model predictive heuristic control: applications to industrial processes. Automatica **14**, 413–428 (1978)
2. Qin, S.J., Badgwell, T.A.: A survey of industrial model predictive control technology. Control Eng. Pract. **11**, 733–764 (2003)
3. Scattolini, R.: Architectures for distributed and hierarchical model predictive control – a review. J. Process Control **19**, 723–731 (2009)
4. Richards, A., How, J.: A decentralized algorithm for robust constrained model predictive control. In: 2004 Proceedings of the American Control Conference, vol. 4265, pp. 4261–4266 (2004)
5. Raimondo, D.M., Magni, L., Scattolini, R.: Decentralized open-loop MPC of nonlinear systems: an input-to-state stability approach. Int. J. Robust Nonlinear Control **17**, 1651–1667 (2007)
6. Zheng, Y., Li, S., Li, N.: Distributed model predictive control over network information exchange for large-scale systems. Control Eng. Pract. **19**, 757–769 (2011)
7. Xing, C., Lei, X., Pei, S.: Decentralized model predictive control of nonlinear continuous systems with contractive constraints. In: 2013 25th Chinese Control and Decision Conference (CCDC) (2010)
8. Camacho, E.F., Bordons, C.: Distributed model predictive control. Control Syst. IEEE **36**, 269–271 (2015)
9. Stewart, B.T., Venkat, A.N., Rawlings, J.B., Wright, S.J., Pannocchia, G.: Cooperative distributed model predictive control. Syst. Control Lett. **59**, 460–469 (2010)
10. Conley, A., Salgado, M.E.: Gramian based interaction measure. In: Proceedings of the 39th IEEE Conference on Decision and Control (2000)
11. Liu, J., Chen, X., Christofides, P.D.: Sequential and iterative architectures for distributed model predictive control of nonlinear process systems. AIChE J. **56**, 2137–2149 (2010)
12. Motee, N., Sayyar-Rodsari, B.: Optimal partitioning in distributed model predictive control. In: Proceedings of the 2003 American Control Conference (2003)
13. Lei, X., Xing, C., Chen, J., Su, H.: GA based decomposition of large scale distributed model predictive control systems. Control Eng. Pract. **57**, 111–125 (2016)
14. Prett, D.M., García, C.E.: Fundamental process control, pp. 401–440 (1990)

Rehabilitation Intelligence

Rehabilitation Training Analysis Based on Human Lower Limb Muscle Model

Chaoyi Zhao[✉], Qiuhao Zhang, Yong Li, Junyou Yang,
Baiqing Sun, and Yina Wang

Shenyang University of Technology, Shenyang 110870, China
1091262968@qq.com

Abstract. Human lower limbs are closely related to human motor ability, so lower limb muscle injury will affect human motor ability. Therefore, we need to establish an effective muscle model to analyze the muscle output in rehabilitation training. Based on the analysis of human anatomy of human lower limb musculoskeletal connection, with calves, quadriceps, hamstrings muscle is a dynamic model of the three muscles, in view of the quadriceps and hamstring muscles work together in the situation of hip flexion and extension, using Crowninshieldt established based on the smallest muscle force of PCSA optimization objective function of solving the two muscles is optimized, And conducted related experiments on general fitness equipment, The BIOPIC multi-channel physiological signal acquisition instrument and Angle sensor are used to collect the muscle emg signal and lower limb position information of the tester. Then, according to the position information collected, the muscle output is calculated by the lower limb muscle model, and compared with the muscle force obtained by the electromyography signal. The results of the two are consistent.

Keywords: Musculoskeletal model · Human dynamics · Muscle force · PCSA

1 Introduction

There are many patients with physical dysfunction in the world. All of these are directly related to the deficiency or defect of human muscle strength. In recent years, the research on the restorative training of human muscle strength has become a hot topic in the world. The rehabilitation robot can train the muscles of patients and restore the motor function of patients. However, it only trains the patients in the hospital or carries out relevant research in the laboratory. It cannot meet the exercise needs of people with mild disease or sub-health.

The current rehabilitation training in the world can be divided into active training and passive training according to the training methods [1]. Active training is mainly based on the intention of patients to actively participate in the rehabilitation training, which is the core of the auxiliary machinery; The passive training robot is mainly based on the equipment, and the patients are only passively trained. Clinical studies in

Natural Science Foundation of Liaoning Province (20180550994).

rehabilitation medicine have shown that rehabilitation training with patients' intention to exercise is more effective for patients' nervous system reconstruction and motor function recovery. Therefore, compared with passive training, active training is more effective for patients' rehabilitation training.

The training posture of the human body and the fitness equipment used in the training also have a great influence on the muscle training. Joris [2] conducted statistical modeling on the lower limb dynamics of human squat and lunge movements. Zhang [3] established an iterative learning algorithm for human musculoskeletal model by artificially induced joints. Katrin [4] studied the force that muscles produce when the arm is stimulated in a neuromusculoskeletal model.

Cullell [5] designed a lower limb orthopedic device to help patients regain lower limb function. Pennicott [6] proposed a weight-supported robot-assisted device controlled by estimation and volitional feedback. Through the real-time observation of the training data, patients can get better rehabilitation training. TEM Lx 2 [7], a lower limb rehabilitation robot developed by Japan's Yaskawa electric company, is mainly aimed at patients with acute lower limb diseases. Its main purpose is to make the patient recover body function as soon as possible, even restore the ability to walk. wang hongbo [8] proposed a robot for the rehabilitation of paraplegics' lower limbs. The end of the rehabilitation robot is mounted on the side of the chair, and the patient can sit in the chair for rehabilitation training. However, patients can not actively participate in the rehabilitation training, which mainly relies on the rehabilitation robot for training. Although researchers have developed a variety of recovery robots, there are few applications on the market. In view of the above problems, in this paper, the musculoskeletal model of human lower limbs was constructed based on general fitness equipment. For joints with multiple muscles acting together, this paper used PCSA parameters to establish an optimization objective function to optimize and analyze the force of each muscle and determine the output state of each muscle.

2   Establishment of Musculoskeletal Model of Human Lower Limbs

In order to analyze the muscle output in lower limb rehabilitation training, it is necessary to establish a lower limb musculoskeletal model to predict the muscle strength of lower limb in rehabilitation exercise [6–8]. First, the lower limbs of the human body are regarded as a two-link model and cartesian coordinate system is established [9], as shown in Fig. 1.

Where, m_1 refers to the mass of the thigh, m_2 to the mass of the calf, m_3 to the mass of the load, l_1 to the length of the thigh, l_2 to the length of the calf; l_{c1} is the length from the center of mass of the thigh to the hip joint, and l_{c2} is the length from the center of mass of the calf to the knee joint; θ_1 and θ_2 are the Angle between the thigh and the calf and the horizontal plane respectively; τ_1 and τ_2 are joint moments of the hip joint and the knee joint respectively. The kinetic equations were established to calculate the joint moments of the hip and knee joints.

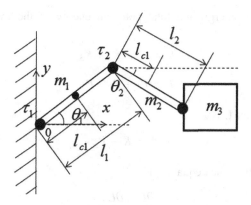

Fig. 1. Two bar diagram

First, we need to calculate the kinetic energy and potential energy of the thigh:
Secondly, we need to calculate the kinetic energy and potential energy of the calf:

$$
\begin{cases}
K_1 = \dfrac{1}{2}m_1 l_{c1}^2 \dot{\theta}_1^2 + \dfrac{1}{2}I_1 \dot{\theta}_1^2 \\
P_1 = m_1 g l_{c1} \sin(\theta_1)
\end{cases}
\tag{1}
$$

$$
\begin{cases}
K_2 = \dfrac{1}{2}m_2 v_2^2 + \dfrac{1}{2}I_2 \dot{\theta}_2^2 \\
P_2 = m_2 g y_2^2 \\
x_2 = l_1 \cos(\theta_1) + l_{c2} \cos(\theta_2) \\
\dot{x}_2 = -l_1 \sin(\theta_1)\,\dot{\theta}_1 - l_{c2} \sin(\theta_2)\,\dot{\theta}_2 \\
y_2 = l_1 \sin(\theta_1) + l_{c2} \sin(\theta_2) \\
\dot{y}_2 = -l_1 \cos(\theta_1)\,\dot{\theta}_1 - l_{c2} \cos(\theta_2)\,\dot{\theta}_2 \\
v_2^2 = \left(\dot{x}_2\right)^2 + \left(\dot{y}_2\right)^2
\end{cases}
\tag{2}
$$

Then calculate the kinetic energy and potential energy of the load:

$$
\begin{cases}
K_3 = \dfrac{1}{2}m_3 v_3^2 \\
P_3 = m_3 g x_3^2 \\
x_3 = l_1 \cos(\theta_1) + l_2 \cos(\theta_2) \\
\dot{x}_3 = -l_1 \sin(\theta_1)\,\dot{\theta}_1 - l_2 \sin(\theta_2)\,\dot{\theta}_2 \\
y_3 = l_1 \sin(\theta_1) + l_2 \sin(\theta_2) \\
\dot{y}_3 = -l_1 \cos(\theta_1)\,\dot{\theta}_1 - l_2 \cos(\theta_2)\,\dot{\theta}_2 \\
v_3^2 = \left(\dot{x}_3\right)^2 + \left(\dot{y}_3\right)^2
\end{cases}
\tag{3}
$$

The total kinetic energy and total potential energy of the two-link mechanical mechanism are:

$$K = K_1 + K_2 + K_3 \tag{4}$$

$$P = P_1 + P_2 + P_3 \tag{5}$$

Lagrange function L is:

$$L = K - P \tag{6}$$

The Lagrange dynamics equation is:

$$\frac{d}{dt}\frac{\partial L}{\partial \dot{\theta}} - \frac{\partial L}{\partial \theta} = \tau \tag{7}$$

According to Eq. (7), the dynamic equations of τ_1 and τ_2 can be obtained:

$$
\begin{aligned}
\tau_1 &= \left(I_1 + m_1 l_{c1}^2 + m_2 l_1^2 + m_3 l_1^2\right) \cdot \ddot{\theta}_1 \\
&\quad + (m_2 l_1 l_{c2} + m_3 l_1 l_2) \cdot \cos(\theta_1 - \theta_2) \cdot \ddot{\theta}_2 \\
&\quad + (m_2 l_1 l_{c2} + m_3 l_1 l_2) \cdot \sin(\theta_1 - \theta_2) \cdot \dot{\theta}_1 \dot{\theta}_2 \cdot \left(\dot{\theta}_1 - 1\right) \\
&\quad + (m_1 l_{c1} + m_2 l_1)g\cos(\theta_1) - m_3 l_1 g \sin(\theta_1)
\end{aligned} \tag{8}
$$

$$
\begin{aligned}
\tau_2 &= \left(I_2 + m_2 l_{c2}^2 + m_3 l_2^2\right) \cdot \ddot{\theta}_2 \\
&\quad + (m_2 l_1 l_{c2} + m_3 l_1 l_2) \cdot \cos(\theta_1 - \theta_2) \cdot \ddot{\theta}_1 \\
&\quad + (m_2 l_1 l_{c2} + m_3 l_1 l_2) \cdot \sin(\theta_1 - \theta_2) \cdot \dot{\theta}_1 \dot{\theta}_2 \cdot \left(\dot{\theta}_2 - 1\right) \\
&\quad + m_2 l_{c2} g \cos(\theta_2) - m_3 l_2 g \sin(\theta_2)
\end{aligned} \tag{9}
$$

Equations (8) and (9) can be used to form the equation of state space:

$$\tau = M(\theta)\ddot{\theta} + V\left(\theta, \dot{\theta}\right) + G(\theta) \tag{10}$$

In Eq. (11), (12) and (13), $M(\theta)$ is the mass matrix of $n \times n$, $V(\theta, \dot{\theta})$ is the centrifugal force and coriander force vectors of $n \times 1$, and $G(\theta)$ is the gravity vector of $n \times 1$. The matrix equations of $M(\theta)$, $V(\theta, \dot{\theta})$ and $G(\theta)$ are as follows:

$$M(\theta) = \begin{bmatrix} I_1 + m_1 l_{c1}^2 + m_2 l_1^2 + m_3 l_1^2 & (m_2 l_1 l_{c2} + m_3 l_1 l_2) \cdot \cos(\theta_1 - \theta_2) \\ (m_2 l_1 l_{c2} + m_3 l_1 l_2) \cdot \cos(\theta_1 - \theta_2) & I_2 + m_2 l_{c2}^2 + m_3 l_2^2 \end{bmatrix} \tag{11}$$

$$G(\theta) = \begin{bmatrix} (m_1 l_{c1} + m_2 l_1)g\cos(\theta_1) - m_3 l_1 g\sin(\theta_1) \\ m_2 l_{c2} g\cos(\theta_2) - m_3 l_2 g\sin(\theta_2) \end{bmatrix} \tag{12}$$

$$V\left(\theta, \dot{\theta}\right) = \begin{bmatrix} (m_2 l_1 l_{c2} + m_3 l_1 l_2) \cdot \sin(\theta_1 - \theta_2) \cdot \dot{\theta}_1 \dot{\theta}_2 \cdot \left(\dot{\theta}_1 - 1\right) \\ (m_2 l_1 l_{c2} + m_3 l_1 l_2) \cdot \sin(\theta_1 - \theta_2) \cdot \dot{\theta}_1 \dot{\theta}_2 \cdot \left(\dot{\theta}_2 - 1\right) \end{bmatrix} \tag{13}$$

3 Calculation of Lower Limb Muscle Force

Human physiological structure is complex, the establishment of the human body model is the basis of the study of human activities, human activity has the characteristics of complexity, integrity and diversity, is unmatched by any precision machine, thus directly for complex mechanical analysis of the human body is very difficult, must be in the model is simplified to the human body. It is a common practice to simplify the joints of the human body into hinges and treat the bones as rigid bodies. Therefore, the human body can be seen as a system composed of multiple rigid bodies connected by several hinges [10, 11].

Therefore, it is necessary to evaluate the motor ability of human lower limbs according to muscle strength in this paper, so as to predict the motor state of human body when muscle weakness occurs during rehabilitation training according to muscle strength [12].

Based on the analysis of the main exertion muscles of the lower limbs during human training, it can be determined that the main exertion muscles of the thigh are quadriceps and hamstring, and the main muscle of the calf is gastrocnemius [13].

This is shown in Fig. 2, it is simplified and geometrized for analysis purposes. Some assumptions need to be made during the geometric transformation to improve the accuracy of the model. Hypothesis 1: bones are rigid and do not bend. Muscles are represented by muscle tension lines. Hypothesis 2: during movement of the knee joint, the patella is always on the same level with the center of rotation of the knee joint.

In Fig. 3, F_1 represents the strength of the gastrocnemius muscle; f_1 represents the force component of the gastrocnemius muscle perpendicular to the tibia; α_1 and θ_3 represent the included Angle of the gastrocnemius and gastrocnemius, respectively, between the attachment point of the femur and the center of rotation of the knee joint, and the included Angle of the normal perpendicular to the level of the tibia. The line segment CD represents the tibia, and the ID represents the gastrocnemius during the training. The line segment CI represents the distance of the gastrocnemius muscle from the attachment point of the femur to the center of rotation of the knee joint. Since it is

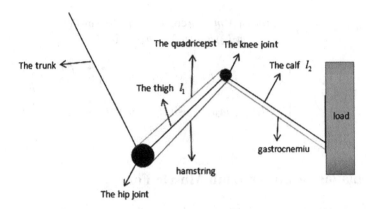

Fig. 2. Simplified figure of musculoskeletal model of human lower limbs

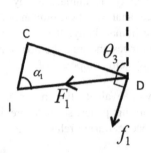

Fig. 3. Geometric diagram of the position relationship between the triceps and the bone

assumed that the contact point and the center of rotation of the knee joint are always on the same horizontal plane, the points I and C are placed on the same horizontal plane after geometric transformation. According to the simple geometric analysis of the simplified figure, the following equations can be listed.

Muscle strength of the gastrocnemius:

$$l_{ID}^2 = l_{CI}^2 + l_{CD}^2 - 2 * l_{CD} * l_{CI} * COS\theta_3 \tag{14}$$

$$f_1 = \frac{\tau_2}{l_{CD}} \tag{15}$$

$$F_1 = \frac{f_1}{\sin(\pi - \theta_3 - \alpha_1)} \tag{16}$$

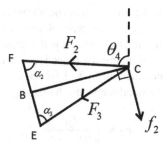

Fig. 4. Geometry of hamstring, quadriceps and bone position

In Fig. 4, F_2 and F_3 shares the output size of quadriceps and hamstrings, f_2 said quadriceps and hamstrings in perpendicular to the direction of the femur in the knee joint action on the component of the size of the quadriceps muscle and α_2 and α_3 respectively with quadriceps in hip attachment points to the center of rotation of the hip line and hamstring in the hip joint Angle of attachment point to the center of rotation of the hip line, θ_4 is the Angle between the femur and the normal perpendicular to the horizontal plane. The line segment BC represents the femur, while FC and EC represent the quadriceps and hamstrings during the training process. The line segments BF and BE represent the distance between the quadriceps and hamstring respectively from the attachment point of the hip joint to the center of rotation of the hip joint. Since it is assumed that the contact point and the center of rotation of the hip joint are always on the same horizontal plane, the three points B, E and F are placed on the same horizontal plane after geometric change. According to the simple geometric analysis of the simplified figure, the following equations can be listed.

The resultant force component of the quadriceps and hamstring muscles is:

$$f_2 = \frac{\tau_1}{l_{BC}} \tag{17}$$

Because of the lower extremities in the model of quadriceps and hamstring Shared a torque, so based on PCSA (physiological cross - sectional area) of the smallest muscle force optimization objective function [14]:

$$J = \min \sum_{i=1}^{n} \left(\frac{F_i}{A_i}\right)^3 \tag{18}$$

Where, F_i is the muscle force of the ith muscle, A_i is the cross-sectional area of the ith muscle, and n is the total number of muscles.

In addition, it also needs to satisfy the kinetic equation:

$$\sum_{i=1}^{m} F_i \times r_i = T \tag{19}$$

$$0 \leq F_i \leq \sigma_i \cdot PCSA_i \tag{20}$$

In the above equation, r_i is the force arm of the ith muscle, and T is the joint torque. F is unknown muscle force, PCSA is the physiological cross-cutting area of muscle, and n is the number of unknown muscle force. σ is the ultimate muscle tension, and I'm going to take 61 N/cm^2. By optimizing the objective function, the components of the two muscle forces can be calculated as f_{21} and f_{22} respectively.

The muscle strength of the quadriceps and hamstring muscles is as follows:

$$F_2 = \frac{f_{21}}{\sin(\theta_4 - \alpha_3)} \tag{21}$$

$$F_3 = \frac{f_{22}}{\sin(\pi - \theta_4 - \alpha_2)} \tag{22}$$

4   Lower Limb Training Experiment

The lower limb rehabilitation training was carried out with a sitting kick trainer, and the Angle sensor and BIOPIC multi-channel physiological signal acquisition instrument were used to collect the lower limb joint Angle and electromyographys signals to verify the validity of the musculoskeletal model.

4.1   Establishment of Experimental Platform

The whole experiment was done on a sitting kick trainer. The main instruments used in this experiment are the sitting kick trainer, the Angle sensor and the BIOPIC multi-channel physiological signal acquisition instrument.

As shown in Fig. 5(a), the subject sits on the chair and pushes the pedal back and forth to achieve the goal of rehabilitation training. The training intensity can be adjusted by changing the load weight.

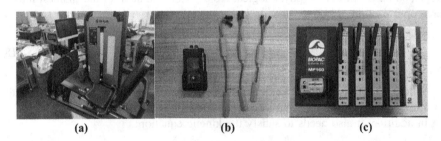

(a)   (b)   (c)

Fig. 5. Experimental platform equipment

The Angle sensor is shown in Fig. 5(b). The Angle sensor is fixed to the hip joint, knee joint and ankle joint of the subject, and the lower limb joint Angle data of the subject during rehabilitation training are recorded through the controller.

The BIOPIC multi-channel physiological signal acquisition instrument is shown in Fig. 5(c). The computer connected to BIOPIC is used to send synchronous start signals to the EMG signal acquisition module, and the EMG signal acquisition module starts to collect signals.

4.2 Experimental Process

As shown in Fig. 6, the subject (male, 85 kg in weight, 185 cm in height, 57 kg in load) put on the device and sat on the kick trainer to start the test. The test included leg extension and leg bend. At the beginning of the experiment, both the Angle sensor and the BIOPIC multi-channel physiological signal acquisition instrument were turned on at the same time, and the subjects started to move, and the leg extension and leg flexion movements were repeated for 5 times to obtain the changes in the Angle of the knee joint and the hip joint, as shown in Fig. 7.

Fig. 6. Lower limb rehabilitation training experiment

4.3 Experimental Results

In Fig. 7, dotted and solid lines represent the Angle of the knee and hip joint, the participants in the lower limb rehabilitation training, two joints of joint Angle change trend is consistent, but the pace of Angle change significantly different, especially the change of knee joint Angle is the largest, and the muscles of the quadriceps and hamstring output is proportional to the angular acceleration of the knee joint, we can speculate quadriceps and hamstring muscle force than gastrocnemius muscle force, according to the Fig. 8 proved consistent with the results of speculation.

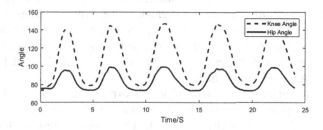

Fig. 7. Lower limb joint Angle data

In Fig. 8, the dotted lines represent the results calculated using the EMG signal to calculate muscle force. The solid line represents the muscle force calculated using the musculoskeletal model in this paper. This part of data is the valid data from the beginning to the end of the synchronization signal, and its x-coordinate unit is also time, and the unit is S. From the perspective of trend, the muscle force calculated by the musculoskeletal model in this paper is basically consistent with the muscle force curve calculated by EMG signal in the literature [15, 16].

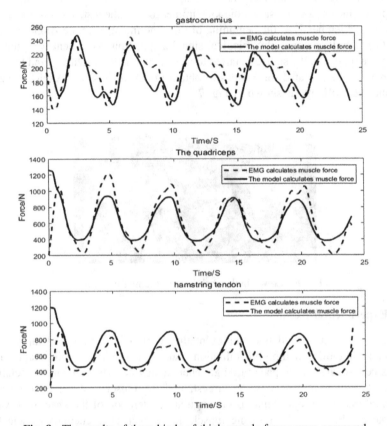

Fig. 8. The results of three kinds of thigh muscle forces were compared

In Fig. 9, the visible and the dot-dash line represent the calf, quadriceps, hamstrings musculoskeletal model to calculate the strength of the error rate, from the numerical point of view, the three muscles of error rate in zero floating up and down, most are not more than ten percent, and calves, quadriceps, hamstrings musculoskeletal model to calculate the strength of the average error rate is 1.6162, 0.7053 and 1.1612 respectively, the average error rate is smaller.

A total of 10 subjects were involved in this experiment, including 5 men and 5 women, all 24–25 years old, in good health, from different regions of the country. The experimental results are basically consistent. Therefore, the results show that the musculoskeletal model and the algorithm to optimize muscle force are feasible.

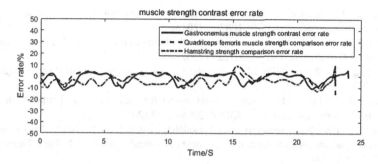

Fig. 9. The error rate of the three thigh muscle forces of the experimenter lower extremity

5 Conclusion

In this paper, the musculoskeletal model of human lower limbs was established on the general training equipment, and the stress state of joints during training was analyzed. For joints with multiple muscles acting together, PCSA method was used to optimize the calculation of muscle force. Finally, the experiment was carried out. The muscle force state obtained by the model was compared with the muscle force obtained by the EMG signal, and the correctness of the model and algorithm was proved. In this experiment, the quadriceps and hamstrings performed more than the gastrocnemius, so the model could be used to train different muscles to produce effective movements. In this paper, the posture information of human body is used to calculate the muscle strength of lower limbs, and the muscle strength is closely related to the health degree of human body. Compared with other methods of calculating muscle strength, it can more intuitively reflect the movement posture and health degree of human body. It lays a certain foundation for the follow-up research of rehabilitation training.

References

1. Da-Qian, W.: Application of lower limb exoskeletons rehabilitation robots in rehabilitation treatment of activity limited knee joint. Chin. J. Tissue Eng. Res. **16**(04), 597 (2012)
2. De Roeck, J.: Statistical modeling of lower limb kinetics during deep squat and forward lunge. Front. Bioeng. Biotechnol. **8**, 233 (2020)
3. Zhang, Q.: Artificially induced joint movement control with musculoskeletal model-integrated iterative learning algorithm. Biomed. Signal Process. Control **59**, 101843 (2020)
4. Stollenmaier, K.: Predicting perturbed human arm movements in a neuro-musculoskeletal model to investigate the muscular force response. Front. Bioeng. Biotechnol. **8**, 308 (2020)
5. Cullell, A, Moreno, J.C, Rocon, E.: Biologically based design of an actuator system for a knee-ankle-foot orthosis. IOS Press (2014)
6. Pennycott, A.: Towards more effective robotic gait training for stroke rehabilitation: a review. NeuroEng. Rehabil. **9**, 65 (2012)
7. Okada, S.: TEM: a therapeutic exercise machine for the lower extremities of spastic patients. Adv. Robot. **14**(7), 597–606 (2001)

8. Yang, K.: Structural design and modal analysis of exoskeleton robot for rehabilitation of lower limb. J. Phys.: Conf. Ser. **1087**(6) (2018)
9. Leardini, A.: A model for lever-arm length calculation of the flexor and extensor muscles at the ankle. Gait Posture **15**(3), 220–229 (2002)
10. Charlton, I.W.: Repeatability of an optimised lower body model. Gait Posture **20**(2), 213–221 (2004)
11. Rajagopal, A.: Full-body musculoskeletal model for muscle-driven simulation of human gait. IEEE Trans. Biomed. Eng. **63**(10), 2068–2079 (2016)
12. Buchanan, T.S.: Neuromusculoskeletal modeling: estimation of muscle forces and joint moments and movements from measurements of neural command. J. Appl. biomechanics **20**, 367–395 (2004)
13. Hamilton, N., Luttgens, K.: Kinesiology: Scientific Basis of Human Motion. Brown & Benchmark (2011)
14. Bolsterlee, B.: The effect of scaling physiological cross-sectional area on musculoskeletal model predictions. J. Biomech. **48**(10), 1760–1768 (2015)
15. Ma, Y., Xie, S., Zhang, Y.: A patient-specific EMG-driven neuromuscular model for the potential use of human-inspired gait rehabilitation robots. Pergamon Press, Inc. (2016)
16. Shao, Q.: An EMG-driven model to estimate muscle forces and joint moments in stroke patients. Comput. Biol. Med. **39**(12), 1083–1088 (2009)

sEMG-Based Hand Gesture Classification with Transient Signal

Yue Zhang[1]([✉]), Jiahui Yu[2], Dalin Zhou[2], and Honghai Liu[2]([✉])

[1] College of Computer Science and Technology, Zhejiang University of Technology, Hangzhou, People's Republic of China
zhangyuemessi@163.com
[2] School of Computing, University of Portsmouth, Portsmouth, UK
{jiahui.yu,dalin.zhou,honghai.liu}@port.ac.uk

Abstract. Surface electromyography (sEMG) can provide a novel control method for human machine interface (HMI) with the improvement of signal decoding technology. sEMG-based hand gesture recognition is the key part of HMI control strategy. However, unstable and complex daily used scenarios hinder the further development of sEMG-based control strategy. In this paper, we concentrate on the data preprocessing part. Three different signal segments were extracted including the transient signal segments between gestures, standard signal segments and stationary signal segments which is smaller than standard segment. By setting up several experiments to analyze and evaluate the classification performance with these transient information. Our research found that transient signal segments can reflects more effective information than the stationary signals in inter-subject scenes. It gained more classification accuracy and stability. In addition, it also performance better in other two scenes in ten hand gesture recognition in intra-session and inter-session.

Keywords: sEMG · Feature extraction · Traditional classifier · Transient signal

1 Introduction

Surface electromyography is a bioelectrical signal that represents the impulses of motor units during skeletal muscle fiber excitement and contraction. It is a comprehensive temporal and spatial characteristic map of electrical activity of complex subepidermal muscles, which can be extracted from the surface of the skin by recorded electrodes. There are different degrees of correlation between sEMG signals and muscle activity and function, which reflects neuromuscular activity to a certain extent (see [1]).

Given the characteristic that sEMG signals are easy to collect, non-invasive, and bionic, many researchers have devoted themselves to studying the essential characteristics of sEMG signals, revealing the physiological mechanism that they

© Springer Nature Singapore Pte Ltd. 2020
J. Qian et al. (Eds.): ICRRI 2020, CCIS 1336, pp. 401–412, 2020.
https://doi.org/10.1007/978-981-33-4932-2_29

generate and the muscle motor systems they reflect. At present, sEMG is widely used in the diagnosis and rehabilitation of neuromuscular diseases including muscle physiology, muscle metabolism, rehabilitation medicine, sports and other fields (see [2]). In addition, sEMG can also be used to provide a safe, non-intrusive control method for advanced prosthesis control, functional electrical stimulation and other advanced HMI (see [3] and [4]). The sEMG-based gesture recognition studied in this paper is a key part of the control of multifunctional dexterous prosthetic hand. Also, it is an important research direction in the field of rehabilitation and intelligent control.

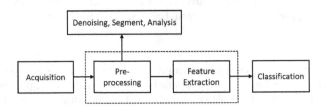

Fig. 1. Signal processing flow

Currently, gesture recognition based on pattern recognition has been regarded as the optimizing approach among different kind of strategies (see [5]). A general pattern recognition system can be divided into four parts: data collection, data preprocessing, feature extraction and classification, as shown in Fig. 1. Each part has a significant impact on the results of gesture recognition. In data collection, we always use an advanced commercial equipment to extract weak effective signals. Data preprocessing is always combined with the part of feature extraction (see [6]), it plays a vital role in feature extraction. The raw signal from acquisition device is often accompanied by some unstable noise which hinder the classification performance. Data preprocessing can remove the noise by some software filtering methods, such as high-pass filtering or low-pass filtering. In addition, this part can also extract and segment the needed part of signal. Feature extraction plays a critical role to the classification performance (see [7] and [8] and [9]). It obtains the characteristic parameters of each analysis window to characterize the characteristic vector of the EMG signal. So far, a large number of sEMG features have been investigated for hand gesture classification, and they can be summarized into four types: time domain (TD), frequency domain (FD), time-frequency domain (TFD) and parametric model analysis. Phinyomark et al. (see [10]) compared thirty-seven TD features and FD features, the experimental result demonstrated that the top three sEMG features could reach the accuracy more than 80%. Besides that, it reveals that most of the TD features were redundant and better than the FD features. Classification is used to build a model which can map the characteristic vector to different motions. Generally, some classic supervised classification methods are used in gesture recognition, such as linear discriminant analysis (LDA), support vector machine (SVM), Naive Bayes and K nearest neighbor (KNN).

Although myoelectric pattern recognition approach can achieve good performance in special laboratory environment, but failed in daily life due to the changes between training data and testing data generated by muscle fatigue (see [7]), electrode shifts (see [8]), and arm position changes (see [9]). To overcome the influence of these external factors, some methods were proposed. Deep learning (DL) has become popular in recent years and has achieved outstanding performance in many fields, such as image processing, speech recognition and natural language processing (see [11]). Convolutional neural network (CNN) and recurrent neural network (RNN), two classic network architectures of deep learning, have been applied to the gesture classification of sEMG. Manfredo et al. (see [12]) use a simple CNN architecture which is combined by four convolutional layers, one fully connected layer and a softmax function to classify the gestures on DB1 in Ninapro database. Finally, it obtains 4.53% higher accuracy than the classical classification methods. Geng et al. (see [13]) applied a deep CNN to a high-density EMG, and achieved an accuracy of 89.3% with a single frame and 99.0% with over 40 frames for 8 gestures recognition. Adaptive learning can enhance the robust of algorithm by adjusting the parameters of training model when external factors changes. Liu J. (see [14]) proposed an adaptive unsupervised classifier based on support vector machine and incremental learning method. This classifier considers the real-time changes between testing data and training data when predict the classification label. Then some adjustments on these difference would be added to the classification model in an unsupervised manner. Thus, continuously updating the model parameters makes the classifier adaptive to the changes. Chen et al. (see [15]) extend two off-line pattern recognition methods, linear discriminant analysis and quadratic discriminant analysis, to self-enhancing methods which continuously updating the class mean vectors, the class covariance and the pooled covariance using the testing data respectively.

In general, researchers always place their focus point on feature extraction and classification. They regard the signal of stationary section of a gesture as the most representative data. However, when capturing sEMG data of a repeated gesture in different session, the experimental volunteer completes the motion with different force because of fatigue or force information missing. Thus, the different session of stationary section exists some unstable factors. In this paper, we will extract transient signal segments when a hand gesture changes as the input data. Several TD features and supervised classifications will be used to compare with traditional processes.

2   Materials and Methods

2.1   Subjects

Eight able-bodied subjects are employed in to investigate the classification performance with the transient signal segments of sEMG signals. (1 female, 7 males, age: 25 ± 5, height: 175 ± 10 cm, weight: 65 ± 10 kg). Subjects have no previous history of neuropathies or traumas to the upper limbs and fingers. They were obtained full informed consent according to the Declaration of Helsinki. All subjects receive the relevant guidelines before capturing sEMG data during the entire experiment. The data capture cycle was one week.

2.2   Apparatus

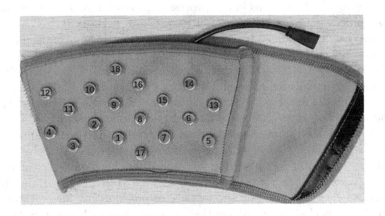

Fig. 2. The electrode sleeve for sEMG signal recording, 16 monopolar channels record the differential voltage between electrode 1–16 and electrode 17 and 18 provides the bias voltage to remove DC offset.

To capture the sEMG signal on the forearm, a multi-channel sEMG acquisition system, developed by our research group (see [16]), was employed to record the multi-channel sEMG signals. This device supports up to 16 bipolar EMG channels, 5000 amplification gains, 1 kHz sampling frequency and 12 bits ADC resolution. To remove the motion artefacts, a band-pass filter with cutoff frequencies at 20 Hz and 500 Hz is integrated in the device. In addition, a notch filter with the fundamental frequency at 50 Hz is also deployed in hardware to remove the power frequency signal from the power line. The noise of each EMG channel was less than 1 μV. The entire device was powered by a 3.3 V rechargeable lithium battery, supporting continuous signal recording for up to ten hours.

This acquisition system use an electrode sleeve as signal conducting and recording equipment, as seen Fig. 2, which embedded 18 electrodes in Zig configuration on an elastic fabric sleeve (see [17]). The diameter of each electrode is 12 mm, 25 mm vertical 30 mm vertical and horizontal distances between two electrodes. 1 to 16 represent 16 independent channels, one reference electrode marked as 17 and one bias electrode marked as 18. All 16 channels are recorded synchronously.

2.3   Acquisition Protocol

Before the experiment, each subject should wear the electrode sleeve in the right way. They were required to be familiar with the experimental process and

Fig. 3. The scenes for sEMG data recording.

practice the pre-defined hand gestures for several minutes. The selected of hand gestures in this experiment refer to the NinaPro dataset (see [18]) and CSL-HDEMG database (see [19]) which are the two popular sEMG databases. All the hand gestures can be seen in Fig. 4. Among them, the first row contains five finger-related gestures, they are (1) tip pinch, (2) flexion of ring and little finger, thumb flexed over ring and little finger, (3) flexion of middle, (4) ring and little finger, middle and ring flexion, and (5) thump up. The second row is wrist-related gestures, includes (1) hand close, (2) hand open, (3) wrist deviation, (4) wrist extension and (5) wrist flexion. Each subject was asked to complete the action with norm strength (Fig. 3).

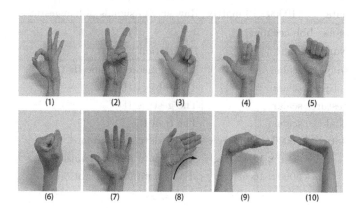

Fig. 4. Ten gestures were used for this experiment. (1) tip pinch, (2) flexion of ring and little finger, thumb flexed over ring and little finger, (3) flexion of middle, (4) ring and little finger, middle and ring flexion, and (5) thump up, (6) hand close (hc), (7) hand open (ho), (8) wrist radial deviation (wrd), (9) wrist extension (we), (10) wrist flexion (wf).

Throughout the acquisition process, each subject wore the electrode sleeve on the left forearm and keep a standard sitting position in comfortable on the chair, as seen in Fig. 2. Ten hand gestures were performed in sequence and each gesture maintained for at least 12s. In order to avoid muscle fatigue, subjects were asked to relax their hand for at least 10s between two adjacent gestures in a session. After finishing the ten hand gestures collected in short period, the subjects were asked to have a rest for half an hour. In this time period, subject can move around the office without removing the sleeve. The second session and third session were carried out in the same way. So, we separately gained three sessions from each subject and all sessions have the same electrode distribution. After three days, another three sessions were carried out to collect sEMG signals from each subject in the same way. It was worth pointing out that no predefined contraction force or elbow angle were applied in the experiment to mimic the real application of a sEMG based HMI as much as possible, although it was well known that muscular contract force and arm position could influence the robustness of patter recognition system (see [20]).

2.4   Data Preprocessing

Table 1. The format of EMG database.

aaa-ccc.mat		
Name	Type	Description
subject	scalar	The subject ID
group	scalar	Gesture group ID
data	110000×16 matrix	Raw sEMG signals of five gestures
ss-aaa-bbb-ccc.mat		
Name	Type	Description
shape	scalar	The segmented shape ID
subject	scalar	The subject ID
gesture	scalar	The gesture ID
trial	scalar	The trail ID
data	10000 × 16 matrix	sEMG data for standard
	4000 × 16 matrix	sEMG data for one transient
	6000 × 16 matrix	sEMG data for one stationary

According to the acquisition protocol, each session contains ten hand gestures and each gesture maintained for 12s. To evaluate the performance of transient segments of sEMG signals in classification, we segment the raw data into three shapes: standard signal segments, transient signal segments, stationary signal segments. First, we extract standard section part. Only the later 10s of 12s signals of each gesture were segmented and labeled for classification. It was to exclude the transient state between two gestures. Thus, 10s stationary sEMG signal containing 10000 frames were extracted for further processing and analysis. Then, the stationary section has the same segmented methods except the effective segmentation area is in the middle of standard section. Only 12s to 18s signals of each gesture were segmented and labeled for classification. At last, transient segment part was processed. There is a movement process when the gesture changes from a rest state to a specific action and this process will hold for 1s or 2s in the earlier part of 12s signals. To collected enough transient information during the changes between gestures, We defined the earlier 4s of 12s signals as the transient segment. Only 8s to 12s signals of each gesture were segmented for training and testing.

The whole dataset was organized in Matlab format with architecture as described in Table 1. The files titled in the format of aaa-ccc.mat and ss-aaa-bbb-ccc.mat indicated the raw and preprocessing data, respectively, where aaa was the subject ID, BBB was the gesture ID, ccc was the session ID and ss represent the three segmented shapes: standard, transient and stationary. For example, 001-002.mat represent the raw sEMG data captured from subject 1 in the second session, and 02-003-004-005.mat contained the transient sEMG data of gesture 4, captured from subject 3 in the fifth session.

2.5   Feature Extraction and Classification

TD features outperformed most feature extraction methods. Moreover, it is low calculation. In this paper, we choose four TD features and an AR model with four coefficients as the evaluation features. In this scheme, each analysis window is equal to 300 ms (300 samples in 1 kHz sampling) and each incremental window is set to 100 ms. The selected four TD features are: Modified mean absolute value (MAV), Waveform length (WL), Root Mean Square (RMS) and Average amplitude change (AAC).

For the classification of the sEMG signals, three traditional classification methods were implemented in MATLAB tool: SVM, LDA and KNN. Initially, the SVM classification with Radial Basis Function (RBF) kernel were applied and the gamma of kernel function is set to 0.125.

3   Experiment Setup

To verify the classification with transient signal segments can acquire similar or superior performance than the other two, three types of experiment and evaluation method as listed below.

Table 2. A comparison of three segmented shapes with different classification algorithms in inter-session evaluation.

Method	Transient			Stationary			Standard		
	KNN	LDA	SVM	KNN	LDA	SVM	KNN	LDA	SVM
WL	0.7060	0.6698	0.6585	0.6935	0.6882	0.7220	**0.6935**	0.6882	0.7224
RMS	0.6921	0.6771	0.6621	0.6826	0.7075	0.7298	0.6900	0.6942	0.7213
MAV	**0.7132**	0.6985	0.6258	**0.6991**	0.7103	0.6504	0.6689	0.7116	0.6962
AR4	0.6744	**0.7895**	**0.7732**	0.6391	**0.8238**	**0.7943**	0.6528	**0.8252**	**0.8074**
AAC	0.6768	0.6875	0.6620	0.6623	0.7164	0.7097	0.6706	0.6953	0.7295

[1] Abbreviations: Waveform length (WL), Root mean square (RMS), Modified mean absolute value (MAV), Auto-regressive coefficients (AR4), Average amplitude change (AAC).

Table 3. A comparison of three segmented shapes with different classification algorithms in inter-subject evaluation.

Method	Transient			Stationary			Standard		
	KNN	LDA	SVM	KNN	LDA	SVM	KNN	LDA	SVM
WL	0.5743	0.5635	0.5554	0.5397	0.5175	0.5774	0.5406	0.5311	0.5565
RMS	0.6143	0.5413	0.5946	0.5362	0.5150	0.5746	0.5486	0.5246	0.4981
MAV	**0.5983**	0.5494	0.5524	**0.5465**	0.5163	0.5812	**0.5558**	0.5245	0.5185
AR4	0.4865	0.5612	0.6211	0.4751	**0.5621**	**0.5958**	0.4808	**0.5645**	**0.5823**
AAC	0.5742	**0.6636**	**0.6358**	0.5396	0.5175	0.5689	0.5389	0.5311	0.5102

3.1 Intra-session Strategy

To evaluate the general classification performance, data within one session was used in this experiment. All subjects are independent, so we choose one subject's data to repeat the procedure. Each session will be divided into training data and testing data. The average classification accuracy across six tests was used to evaluate the performance of the three segmented shapes.

3.2 Inter-session Strategy

Three sessions collected in the same electrode distribution were used for training and the remaining three sessions collected in three days later for testing. Then, the later three sessions for training and earlier three sessions for testing. Although the training and testing data were collected from the same subject, a large gap will exist between them due to electrode shift, etc. The average classification accuracy across eight tests was used to evaluate the performance of the three segmented shapes.

3.3 Inter-subject Strategy

A single subject's data was chosen to testing the classifier and the remaining data as the training dataset. In the current case, eight training and testing procedures will be executed, the data of one of eight subjects were used as the testing data in turn, and the remaining data as the training data. The average classification accuracy across eight tests was used to evaluate the performance of the three segmented shapes.

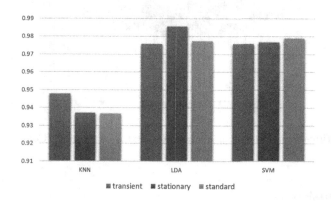

Fig. 5. The classification accuracy between three segemented shapes.

4   Results and Discussion

4.1   Classification Accuracy Performance

Three different experiments were implemented in this paper to evaluate the performance of different preprocessing methods. In the intra-session part, a combination of feature extraction and classification methods were applied to the three segmented shapes. The result can be summarized as a histogram in the Fig. 5. The average accuracy is over 90%. LDA can gained more accuracy compared to two other classifications. The transient signal segments between two gestures hidden some important information of muscle contraction.

In the part of inter-session, the experimental results demonstrated that the classification with transient signal segments are similar to the other two, as listed in Table 2, the highest accuracy that transient segments acquired is 78.95% with the LDA and AR4. In addition, LDA outperformed the other two classifier in three segmented shapes, it obtains the highest 78.95%, 82.38% and 82.52%, respectively, in the three shapes. AR4 can achieve better characterization of feature vector and get better classification performance with different classifier.

As seen in Table 3, the classification accuracy of inter-subject can be seen in the table. The accuracy of transient signal segment performed best. It acquired 66.36% with LDA and AAC, which is 6% higher than the top accuracy in stationary signal segment and standard signal segment. Besides, the average accuracy of transient is around 55% to 60%, which illustrate the stability of transient signal segment in inter-subject. Among all the features, AR4 and AAC performed better.

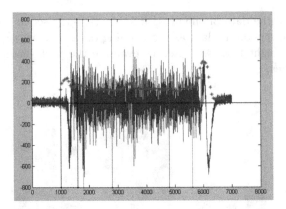

Fig. 6. The three segmented shapes in raw and feature data plot. (Color figure online)

4.2 Discussion

As we can see in the Fig. 6, raw data signals of one gesture was plotted on the figure, blue curve represent the raw signal, the red dotted line represent the RMS feature extracted from this signal with the length of analysis window is 100, and shift window is 50. In general, we extract the effective signal between green line, which is standard and steady. Sometimes, real-time performance is required, so signal segment between red lines will be extract as the whole representative of the motion. The signal segment between two black lines in the left is the transient signal which occurs along with hand gestures changes.

In intra-session, the sEMG data is offline, any popular machine learning classifier can extract the hidden model between the training data and theirs labels. However, when external environment changes, model extracted from the stationary signal segments will be expired. Compared to the stationary signal, Transient signal is always related to the changes, which decrease the influences by external factors. Certainly, the classification accuracy with transient signal segments is not very well for daily use, some essential characteristics should be discovered.

5 Conclusion

In this paper, we discuss the transient signal segments acquired when gesture changes. To reveal the useful information hidden in the transient signal, we designed some comparative experiments. Three kinds of signal segments were extracted as the representative sector of hand gestures. Traditional feature extraction methods including WL, AAC, AR4, RMS, MAV and classification methods are combined to evaluate the proposed methods. The results show that transient signal segments can achieve satisfactory result in single session. In addition, transient signal segments can gain more stable representative information between different subjects. It acquired better classification performance in the part of inter-subject. Two referenced segmented shapes, stationary and standard, are all extracted from the steady part of one hand gesture, they can get almost the same performance in three experiments, which illustrate that some representative small stationary signal segment can get better performance in accuracy and calculation cost. In the future, we plan to fit the transient sectors by some regression methods for promoting the robustness of long-term sEMG signals.

References

1. Jiang, N., Dosen, S., Muller, K., et al.: Myoelectric control of artificial limbs-is there a need to change focus? IEEE Sig. Process. Mag. **29**(5), 150–152 (2012)
2. Wehner, T., Vogt, S., Stadler, M.: Task-specific EMG-characteristics during mental training. Psychol. Res. **46**(4), 389–401 (1984)
3. Castellini, C., Smagt, P.V.D.: Surface EMG in advanced hand prosthetics. Biol. Cybern. **100**(1), 35–47 (2009)
4. Cipriani, C., Segil, J.L., Birdwell, J.A., et al.: Dexterous control of a prosthetic hand using fine-wire intramuscular electrodes in targeted extrinsic muscles. IEEE Trans. Neural Syst. Rehabil. Eng. Publ. IEEE Eng. Med. Biol. Soc. **22**(4), 828–836 (2014)
5. Chu, J.U., Moon, I., Mun, M.S.: A real-time EMG pattern recognition system based on linear-nonlinear feature projection for a multifunction myoelectric hand. In: International Conference on Rehabilitation Robotics, pp. 295–298. IEEE Xplore (2006)
6. Englehart, K., Hudgins, B.: A robust, real-time control scheme for multifunction myoelectric control. IEEE Trans. Biomed. Eng. **50**(7), 848–854 (2003)
7. Tkach, D., Huang, H., Kuiken, T.: Study of stability of time-domain features for electromyographic pattern recognition. J. Neuroeng. Rehabil. **7**(21), 13 (2010)
8. Jacobs, I.S., Bean, C.P.: Fine particles, thin films and exchange anisotropy. In: Rado, G.T., Suhl, H. (eds.) Magnetism, vol. III, pp. 271–350. Academic, New York (1963)
9. Hargrove, L., Englehart, K., Hudgins, B.: A training strategy to reduce classification degradation due to electrode displacements in pattern recognition based myoelectric control. Biomed. Signal Process. Control **3**, 175–180 (2008)
10. Phinyomark, A., Phukpattaranont, P., Limsakul, C.: Feature reduction and selection for EMG signal classification. Expert Syst. Appl. **39**(8), 7420–7431 (2012)
11. Ciresan, D.C., et al.: High-performance neural networks for visual object classification. arXiv preprint arXiv:1102.0183 (2011)
12. Atzori, M., Cognolato, M., Müller, H.: Deep learning with convolutional neural networks applied to electromyography data: a resource for the classification of movements for prosthetic hands. Front. Neurorobotics **10**, 9 (2016)
13. Geng, W., et al.: Gesture recognition by instantaneous surface EMG images. Sci. Reports **6**, 36571 (2016)
14. Liu, J.: Adaptive myoelectric pattern recognition toward improved multifunctional prosthesis control. Med. Eng. Phys. **37**, 424–30 (2015)
15. Chen, X., Zhang, D., Zhu, X.: Application of a self-enhancing classification method to electromyography pattern recognition for multifunctional prosthesis control. J. Neuroeng. Rehabil. **10**(1), 44 (2013)

16. Fang, Y., Zhu, X.Y., Liu, H.H.: Development of a surface EMG acquisition system with novel electrodes configuration and signal representation. In: 6th International Conference, ICIRA 2013, pp. 405–414 (2013)

17. Fang, Y., Liu, H.H.: Robust sEMG electrodes configuration for pattern recognition based prosthesis control. In: 2014 International Conference, ICSMC 2014, pp. 2210–2215 (2014)

18. Atzori, M., et al.: Electromyography data for non-invasive naturally-controlled robotic hand prostheses. Sci. Data **1**(1), 1–13 (2014)

19. Amma, C., et al.: Advancing muscle-computer interfaces with high-density electromyography. In: Proceedings of the 33rd Annual ACM Conference on Human Factors in Computing Systems (2015)

20. Geng, Y., Zhou, P., Li, G.: Toward attenuating the impact of arm positions on electromyography pattern-recognition based motion classification in transradial amputees. J. NeuroEng. Rehab. **9**(74) (2012). http://www.jneuroengrehab.com/content/9/1/74/abstract

Improving Gesture Recognition by Bidirectional Temporal Convolutional Netwoks

Haoyu Chen[1(✉)], Yue Zhang[1], Dalin Zhou[2], and Honghai Liu[2(✉)]

[1] College of Computer Science and Technology, Zhejiang University of Technology,
Hangzhou, China
chyhave@163.com, zhangyuemessi@163.com
[2] School of Computing, University of Portsmouth, Portsmouth, UK
dalin.zhou@port.ac.uk, honghai.liu@port.ac.uk

Abstract. Surface electromyography (sEMG) based gesture recognition as an important role in Muscle-Computer interface has been researched for decades. Recently, deep learning based method has had a profound impact on this field. CNN, RNN and RNN-CNN based methods were studied by many researchers. Motivated by Bidirectional Long short-term memory (Bi-LSTM) and Temporal Convolutional Networks (TCN), we propose 1D CNN based networks called Bidirectional Temporal Convolutional Networks (Bi-TCN). The positive order signal and reverse order sEMG signal are feed to our networks to learn the different representation of the same sEMG signal. We evaluate proposed networks on two benchmark datasets, Ninapro DB1 and DB5. Our networks yields 90.74% prediction accuracy on DB1 and 90.06% prediction accuracy on DB5. The results demonstrate our networks is comparable to the state-of-the-art works.

Keywords: sEMG · Deep learning · CNN · TCN · LSTM · Gesture recognition

1 Introduction

Human-Computer interface system requires natural and intuitive method of interaction and can't cause too much confusion for the human users. Surface Electromyography (sEMG) naturally becomes an important role for muscle-computer interaction, which can be easily collected by non-invasion electrodes and contain abundant information about human intention of movement. Therefore, sEMG based gesture recognition has been widely researched in many fields, such as Virtual Reality (VR) interaction [27], Assistive system [9], etc.

Gesture recognition as a pattern recognition, supervised machine learning methods are widely researched, such as Support Vector Machine (SVM) [2,17], K-nearest Neighbor (KNN) [15], Linear Discriminant Analysis (LDA) [2], Hidden Markov Model (HMM) [7], Random Forest (RF) [20], etc. These conventional

© Springer Nature Singapore Pte Ltd. 2020
J. Qian et al. (Eds.): ICRRI 2020, CCIS 1336, pp. 413–424, 2020.
https://doi.org/10.1007/978-981-33-4932-2_30

methods require extracting features before classifying. Thus, a plenty of hand-craft features were researched.

In recent years, Deep learning has been demonstrated remarkable capabilities in many fields, such as image classification, natural language processing (NLP) and object detection, etc. Similarly, deep learning based methods also give gesture recognition a new perspective. Unlike traditional machine learning method which requires a mount of human-selected, handcraft features. Deep learning based method can learn to extract appropriate feature by itself. Specifically, an increasing number of Convolutional Neural Networks (CNN) or Recurrent Neural Networks (RNN) based methods were researched to improve gesture recognition. More recently, 1D CNN have been considered the more appropriate approach to address multi-variate time sequence such as sEMG and EEG.

In this work we proposed a TCN [5] based networks named Bi-TCN. Unlike CNN and CNN-RNN [22,26] based method, TCN consider sEMG based gesture recognition as a multivariate time sequence classification and use 1D casual convolution. Motivated by Bi-LSTM, proposed networks also requires different order of sEMG signal. In the recognition period, the positive order of sEMG signal and reverse order of the same sEMG signal are feed to our networks. We validation our method on two benchmark datasets, Ninapro DB1 [3] and DB5 [18]. The results demonstrate our method is comparable to state-of-the-art works.

The organization of this paper is given as follow: Sect. 2 provides an overview of the related gesture recognition approaches. In Sect. 3, we provide the detail about proposed Bi-TCN architecture. The description of datasets is given in Sect. 4. In Sect. 5, we describe the detail of the experiment. The results and discussion are presented in Sect. 6. Finally, we draw conclusions for this paper and give future work in Sect. 7.

2   Related Work

Same as other classification task, sEMG based Gesture recognition can be divided into two categories. The first is traditional machine learning based method and the second is deep learning based method.

For conventional machine learning method, feature extracting is required before classifying. There are three categories of these features: Time domain, Frequency domain and Time-Frequency domain. Atzori [3] extract Root mean square (RMS), Histogram (HIST), marginal Discrete Wavelet Transform (mDWT) as features. By combining these features with different conventional classifiers, the results shows that random forest with combined features perform the best. Ali H. Al-Timemy [1] use time domain-auto regression (TD-AR) feature and SVM to classify sEMG. They achieve 98% accuracy over ten intact subjects for the classification of 15 classes finger movement and 90% accuracy over six amputee subjects with 12 classes finger movement.

Recently, deep learning based method was widely studied. For most Deep Learning (DL) based methods, handcraft features is no more required. Hongfeng

Chen [6] *et al.* proposed a new feature extracted by CNN, called CNNfeat. By comparing CNNfeat with 25 traditional features, the result demonstrated that CNNfeat outperform the traditional features. Wentao Wei [25] *et al.* use "divide-and-conquer" strategy to split sEMG image into equal-sized stream. Then, they use CNN learns representation of each stream and feed these learned feature to a fusion classification networks. In order to learn to extract spatial-temporal information of SEMG, Yu Hu *et al.* [13] combine CNN and RNN as a new classification model. The average accuracy achieved on Ninapro DB1 and DB2 are 87.2% and 82.2%, respectively. Elahe Rahimian *et al.* [19] proposed 1D CNN based architecture named XceptionTime, they achieve SOTA performance with 92.3% prediction accuracy on Ninapro DB1.

On the other hand, some researchers combine traditional features with deep learning method to learn multi-view classification. Wentao Wei *et al.* [24] combine classical sEMG features to train a multi-view classifier. They select 3 classical handcraft features and use these features to train their multi-view networks. Finally, they achieve 88.2%, 83.7%, 64.3%, 90%, 64%, 88.3% prediction accuracy on Ninapro DB 1, 2, 3, 5, 6, 7 without IMU, respectively.

3 Proposed Method

In many existing works [24, 25], sEMG based gesture recognition was regarded as image classification task. These methods may be not appropriate for the loss of temporal information hidden in the sEMG signal. To consider the temporal information in sEMG signal, some researchers combine LSTM with CNN, like [22, 26]. By this way, the networks can learn the spatial-temporal information of the sEMG signal. But this method will lead to parameter explosion and pressure on real-time calculation. To address above problem, a new networks architecture are presented name Temporal convolutional networks (TCN). TCN [5] has been proved to be a more appropriate method to address multivariate time series classification task by using a new technology called casual convolution and 1D CNN.

Temporal Convolutional Networks based on two principles: The first is the future information can not be leaked to the past, and the second is that neural network produces an output of the same length as the input. To satisfy these two principles, casual convolution is introduced. More detail, casual convolution can be implemented simply by padding the time sequence and discarding the over-padding value at the end after convolution operation. Padding value can be calculated as follow:

$$padding = (k - 1) * d \tag{1}$$

where k is kernel size, d is dilation factor.

However, the long sequence challenges the simple casual convolution. We need stack casual convolution for many layers to look back history for very long time sequences. However, this operation will leads to a large increase in parameters and over-fitting. To solve this problem, the dilated convolution is introduced which can be calculated as follow:

$$f(s) = (x *_d f)(s) = \sum_{i=0}^{k-1} f(i) \cdot x_{s-d_i} \tag{2}$$

Where k is kernel size,d is dilation factor,$s - d_i$ is the direction of the past.

To summarized, the temporal convolution is equal to 1D casual CNN + 1D dilation CNN. By using these strategies, TCN can extract information of multi-variate time sequences with a few layer, meanwhile, it will not destroying temporal attribute. The basic Temporal Convolution Networks is showed in Fig. 1.

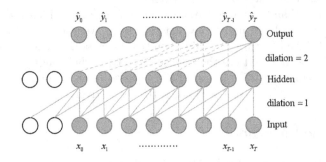

Fig. 1. Temporal convolutional networks.

In this work, motivated by Temporal Convolutional Networks (TCN) and Bidirectional Long Short-Term Memory (Bi-LSTM), we propose Bidirectional Temporal Convolutional Networks (Bi-TCN) which contains a pair parallel TCN as showed in Fig. 2. The structure of two TCN is the same but the input is different. Like Bi-LSTM, one of the TCN's input is normal sEMG signal, and the other's input is reverse order of the same sEMG signal.

The first layer of proposed TCN is Batchnorm. Then, followed by four casual convolution layers, the dilation rate of these convolution is set to 1, 1, 2, 2 respectively and the kernel size of all convolutional layers is 3. After each casual convolution layer, we apply PReLU activation function, Bachnorm and Dropout. In this work, we set the possibility of Dropout layer to 0.5. In the first two casual convolution layers, we apply SE-Module [12]. Basic structure of SE-Module is showed in Fig. 3. This module can make model pay more attention on key channels. In the last two convolutional layers, we apply residual connect [11], which solves the degradation problem of deep neural networks very well by allowing layers to learn modifications to the identity mapping. The kernel size of residual layers is set to 1 with padding zero. After four casual convolutional layers, we summed the feature according to time dimension. The final layer of proposed TCN is one layer MLP which predicts the classification result. All parameters of model are initialized by using Xavier method [10].

Our method requires two stages to train. In the first stage, we train two TCN separately. Firstly, sEMG signal are feed to one TCN. Then, the extracting

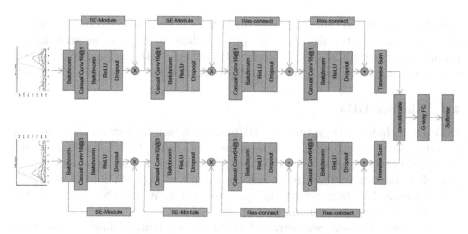

Fig. 2. Bidirectional temporal convolutional networks.

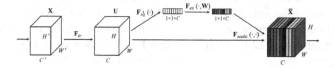

Fig. 3. SE-Module.

feature are summed according to time dimension. Finally, the summed feature are feed to on layer fully connected to train and classification. Similarly, we train the other TCN with the reverse order sEMG signal. After the first stage, we discard the fully connected layer of both two TCN and freeze its parameters.

In the second stage of train, the positive and reverse order sEMG signal are simultaneously feed to mentioned TCN. Then, the extracting feature of two TCN are summed according to time dimension and concatenated. Finally, the summed feature are feed to final fully connected layer to train. In this way, our proposed networks combine two views of the sEMG signal.

4   Datasets

To validate our proposed model, a benchmark database Ninapro is used in this paper. The Ninapro database have 7 datasets, we select the first and the fifth dataset as our validation datasets.

4.1   Ninapro DB1

The first Database of Ninapro is recorded by Otto Bock MyoBock 13E200 with 10 wireless electrodes (channels) at a sampling rate of 100 Hz. DB1 (Ninapro database1) [3] comprise 52 gestures from 27 intact subjects, each movement include 10 repetition, and each repetition lasted 5 s. After recording, a generalized likelihood ratio algorithm is applied offline to correct the stimulation. As previous works [3,24,25], we apply 1 Hz butter-worth low-pass to filter this signal, and use repetition 1, 3, 4, 6, 8, 9, 10 as train set, and remain as test set.

4.2   Ninapro DB5

DB5 [18] include same gesture as DB1 from 10 intact subjects, each movement include 6 repetitions. The acquired equipment of DB5 is double MYO band. One of the MYO band is placed close to elbow, and the other is placed close to hand. The MYO band have 8 electrodes and the sample ratio of MYO band is 200 Hz. We also apply 1 Hz butter-worth low-pass to filter it. We use the repetition of 1, 3, 4, 6 for training, and remain for testing.

For above datasets, we extract features by applying sliding window with 200 ms window size, because the time of classification for real-time recognition should below 300 ms [14]. Besides, The window step of DB1 is 10 ms and of DB5 is 100 ms. Like many previous works [4,19,24], we discard the rest gesture. Additionally, the gesture in above two datasets can be divide into three categories. Exercise A: 12 basic movements; Exercise B: 8 isometric and isotonic hand configurations and 9 basic movements of the wrist; Exercise C: 23 grasping functional movements. Figure 4 show a part of these gestures. The other information of above datasets is also showed in Table 1.

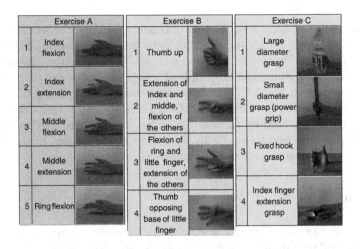

Fig. 4. Gestures.

5   Experiment Setup

Because of different sampling ratio of two datasets, there are different setting during training model. For the DB1, the filter number of convolutional layers is

Table 1. Specification of used Databases.

Database	DB1	DB5
Number of gestures	52	52
Number of trials	10	6
Intact subjects	27	10
Number of channel	10	16
Sampling ratio	100 Hz	200 Hz
Repetition for training	1,3, 4, 6, 8, 9, 10	1, 3, 4,6
Repetition for testing	2, 5, 7	2, 5

set to 16, 32, 64, 64, respectively. The final FC layer of Bi-TCN is comprised by one MLP layer. Batchnorm is applied before final MLP. For the DB5, the filter number of convolutional layers is set to 32, 64, 128, 128, respectively. The final FC layer of Bi-TCN is comprised by two MLP layer. Output size of first MLP layer is 128. Similarly, batchnorm is applied before each MLP layer. The reason is that the length of DB5 data is twice than DB1's which require larger capacity of networks. We use Adam as our optimizer and ReducLRPlateau as learning scheduler during both training stages. The parameters of ReduceLRPlateau is set as: $model =' min', factor = 0.5, patient = 5, eps = 1e-8$. In all stages, the training epoch for DB1 is 50, and for DB5 is 80. In the first stage, the learning rate is set to 0.01. In the second stage, the initial learning rate is set to 0.001. We use CrossEntropyLoss as our loss function, which can be represented as follow:

$$H(p,q) = -\sum \sum_{i=1}^{k} p(x_i)log(q(x_i)) \tag{3}$$

We also used label-smoothing [16] for slowing down over-fitting. The label-smoothing can be calculated as follow:

$$q_i = \begin{cases} 1-\varepsilon & if \ i = y \\ \varepsilon/(K-1) & otherwise \end{cases} \tag{4}$$

Where ε is a hyper parameters, K is number of gestures. We set ε equal to 0.1.

RBF-SVM, LDA and Random Forest are selected as conventional classifiers. Traditional machine learning requires extracting feature before training and testing. Root mean square (RMS) is selected as our validation feature.

The gesture recognition accuracy is calculated as given below:

$$Accuracy = \frac{Correct \ classification}{Total \ test \ sample} * 100\% \tag{5}$$

We run all experiments for 5 times and report the average result. All experiments were implemented in workstation with INTEL 9700k processor, 16 GB RAM and 2070super, 8 GB GPU.

6   Result and Discussion

The classification results can be seen in Fig. 5, SVM with RBF kernel yields the best performance among the traditional classifiers, with 81.05% prediction accuracy for DB1 and 71.22% prediction accuracy for DB5, but it's still inferior to the performances of deep learning method. Our proposed TCN yield 90.16% ± 0.33% recognition accuracy on DB1 and 88.00% ± 0.34% on DB5. Comparing with proposed TCN, Bi-TCN improves average accuracy from 90.16% to 90.74% ± 0.26% on DB1, with a increasing of 0.58%. On DB5, it improves average accuracy from 88.00% to 90.06% ± 0.15%, with a increasing of 2.06%.

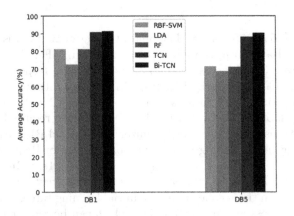

Fig. 5. Performances comparsion of different classifier.

The train loss and test loss of Bi-TCN can be seen in Fig. 6 and Fig. 7. From the results, we can clearly see the loss of Bi-TCN are lower than proposed TCN. In the confusion matrix, some incorrect classification of proposed TCN are correctly classified by Bi-TCN. This phenomenon demonstrate that through combine positive order and reverse order of the same sEMG signal, our Bi-TCN can extract more information hidden in the signal than proposed TCN. On the other hand, as showed in the confusion matrix, the misclassification occurs more often between adjacent classes. This is because the gesture of adjacent classes are more similar and the sEMG signal of these gesture are also similar, which makes classification more difficult. That's phenomenon indicates we should pay more attention to the classification of similar gesture to further improve the accuracy of gesture recognition.

We also compare our method with related works. The result can bee seen in the Table 2, our method achieve best result on DB5. For DB1 our method's prediction accuracy is lower than SOTA, but it's worth noting that our TCN only have 32k parameters and Bi-TCN have 72k parameters, while SOTA method have 410k parameters. The less parameters means less computation power needed, which will improve real-time performance. In [23] which also use TCN

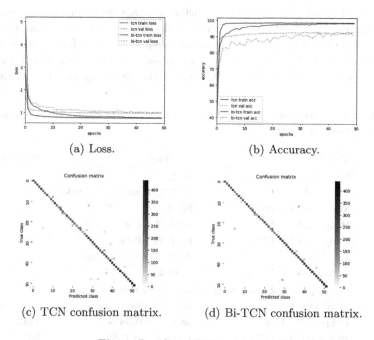

(a) Loss.

(b) Accuracy.

(c) TCN confusion matrix.

(d) Bi-TCN confusion matrix.

Fig. 6. Results of DB1 subject 1.

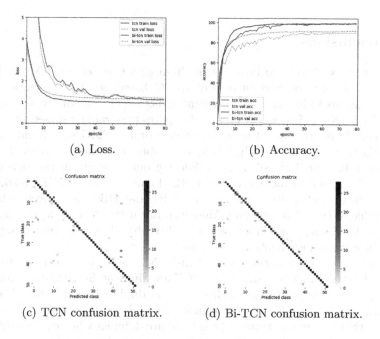

(a) Loss.

(b) Accuracy.

(c) TCN confusion matrix.

(d) Bi-TCN confusion matrix.

Fig. 7. Results of DB5 subject 1.

framework, they achieve 89.76% recognition accuracy on 53 movements but with 300 ms window. It's worth know that with decreasing of window size, the accuracy will also decrease, thus, our proposed method is outperformance than general TCN methods. On DB5, our method achieve 90.06% recognition accuracy on 52 movement, while SOTA get 90% prediction accuracy on 41 movements.

Table 2. Gesture recognition accuracy of the proposed TCN and Bi-TCN compared with the accuracy of existing works.

Database	Model	Accuracy	Parameters
DB1	Multi-view [24]	88.2%	-
	TCN [23]	89.76%(53 movements, 300 ms)	85k
	XceptionTime [19]	92.3%	413k
	Proposed TCN	**90.16% ± 0.33%**	**32k**
	Proposed Bi-TCN	**90.74% ± 0.26%**	**72k**
DB5	TL CNN [8]	68.98%(18 movements)	-
	LCNN [26]	71.66%	-
	Stacking Ensemble Learning [21]	72.09%(40 movements)	-
	Multi-view [24]	90%(41 movements)	-
	Proposed TCN	**88.00% ± 0.34%**	**121k**
	Proposed Bi-TCN	**90.06% ± 0.15%**	**257k**

7   Conclusions

In this paper, we proposed Bidirectional Temporal Convolutional Network (Bi-TCN) to improve the performance of sEMG based gesture recognition. Our proposed networks is consist of two same structure of Temporal Convolutional Networks(TCN). This new TCN method requires two training stages. In the first stage, two TCN are trained by using reverse order sEMG signal and positive order sEMG signal, respectively. Then, we combine the feature extracted by two TCN to train a final classifier. By validating our network on two benchmark dataset, the results demonstrate our method can achieve 90.74% ± 0.26% and 90.06% ± 0.15% recognition accuracy on DB1 and DB5 respectively which is comparable to the SATO method. Moreover, our networks have less parameters than previous' which will improve the ability of real-time gesture recognition and reduce the requirements on equipment.

Despite of the good performance of Bi-TCN, it is difficult to ensure that the features learned by TCN and Reverse-TCN are different. So our future work will focus on find appropriate approaches to avoid duplication of learned features. Furthermore, we also will consider utilizing Graph Neural Networks (GNN). The reason is that the muscle groups of the hand are different when execute different gestures and this relation of the muscle groups can be regarded as a relation graph.

References

1. Al-Timemy, A.H., Bugmann, G., Escudero, J., Outram, N.: Classification of finger movements for the dexterous hand prosthesis control with surface electromyography. IEEE J. Biomed. Health Inform. **17**(3), 608–618 (2013)
2. Alkan, A., Günay, M.: Identification of EMG signals using discriminant analysis and SVM classifier. Expert Syst. Appl. **39**(1), 44–47 (2012)
3. Atzori, M., et al.: Electromyography data for non-invasive naturally-controlled robotic hand prostheses. Sci. Data **1**(1), 1–13 (2014)
4. Atzori, M., et al.: Building the ninapro database: a resource for the biorobotics community. In: 2012 4th IEEE RAS & EMBS International Conference on Biomedical Robotics and Biomechatronics (BioRob), pp. 1258–1265. IEEE (2012)
5. Bai, S., Kolter, J.Z., Koltun, V.: An empirical evaluation of generic convolutional and recurrent networks for sequence modeling. arXiv preprint arXiv:1803.01271 (2018)
6. Chen, H., Zhang, Y., Li, G., Fang, Y., Liu, H.: Surface electromyography feature extraction via convolutional neural network. Int. J. Mach. Learn. Cybernet. **11**(1), 185–196 (2020)
7. Chiang, J., Wang, Z.J., McKeown, M.J.: A hidden Markov, multivariate autoregressive (HMM-mAR) network framework for analysis of surface EMG (sEMG) data. IEEE Trans. Sig. Process. **56**(8), 4069–4081 (2008)
8. Côté-Allard, U., et al.: Deep learning for electromyographic hand gesture signal classification using transfer learning. IEEE Trans. Neural Syst. Rehabil. Eng. **27**(4), 760–771 (2019)
9. Fall, C.L., et al.: Wireless sEMG-based body-machine interface for assistive technology devices. IEEE J. Biomed. Health Inform. **21**(4), 967–977 (2016)
10. Glorot, X., Bengio, Y.: Understanding the difficulty of training deep feedforward neural networks. J. Mach. Learn. Res. **9**, 249–256 (2010)
11. He, K., Zhang, X., Ren, S., Sun, J.: Deep residual learning for image recognition. In: Proceedings of the IEEE Conference on Computer Vision and Pattern Recognition, pp. 770–778 (2016)
12. Hu, J., Shen, L., Sun, G.: Squeeze-and-excitation networks. In: Proceedings of the IEEE Conference on Computer Vision and Pattern Recognition, pp. 7132–7141 (2018)
13. Hu, Y., et al.: A novel attention-based hybrid CNN-RNN architecture for sEMG-based gesture recognition. PloS One **13**(10) (2018)
14. Hudgins, B., Parker, P., Scott, R.N.: A new strategy for multifunction myoelectric control. IEEE Trans. Biomed. Eng. **40**(1), 82–94 (1993)
15. Kamavuako, E.N., et al.: Surface versus untargeted intramuscular EMG based classification of simultaneous and dynamically changing movements. IEEE Trans. Neural Syst. Rehabi. Eng. **21**(6), 992–998 (2013)
16. Müller, R., Kornblith, S., Hinton, G.E.: When does label smoothing help? In: Advances in Neural Information Processing Systems, pp. 4696–4705 (2019)
17. Oskoei, M.A., Hu, H.: Support vector machine-based classification scheme for myoelectric control applied to upper limb. IEEE Trans. Biomed. Eng. **55**(8), 1956–1965 (2008)
18. Pizzolato, S., et al.: Comparison of six electromyography acquisition setups on hand movement classification tasks. PloS One **12**(10) (2017)

19. Rahimian, E., et al.: Xceptiontime: independent time-window xceptiontime architecture for hand gesture classification. In: ICASSP 2020–2020 IEEE International Conference on Acoustics, Speech and Signal Processing (ICASSP), pp. 1304–1308. IEEE (2020)
20. Scheme, E., Englehart, K.: Electromyogram pattern recognition for control of powered upper-limb prostheses: state of the art and challenges for clinical use. J. Rehabil. Res. Dev. **48**(6) (2011)
21. Shen, S., Kang, G., Chen, X.-R., Yang, M., Wang, R.-C.: Movements classification of multi-channel sEMG based on CNN and stacking ensemble learning. IEEE Access **7**, 137489–137500 (2019)
22. Tong, R., Zhang, Y., Chen, H., Liu, H.: Learn the temporal-spatial feature of sEMG via dual-flow network. Int. J. Humanoid Rob. **16**(04), 1941004 (2019)
23. Tsinganos, P., et al.: Improved gesture recognition based on sEMG signals and TCN. In: ICASSP 2019–2019 IEEE International Conference on Acoustics, Speech and Signal Processing (ICASSP), pp. 1169–1173. IEEE (2019)
24. Wei, W., et al.: Surface-electromyography-based gesture recognition by multi-view deep learning. IEEE Trans. Biomed. Eng. **66**(10), 2964–2973 (2019)
25. Wei, W., et al.: A multi-stream convolutional neural network for sEMG-based gesture recognition in muscle-computer interface. Pattern Recogn. Lett. **119**, 131–138 (2019)
26. Wu, Y., Zheng, B., Zhao, Y.: Dynamic gesture recognition based on LSTM-CNN. In: 2018 Chinese Automation Congress (CAC), pp. 2446–2450. IEEE (2018)
27. Yoo, J.W., Lee, D.R., Sim, Y.J., You, J.H., Kim, C.J.: Effects of innovative virtual reality game and EMG biofeedback on neuromotor control in cerebral palsy. Biomed. Mater. Eng. **24**(6), 3613–3618 (2014)

Correction to: Research on Short-Term Traffic Flow Forecast Based on Improved GA-Elman

Qi Li and Hongliang Wang

Correction to:
Chapter "Research on Short-Term Traffic Flow Forecast Based on Improved GA-Elman" in: J. Qian et al. (Eds.):
Robotics and Rehabilitation Intelligence, **CCIS 1336,**
https://doi.org/10.1007/978-981-33-4932-2_13

In the originally published chapter 13 the affiliation information of the authors was incorrect. The affiliation of the authors has been corrected as "School of Computer and Communication Engineering, Liaoning Petrochemical University".

The updated version of this chapter can be found at
https://doi.org/10.1007/978-981-33-4932-2_13

© Springer Nature Singapore Pte Ltd. 2021
J. Qian et al. (Eds.): ICRRI 2020, CCIS 1336, p. C1, 2021.
https://doi.org/10.1007/978-981-33-4932-2_31

Author Index

Printed in the United States
by Baker & Taylor Publisher Services